Cyber Physical System 2.0

The book covers the emerging communication and computational technologies for future cyber-physical systems and discusses the security of in-vehicle communication protocols using automotive embedded systems, presenting an in-depth analysis across various domains, such as manufacturing, transportation, health-care, and smart cities.

This book:

- Discusses how communication and computing co-design provides dynamic adaptability and centralized control.
- Presents the convergence of physical and digital realities within the metaverse and multiverse, setting the stage for the future of cyber-physical-social systems (CPSS).
- Presents emerging communication and computational technologies, such as 6G, software-defined networking, cloud computing, blockchain, artificial intelligence, machine learning, virtual reality, and blockchain, for the design and implementation of cyber-physical systems.
- Explores advanced topics such as security and privacy in industrial CPS, strategies for protecting serial industrial networks, and enhancing firmware update security in automotive systems.

It is primarily written for senior undergraduates, graduate students, and academic researchers in the fields of electrical engineering, electronics and communication engineering, computer science and engineering, and information technology.

Cyber Physical System 2.0

Communication and Computational Technologies

Edited by Amitkumar Vidyakant Jha
and Bhargav Appasani

CRC Press
Taylor & Francis Group
Boca Raton London New York

CRC Press is an imprint of the
Taylor & Francis Group, an **informa** business

Designed cover image: © Canva

First edition published 2025
by CRC Press
2385 NW Executive Center Drive, Suite 320, Boca Raton FL 33431

and by CRC Press
4 Park Square, Milton Park, Abingdon, Oxon, OX14 4RN

CRC Press is an imprint of Taylor & Francis Group, LLC

ISBN: 978-1-032-61463-2 (hbk)
ISBN: 978-1-032-90825-0 (pbk)
ISBN: 978-1-003-55999-3 (ebk)

DOI: 10.1201/9781003559993

Typeset in Sabon
by Apex CoVantage, LLC

Contents

Preface

In recent years, cyber-physical system (CPS) has emerged as a transformative force, blending the physical and digital worlds to create interconnected frameworks that revolutionize industries, enhance efficiency, and redefine human interaction with technology. This edited book, *Cyber Physical System 2.0: Communication and Computational Technologies*, delves into the intricacies of CPS 2.0, focusing on communication and computational technologies that underpin its development and adoption.

The first chapter gives an insightful introduction to CPS 2.0, tracing its evolution from its cybernetic roots to its current state and beyond.

Communication lies at the heart of CPS, serving as the conduit through which data flows and decisions are made. Chapter 2 delves into the co-design of communication and computing under imperfect channel conditions, offering valuable insights into optimizing CPS performance in real-world environments. This chapter mainly focuses on the joint design of communication and computing in CPS, considering factors such as communication delay and packet loss when designing scheduling policies.

The integration of software-defined networking (SDN) into CPSs heralds a new era of flexibility and adaptability. Chapter 3 examines the benefits, challenges, and opportunities of SDN in CPSs, shedding light on its potential to revolutionize network management and enhance system resilience. It primarily underscores the importance of dynamic network management in enhancing system resilience and efficiency, while also addressing the imperative of maintaining high-performance communication in real-time CPS operations.

As CPSs evolve, they transcend traditional boundaries, giving rise to cyber-physical-social systems (CPSS) that integrate social dynamics into system design. Chapter 4 explores this transformative landscape, envisioning the emergence of metasystems that blur the lines between physical and virtual worlds.

From smart health-care systems in smart cities to cloud-enabled smart manufacturing processes, CPSs are reshaping industries and societies worldwide. Chapters 5 and 6 delve into the implementation, challenges, and solutions of CPSs in health-care and manufacturing domains, offering valuable insights into their real-world applications.

The architecture of CPSs is as diverse as the systems themselves, ranging from open architectures to AI-driven models and resilient frameworks. Chapters 7 and 8 delve into the architectural intricacies of CPSs, presenting innovative approaches to design, implementation, and optimization.

Chapter 9 provides a CPS architecture for bridge monitoring concerning civil infrastructures that are subjected to aging and deterioration. To represent a CPS approach parallel with physics-informed decision analysis research, this chapter portrays a model-driven structural health monitoring paradigm based on the numerical, experimental, and field lessons learned in the past decade. This chapter describes a CPS outlook from a civil infrastructure lens through model calibration protocols and reliability-based decision analyses with the help of a few case studies.

Understanding the fundamental principles of security and privacy becomes imperative as we navigate through the intricate web of physical and digital components that constitute industrial CPS (ICPS) environments. Chapter 10 sets the stage by elucidating the potential consequences of security breaches and privacy violations, emphasizing the critical need for proactive countermeasures.

In ICPS, the security of non-IP-based networks is again an important research direction. There is a need to improve the defense of serial (non-IP) industrial networks within CPSs. Chapter 11 delineates these networks' evolution and vulnerabilities, especially in the face of escalating cyber threats increased by the advent of Industry 4.0 and the integration of operational technology with information technology. This chapter comprehensively analyzes strategies to mitigate risks and enhances network resilience against cyber intrusions, reinforcing the necessity of a multifaceted approach. The chapter encompasses technologies to safeguard cyber infrastructure by proposing protective measures, including but not limited to network segmentation, authentication protocols, and real-time monitoring.

In Chapter 12, "Enhancing the Security of Firmware Over-the-Air Updates in Automotive Cyber-Physical Systems," the focus shifts to the automotive sector, where traditional vehicles metamorphose into "smart cars." With this evolution comes the imperative of securing firmware updates, crucial for addressing emerging cyber threats in modern vehicles. The chapter offers an overview of security update processes, emphasizing their importance in modern systems.

Critical energy infrastructure has undergone significant changes in the past decades, which has made these systems more vulnerable to breakdowns due to the uncertainty and variability in operation. Dealing with the ever-increasing challenge posed by such cyber-physical events requires improving the resiliency of our critical energy infrastructure. Recent advances in improving the resilience of CPSs, especially with regard to energy systems, are presented in the last chapter of the book. In this chapter, the methods presented range from distributionally robust optimization to autonomous and coordinated control, reinforcement learning–based resilient control, and topology reconfiguration in inter-system resilient control.

As we embark on this journey through the world of CPS 2.0, we invite you to explore the rich tapestry of technologies, challenges, and opportunities that define this rapidly evolving field. Whether you are a seasoned professional, a researcher, or an enthusiast, *Cyber Physical System 2.0: Communication and Computational Technologies* offers invaluable insights into the past, present, and future of CPSs. Let us embark on this journey together as we unravel the mysteries of CPS 2.0 and chart a course toward a smarter, more connected future.

Amitkumar Vidyakant Jha
Bhargav Appasani

Foreword

In the era of rapid technological advancement, the fusion of the physical and digital realms has catalyzed a paradigm shift leading to the emergence of the cyber-physical system (CPS). This book discusses the evolution of CPS to 2.0, guiding us through the intricate landscape of this transformative domain.

The book begins with an introduction to the evolution of CPS to its current state. The interplay of communication and computational technologies that underpin the fabric of CPS 2.0 is explored. Communication, the lifeblood of CPS, emerges as a central theme, illuminating pathways for optimizing performance under real-world constraints. Each chapter presents state-of-the-art results that will reshape network management and system security for CPS.

CPS 2.0 extends beyond technological integration, giving rise to cyber-physical-social systems (CPSS) that intertwine social dynamics with computational frameworks. As CPS architectures blur the physical and virtual boundaries, we face societal implications and ethical considerations inherent in this evolution.

The transformative potential of CPS is having significant impact on a wide range of real-world applications, from smart health-care systems to cloud-enabled manufacturing processes. To achieve the full potential of CPS and enable their cost-effective deployments, we must address several design and implementation challenges associated with CPS 2.0.

Security and privacy emerge as paramount concerns in the industrial CPS (ICPS) landscape. To secure and protect ICPS, we must develop novel, scalable, and robust anomaly detection techniques and firmware security protocols. As we confront escalating cyber threats, the imperative of safeguarding critical infrastructures becomes ever more pressing, demanding multifaceted approaches to resilience and risk mitigation.

As the number of cyberattacks continues to increase, CPS resilience emerges as a cornerstone of operational continuity and public safety. From health-care systems to critical energy infrastructures, the quest for resilient CPSs should consider theoretical insights with practical design iterations, bridging the chasm between theory and real-world applications.

This timely volume serves as an authoritative reference and a guidebook for the rapidly evolving landscape of CPS, and it is highly recommended to those interested in this domain.

Sherali Zeadally, FBCS, FIET
University Research Professor
University of Kentucky Alumni Association Endowed Professor
College of Communication and Information
University of Kentucky

About the editors

Amitkumar Vidyakant Jha received his MTech degree from the Indian Institute of Industrial Technology, Guwahati, India, and his PhD degree from the Kalinga Institute of Industrial Technology (KIIT), Bhubaneswar, India. He has working as an assistant professor at KIIT since 2015. He has authored more than 50 articles in international journals and conference proceedings. He has also authored four book chapters. He has edited a book titled *Metamaterials for Microwave and Terahertz Applications: Absorbers, Sensors and Filters*, which is available at DOI. 10.52305/APHY8244. His research interests include smart grid, communication network, optimization, etc. He is a reviewer of several journals, such as *e-Prime*, *IEEE Access*, *IEEE System Journal*, *IEEE IoT*, *Wireless Personal Communication*, *International Journal of Electrical Power and Energy Systems*, *Energies*, *Mathematics*, etc., and he is the editor of the *Journal of Computer Networks and Communications*. He is also a member of many professional organizations, such as the Indian Science Congress, the International Association of Engineers, the World Leadership Academy, etc.

Bhargav Appasani received his PhD (Engg.) degree from Birla Institute of Technology, Mesra, India. He is currently an associate professor with the School of Electronics Engineering, KIIT University, Bhubaneswar, India. He has published more than 140 articles in international journals and conference proceedings. He has also published six book chapters with reputed international publishers. He has also authored a book with a publication of international repute and is currently editing two more books. He also has a patent filed to his credit. He is an academic editor of the *Journal of Electrical and Computer Engineering* (Hindawi) and a reviewer for *IEEE Transactions on Smart Grid*, *IEEE Transactions on Antennas and Propagation*, and *IEEE Access*. Dr. Appasani has taught many courses, such as machine learning, data structure algorithm, microwave, control system, etc., for several years. His research interests include optimization, AI, metamaterials, communication systems, etc.

Contributors

Veronica Adetola
Pacific Northwest National Laboratory
Richland, WA, United States

F. Aguayo-González
Department of Design Engineering
University of Seville
Polytechnic School
Seville, Spain

Mobayode O. Akinsolu
Faculty of Arts, Computing, and Engineering
Wrexham University
Wales, United Kingdom
and
Faculty of Natural and Applied Sciences
Lead City University
Ibadan, Oyo State, Nigeria

Folahanmi T. Akinsolu
Faculty of Basic Medical and Health Sciences
Lead City University
Ibadan, Oyo State, Nigeria

M. J. Ávila-Gutiérrez
Department of Design Engineering
University of Seville
Polytechnic School
Seville, Spain

Rodrigo Rosetti Binda
Cybersecurity Architecture
Vale SA
Vitória, ES, Brazil

Filipe Andersonn Teixeira da Silveira
Industrial Technology Architecture
Vale SA
Vitória, ES, Brazil

Telmo Fernández De Barrena
Department of Data Intelligence for Energy and Industrial Processes
Fundación Vicomtech
Basque Research and Technology Alliance (BRTA)
Mikeletegi, Donostia-San Sebastian, Spain

Ralf Luis de Moura
Industrial Technology Architecture
Vale SA
Vitória, ES, Brazil

Chekwube Ezechi
Faculty of Natural and Applied Sciences
Lead City University
Ibadan, Oyo State, Nigeria

Paul Fanning
University College Dublin
Dublin, Co Dublin, Ireland

Maria Q. Feng
Columbia University
New York, NY, USA

Juan Luis Ferrando
Department of Data Intelligence for Energy and Industrial Processes
Fundación Vicomtech
Basque Research and Technology Alliance (BRTA)
Mikeletegi, Donostia-San Sebastian, Spain

Tianyi Gao
NA

Ander García
Department of Data Intelligence for Energy and Industrial Processes
Fundación Vicomtech, Basque Research and Technology Alliance (BRTA)
Mikeletegi, Donostia-San Sebastian, Spain

K. Hemant Kumar Reddy
School of Computer Science and Engineering
VIT-AP University
India

Nika Hosseini
Optical Zeitgeist Laboratory
INRS
Montreal, Canada

Yuchen Jiang
Harbin Institute of Technology
Harbin, China

Soumya Kundu
Pacific Northwest National Laboratory
Richland, WA, United States

Hao Luo
Harbin Institute of Technology
Harbin, China

Martin Maier
Optical Zeitgeist Laboratory
INRS
Montreal, Canada

Asmita Manna
Department of Computer Engineering, Pimpri Chinchwad College of Engineering
Pune, India

Brenda Aurora Pires Moura
Computing Science
Universidade Federal do Espirito Santo Vitória
ES, Brazil

Sayak Mukherjee
Pacific Northwest National Laboratory
Richland, WA, United States

Sai Pushpak Nandanoori
Pacific Northwest National Laboratory
Richland, WA, United States

Nawaf Nazir
Pacific Northwest National Laboratory
Richland, WA, United States

Isaac O. Olalere
Department of Industrial Engineering
University of South Africa (UNISA) Florida Campus, Roodepoort
Johannesburg, Gauteng, South Africa

Ekin Ozer
University College Dublin
Dublin, Co Dublin, Ireland

Rachana Y. Patil
Department of Computer Engineering
Pimpri Chinchwad College of Engineering
Pune, India

Yogesh H. Patil
D. Y. Patil College of Engineering
Akurdi, Pune, India

Yifei Qiu
Department of Electronics and Information Engineering
Harbin Institute of Technology (Shenzhen)
Shenzhen, China

Manjiri Ranjanikar
Department of Computer Engineering
Pimpri Chinchwad College of Engineering
Pune, India

Manjula Gururaj Rao
Department of Information Science and Engineering
NMAM Institute of Technology, Nitte (Deemed to be University)
Nitte, Karkala, India

Diptendu Sinha Roy
Department of Computer Science and Engineering
National Institute of Technology Meghalaya
Meghalaya, India

Wilson Sakpere
Faculty of Natural and Applied Sciences
Lead City University
Ibadan, Oyo State, Nigeria

Abimbola Sangodoyin
School of Computer Science
University of Lincoln
Brayford Pool, Lincoln, United Kingdom

Minoo Soltanshahi
Optical Zeitgeist Laboratory
INRS
Montreal, Canada

Serdar Soyoz
Bogazici University
Bebek, Istanbul, Turkiye

Muhammad Usman Tariq
Abu Dhabi University
Abu Dhabi, UAE

Jilun Tian
Harbin Institute of Technology
Harbin, China

Thanh Long Vu
Pacific Northwest National Laboratory
Richland, WA, United States

Ying Wang
Department of Electronics and Information Engineering
Harbin Institute of Technology (Shenzhen)
Shenzhen, China

Tiago Tadeu Wirtti
Industrial Technology Architecture
Vale SA
Vitória, ES, Brazil

Shaohua Wu
Guangdong Provincial Key Laboratory of Aerospace Communication and Networking Technology
Harbin Institute of Technology (Shenzhen)
Shenzhen, China
and
Pengcheng Laboratory
Shenzhen, China

Shimeng Wu
Harbin Institute of Technology
Harbin, China

Qinyu Zhang
Guangdong Provincial Key Laboratory of Aerospace Communication and Networking Technology
Harbin Institute of Technology (Shenzhen)
Shenzhen, China
and
Pengcheng Laboratory
Shenzhen, China

Chapter 1

Introduction to cyber-physical systems 2.0

Evolution, technologies, and challenges

Dr. Muhammad Usman Tariq

LIST OF ABBREVIATIONS

AR	augmented reality
AI	artificial intelligence
CPS	cyber-physical systems
IoT	Internet of Things
ML	machine learning
NSF	National Science Foundation
VR	virtual reality
THz	terahertz

1.1 INTRODUCTION

The phrase "cyber-physical system" (CPS) sounds like a spanking-new buzzword since it is an ongoing topic of conferences, journal articles, and books like this one. According to its historical origins, the prefix *cyber* derives from the ancient Greek term *υ* ("cybernetic"), which originally meant "controlling abilities." Eventually, it became the Latin word *governance*, which became the English term *governing*. This chapter discusses systems in which tangible items and information technology assets are closely interwoven and display a certain level of ongoing coordination [1, 2].

The term *cyber-physical systems*, as well as *CPS*, first came into use in 2006 and has been defined as "systems which merge computation with physical procedures." Embedded computers and networks track and regulate physical procedures via feedback loops, where one step influences the others. The US National Science Foundation (NSF) identified this as an important area of study in 2008. Physical objects are fitted with identifying technologies in the first-generation CPS, whereas actuators and sensors are elements of the second-generation CPS [3]. A feedback procedure involving sensing, physical modernization, decision-making processes, and network compatibility evaluation results in the emergence of the latest third-generation CPS. CPS frequently uses the "cyberizing the physical" and "physicalizing the cyber" aspects.

DOI: 10.1201/9781003559993-1

A cyber-physical system is attributed to a mixture of physical essentials, such as systems, devices, and machinery, integrated and interconnected with computer-based monitoring and control systems. The CPS demonstrates the fusion of the digital and physical worlds, where the physical components are coordinated and controlled by computer networks, systems, and software algorithms [4, 5]. The physical components of a cyber-physical system interact with the digital world via actuators, sensors, and other embedded devices that transmit and collect data to computer systems. These computer systems make decisions, analyze data, and give commands back to the physical components to optimize their performance and control their behavior [6].

CPS can be found in numerous domains, including manufacturing plants, transportation systems, health-care systems, smart grids, and smart cities. Examples of cyber-physical systems include industrial automation systems, medical monitoring devices, and autonomous vehicles [7, 8].

The key features of cyber-physical systems are:

- *Interconnectivity.* CPS components are interconnected through wireless or wired networks, which facilitates coordination and communication between the digital and physical parts of the system [9].
- *Real-time feedback.* CPS perpetually monitors and senses the physical world to facilitate the process of decision-making and control by collecting data and giving real-time feedback to the control system.
- *Control and automation.* CPS utilizes computer-based control systems to control and monitor the behavior of physical components, usually by utilizing closed-loop control and feedback loop mechanisms.
- *Integration of physical and computing elements.* Physical components are combined with computing devices by CPS, such as algorithms and software, to facilitate seamless interaction between the digital and the physical realms [10].
- *Adaptability.* A CPS is designed to adapt to dynamically changing conditions by optimizing their behavior based on collected data and environmental feedback. The applications of cyber-physical systems have diversified and spread across numerous industries. They help improve productivity, efficiency, sustainability, and safety in various domains by enabling the coordination, optimization, and better control of physical processes [11].

This chapter presents a thorough exploration of the evolving landscape of CPS 2.0. This segment serves as a crucial introduction to the evolution of CPS, transcending its traditional boundaries and incorporating more intricate interactions between digital and physical components. It delves into the technological advancements that define CPS 2.0, emphasizing the roles of communication networks, sensor technologies, and artificial intelligence in shaping the intelligence and capabilities of these systems. The integration of 5G and the envisioned 6G plays a pivotal role, highlighted for their strategic contributions to enhancing data transfer speeds, reducing latency, and supporting a vast density of connected devices. Furthermore, the section addresses the challenges accompanying the evolution of CPS 2.0, including the management of extensive data, concerns regarding security and privacy, and ethical considerations. Throughout, the sections provide a comprehensive overview of the key themes shaping the trajectory of CPS 2.0, paving the way for a deeper exploration of this transformative paradigm [12, 13].

1.2 CYBER-PHYSICAL SYSTEMS 1.0

A pivotal point in developing more networked and intelligent systems is represented by CPS 1.0, which is a fundamental milestone in integrating computer and physical processes. The primary objective of CPS 1.0 is to enable connection, automation, and data interchange across several domains, providing a foundation for future developments [14]. The main goal was to develop systems that could perceive, analyze, and react to changes in the external environment to bring about a new wave of automation and efficiency in various sectors [15].

1.2.1 Key components of CPS 1.0

The key components of CPS 1.0 are as follows:

- *Actuators and sensors.* These components constitute the backbone of CPS 1.0. They are in charge of gathering and sending data from the physical environment to the computing systems, and vice versa. Intelligent decision-making is predicated on this reciprocal flow of knowledge [16].
- *Connectivity.* CPS 1.0 strongly focuses on integrating diverse devices via networks, facilitating smooth communication and information sharing. According to [17], collaborative decision-making procedures are built on this interconnection.
- *Centralized control systems.* To interpret incoming data, make judgments, and issue orders to actuators, CPS 1.0 frequently relied on centralized control systems. According to [18], this centralized paradigm offered an organized method for overseeing and managing physical processes.
- *Automation.* CPS 1.0 was characterized by its use of predetermined rules and algorithms to automate repetitive processes. The objectives of this automation were to improve productivity, decrease the need for continual human involvement, and streamline procedures [19].

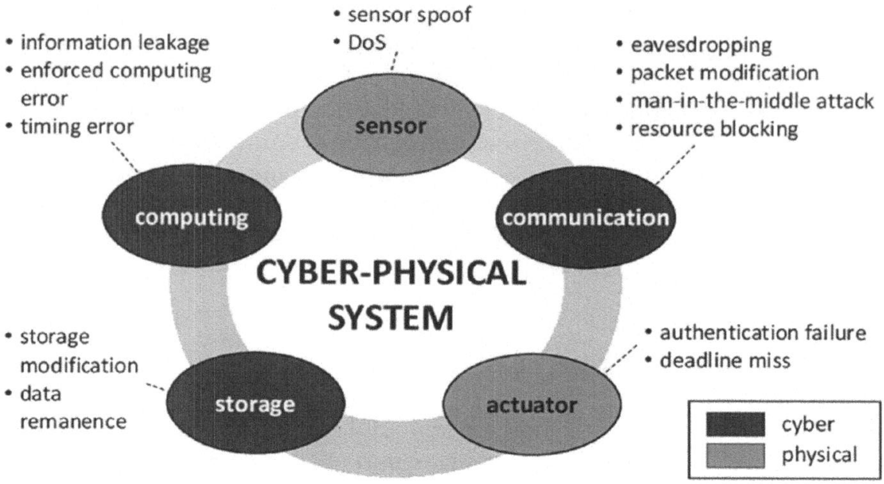

Figure 1.1 Cyber-physical system.

The integration of physical and digital systems is made possible by these components working together in the CPS 1.0 system. To govern the behavior of the physical system, data are gathered from sensors, processed by computational components, and control signals are produced. User interfaces enable people to interact with and monitor the system, whereas communication networks permit information interchange [20].

It is crucial to remember that the architecture and parts of CPS 1.0 can change based on a particular application's needs and the system's specifications. To fit their intended functionality and goals, different CPS 1.0 systems may differ in the complexity and organization of these components, as shown in Figure 1.1.

1.2.2 Challenges associated with CPS 1.0

CPS 1.0 confronts system integration, interoperability, scalability, security, and privacy challenges. Significant obstacles arise from integrating many systems and components, providing smooth communication, and ensuring the security and privacy of sensitive data. These challenges are described as follows:

- *System integration.* CPS 1.0 calls for integrating several components and subsystems, which can be difficult, owing to variations in communication protocols, data formats, and hardware compatibility. A major difficulty was ensuring smooth interoperability and integration between physical and digital components [18].

- *Compatibility*. A difficult task in CPS 1.0 is achieving compatibility across various CPSs and devices. The seamless exchange of data and information between many systems and domains must be improved by incompatibilities in communication protocols, data models, and interfaces.
- *Scalability*. Extending CPS 1.0 systems to more extensive deployments and complicated contexts is difficult. The scalability of CPS has become an issue as the number of interconnected components increases. A major problem was ensuring the systems could manage the increasing demands while maintaining their performance and efficiency.
- *Security and privacy issues*. Due to data interchange and interconnectedness between physical and digital components, CPS 1.0 encountered security and privacy issues. The fusion of numerous networks and systems widened the potential attack surface, leaving CPS open to hacking attacks and illegal access. Significant hurdles were presented to ensure the confidentiality, integrity, and availability of data and safeguard the privacy of sensitive information [5, 21].
- *Low intelligence and limited adaptability*. CPS 1.0 systems frequently lack high levels of intelligence and flexibility. They must help manage complicated and dynamic contexts because their decision-making is based on pre-established rules and algorithms. Adjusting to shifting conditions and unforeseen events was difficult because CPS 1.0 systems were primarily created for static surroundings and predetermined scenarios [22].
- *Retrofitting legacy systems*. One of the challenges of CPS 1.0 was retrofitting legacy systems and infrastructure with CPS capabilities. It is frequently necessary to make adjustments and upgrades and consider compatibility when integrating CPS technology into existing systems, which might be expensive and difficult.
- *Human–machine interaction*. Providing natural and efficient human–machine interaction was a problem for CPS 1.0. The creation of user interfaces and control systems enables seamless human interaction with CPS, understanding of system behavior, and provision of suitable information [17].

To sum up, comprehending CPS 1.0 entails being aware of its fundamental traits (as shown in Figure 1.2), its essential elements, and the technical environment that shaped this early stage of cyber-physical integration. Although CPS 1.0 set the stage for more advanced versions, it was not without problems, especially regarding its lack of intelligence and security flaws. The development of CPS 1.0 and CPS 2.0 will be examined later, emphasizing the breakthroughs and developments that have influenced the current state of linked systems [23].

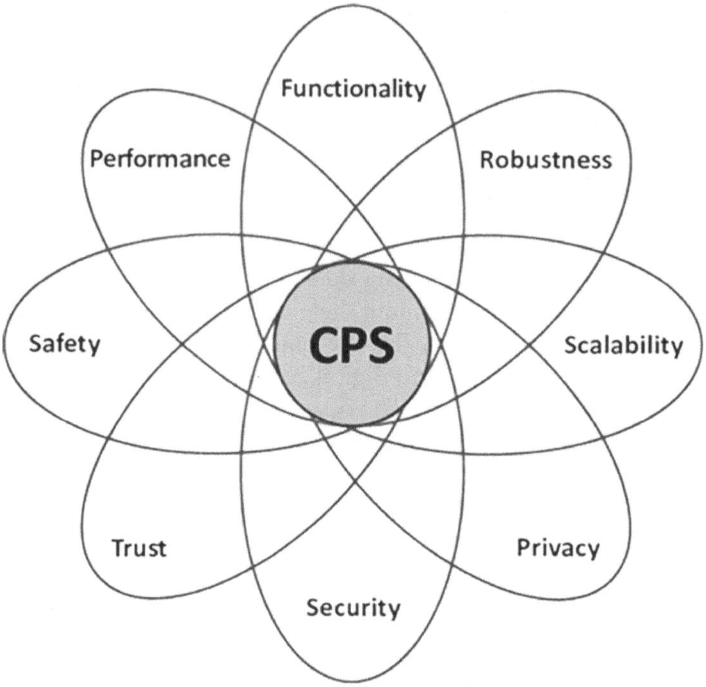

Figure 1.2 CPS attributes.

1.3 CYBER-PHYSICAL SYSTEMS 2.0

CPS 2.0 has emerged as the next frontier in this revolutionary journey, marking a considerable evolution of CPS over the previous ten years. The origins of CPS 2.0 may be found in its initial version, also known as CPS 1.0. The focus of CPS 1.0 was on automation, networking, and data sharing across a range of industries, such as smart infrastructure, manufacturing, and health care [24]. Nevertheless, the need for a more complex and flexible framework emerged when technology developed further, giving rise to CPS 2.0.

1.3.1 Evolution from CPS 1.0 to CPS 2.0

The first phase of the development of cyber-physical systems is represented by CPS 1.0. Although it sets the stage for the fusion of analog and digital components, CPS 1.0 has its constraints to solve these constraints and further improve the capabilities of cyber-physical systems. The improvements and features added in later versions of CPS build on the frameworks created by CPS 1.0. The solutions to these problems paved the way for later

CPS iterations, such as CPS 2.0, which aimed to overcome these restrictions and expand the capabilities and robustness of cyber-physical systems [19, 25]. While CPS 1.0 has already revolutionized many sectors, CPS 2.0 further expands on the concept by employing new technology, fixing imperfections, and capitalizing on emerging patterns.

Modern systems are becoming more intricate, networked, and data-intensive. CPS 2.0 uses cutting-edge technologies, such as AI, machine learning, and edge computing, to manage and improve these complex systems. Because of these developments, CPS 2.0 can manage escalating complexity and make wise decisions in real time [3]. The key differences between CPS 1.0 and 2.0 are summarized in Table 1.1.

Table 1.1 Key differences between CPS 1.0 and CPS 2.0

Feature	*CPS 1.0*	*CPS 2.0*
Security	Limited security measures; vulnerability to cyber threats	Enhanced security protocols, including blockchain, to ensure data integrity and trust
Interoperability	Limited interoperability; challenges in integrating diverse components	Focus on interoperability standards, facilitating seamless integration of diverse technologies
Communication infrastructure	Basic communication protocols, with limited scalability	Advanced communication with 5G technology, supporting high-speed, low-latency data transfer
Technological foundation	Basic integration of sensors, actuators, and basic control systems	Advanced integration of AI, ML, edge computing, blockchain, and 5G technologies
Decision-making capability	Limited intelligence; relies on predefined rules and algorithms	Enhanced decision-making through machine learning, adaptability, and real-time analysis
Scalability	Limited scalability due to technological constraints	Improved scalability to handle the increasing complexity of interconnected systems
Data processing	Centralized processing, with limited real-time capabilities	Edge computing for real-time processing, reducing latency and improving efficiency
Standardization	Lack of standardized protocols, hindering collaboration	Emphasis on standardization to establish common protocols and frameworks
Application domains	Primarily used in basic automation and control applications	Diverse applications, including smart cities, health care, and advanced manufacturing
Adaptability	Limited adaptability to dynamic changes in the environment	High adaptability with machine learning capabilities, enabling systems to learn and evolve

1.3.2 Key features of CPS 2.0

CPS 2.0 is required to fulfill the rising needs of complex, linked systems while overcoming the constraints of existing CPS, embracing emerging technologies, enhancing security and privacy, and improving system performance. CPS 2.0 aims to deliver more intelligent, adaptable, and effective solutions that can respond to these changing needs and open up new possibilities across various industries.

Several essential characteristics can be used to explain the evolution of CPS to CPS 2.0.

- *Expanded connectivity and interoperability.* A vital aspect of CPS 2.0 is enhanced connectivity and interoperability among different systems and parts. It facilitates the use of open designs, standardized protocols for communication, and the effortless integration of cyber and physical components. Consequently, several CPSs may work together, coordinate, and share information more efficiently between different CPSs [26].
- *Advanced autonomy and intelligence.* CPS 2.0 integrates advanced machine learning algorithms, artificial intelligence (AI) techniques, and autonomous decision-making abilities. CPS can employ data to gain insight, adjust to changing conditions, and rapidly make intelligent decisions. Consequently, systems develop that may enhance their performance, anticipate issues, take measures to avoid them, and respond independently to evolving circumstances.

  Thus, the need for CPS 2.0 arises from the desire to be more flexible and efficient, to have adaptive systems that can handle changing conditions and environments, and to overlook unforeseen circumstances.
- *Edge computing and distributed intelligence.* CPS 2.0 uses edge computing, bringing computation and data processing closer to the actual physical components. CPS 2.0 improves real-time reaction, lowers latency, and allows for more effective resource use by putting the processing power closer to the edge. Distributed intelligence further enhances CPS's resilience and fault tolerance by allowing decision-making to be distributed among numerous CPS components.
- *Improved security and privacy.* Conventional CPSs do not have strong security processes, which makes them vulnerable to online threats. Privacy issues may arise because of the need for sufficient safeguarding procedures for private information. As CPS grows increasingly prevalent and networked, protecting the security and privacy of data and operations is essential. Encryption, authentication, access control, and intrusion detection systems are the primary security methods that CPS 2.0 emphasizes. Additionally, it handles privacy issues by enforcing privacy-preserving procedures and adhering to data protection laws [25].

- *Integration of emerging technologies.* Conventional CPSs will likely only partially use cutting-edge technologies or domain integration. Whereas CPS 2.0 hyperlinks with technologies in development to enhance capabilities while opening the door to new applications, it uses additive manufacturing (3D printing), cloud computing, robots, augmented reality (AR), virtual reality (VR), and nanotechnology to increase functionality, efficiency, and interoperability.

Thus, CPS 2.0 incorporates emerging technologies to strengthen and increase the application's capabilities. The chances of the emergence of new use cases for CPS 2.0 are made possible because of the integration of these technologies. An improvement in intelligence, connectivity, and security can be observed in the transition from CPS to CPS 2.0 [27].

By correcting shortcomings, embracing improvements, and utilizing emerging technology, CPS 2.0 strengthens the foundation of conventional CPS. This shift provides creativity, automation, and advancement opportunities across many industries and domains. Regarding connectivity, autonomy, intelligence, security, integration with emerging technologies, and system optimization, CPS 2.0 and traditional CPS differ significantly. CPS 2.0, a more integrated, intelligent, secure, and optimized framework, opens up new opportunities and uses in various industries [28].

1.3.3 Technologies driving CPS 2.0

The development of CPS 2.0 is fueled by the fusion of several cutting-edge technologies that improve the capabilities and functionalities of cyber-physical systems. These innovations have had a significant impact on CPS development. The following are a few of the major technologies that power CPS 2.0:

- *The Internet of Things.* IoT allows for smooth data interchange and communication between analog and digital components by connecting physical devices, sensors, and actuators to the Internet. CPS 2.0 uses IoT technology to improve connectivity, real-time data collection, and system compatibility [29].
- *Artificial intelligence (AI) and machine learning (ML).* These techniques boost the autonomy and intelligence of CPS 2.0. Systems can use these technologies to learn from the data, make wise choices, and adjust to changing environments. These techniques are used for anomaly detection, predictive analytics, autonomous decision-making, and optimization within CPS 2.0.
- *Edge computing.* It enables real-time data processing and decision-making at the source by bringing computational capabilities closer to

the network's edge. Edge computing is used by CPS 2.0 to manage massive amounts of data produced by dispersed sensors and devices, minimize latency, and improve responsiveness. Owing to edge computing, CPS 2.0 systems can function in low-latency, high-bandwidth environments [30, 31].

- *Blockchain*. Blockchain technology offers a decentralized, transparent, and tamper-resistant architecture for safe data sharing, identity management, and trust building. Blockchain can be used by CPS 2.0 to improve security, privacy, and trust in data and transaction exchanges between various system entities. This makes interactions within CPS 2.0 systems secure and auditable.
- *Cloud computing*. Scalable processing resources, online storage, and other services are provided via cloud computing. Cloud computing is used by CPS 2.0 for data storage, distributed processing, and on-demand access to computational resources. Cloud platforms make large-scale data analyses, intricate simulations, and computations that require many resources possible for CPS 2.0.
- *Virtual reality (VR) and augmented reality (AR)*. These technologies offer collaborative, immersive human–machine interaction and visualization with CPS 2.0. These innovations improve the human–machine interface by enabling real-time monitoring and interaction between people and digital and physical components. Applications for CPS 2.0 are made easier to train, maintain, and provide remote assistance via AR and VR.
- *5G networks*. Data transfer in CPS 2.0 systems is made possible by 5G networks, which offer high-speed, low-latency, and dependable connectivity. The improved network capabilities of 5G meet the demands of CPS 2.0, enabling real-time communication, quick responses, and enormous device connectivity [32, 33].

1.3.4 Application of CPS 2.0

The development and capabilities of CPS 2.0 are driven by these technologies and others, such as robotics, additive manufacturing (3D printing), and enhanced sensing technologies. They offer improved connectivity, intelligence, autonomy, security, and efficiency within CPS 2.0, opening new avenues and reshaping various sectors and areas.

Cyber-physical systems 2.0, or *CPS 2.0*, is the term used to describe the next generation of cyber-physical systems, which combine physical items, computer components, and networked communication to produce intelligent and autonomous systems. While CPS 1.0 concentrates on tying together physical and digital systems, CPS 2.0 expands on this concept by introducing cutting-edge tools and capabilities. CPS 2.0 has a wide range of applications and the potential to change many different sectors and businesses [34]. The following are some instances of how CPS 2.0 can be used:

- *Smart cities*. CPS 2.0 can make urban areas more efficient and environmentally friendly. It allows for real-time monitoring and management of many systems, including waste management, transportation, and energy grids. This technique can reduce waste and maximize resource utilization.
- *Industrial automation*. The development of industrial automation is greatly aided by CPS 2.0. CPS 2.0 supports autonomous manufacturing processes, preventive maintenance, and adaptive production systems by fusing physical systems, such as machines and robots, with intelligent software and communication networks. Industrial operations result in improved productivity, safety, and cost savings.
- *Health care*. CPS 2.0 could completely change how health care is delivered. CPS 2.0 offers real-time patient monitoring, individualized treatments, remote health-care services, and effective resource management by integrating medical equipment, patient monitoring systems, electronic health records, and other health-care infrastructures. This can lead to better patient outcomes, lower health-care expenditures, and more rapid and accurate interventions [35].
- *Transportation and autonomous vehicles*. CPS 2.0 is a crucial enabler for intelligent and driverless transportation systems. CPS 2.0 enables real-time data sharing between vehicles, infrastructure, and people by combining sensors, communication networks, and control systems. This increases overall mobility, makes transportation safer and more effective, and reduces traffic congestion.
- *Energy systems*. CPS 2.0 can optimize energy production, distribution, and consumption in energy systems. CPS 2.0 offers real-time monitoring, demand–response management, and energy efficiency optimization by integrating renewable energy sources, smart grids, energy storage systems, and consumer devices. As a result, energy infrastructure becomes more trustworthy and sustainable [36].
- *Agriculture*. Through precision agriculture, CPS 2.0 can transform farming techniques. CPS 2.0, which integrates sensors, drones, autonomous vehicles, and data analytics, offers real-time monitoring of crops, soil conditions, and weather patterns. This enables farmers to manage better irrigation, fertilizer use, insect management, and crop harvesting, resulting in higher yields, lower resource use, and increased sustainability.

These are only a few instances of varied uses of CPS 2.0. A wide range of opportunities for innovation and optimization across several fields is made possible by integrating physical systems with cutting-edge computers, communication, and data analytics capabilities.

1.4 CHALLENGES, SOLUTIONS, AND FUTURE TRENDS

This section provides an overview of the most important aspects for CPS 2.0.

1.4.1 Challenges and solutions

CPS 2.0 has a wealth of prospects and breakthroughs; it also has several problems that must be resolved for successful implementation. Following are some of the main obstacles CPS 2.0 may be faced with in the future:

- *Risks to security and privacy*. As CPS 2.0 systems become more data-driven and linked, there is a higher chance of cyberattacks, data breaches, and privacy violations. Important obstacles that must be solved include ensuring strong security measures, putting encryption, authentication, and access control methods into place, and addressing privacy issues [37].
- *Standards and interoperability*. CPS 2.0 requires the fusion of numerous devices, networks, and technologies. A fundamental difficulty is ensuring compatibility and standardization across many systems, protocols, and communication interfaces. Common frameworks and standards must be created to ensure smooth communication and collaboration between CPS 2.0.
- *Scalability and complexity*. It is anticipated that CPS 2.0 systems will be able to handle large deployments, including a large number of interconnected devices and data sources. Significant hurdles lie in managing the complexity of such systems, maintaining scalability, and managing the enormous amounts of data created. Effective data processing, resource allocation, and system management strategies are required to overcome these issues.
- *Resource management and energy efficiency*. CPS 2.0 systems use much energy and resources. It may be difficult to balance the demands for functionality and performance, energy efficiency, and sustainable resource management. To address this problem, it is necessary to create energy optimization algorithms, including renewable energy sources, and to implement severe resource allocation techniques [1–6, 8–11, 13–16, 18, 20, 21, 23, 24, 26–29, 31, 33–42].
- *Acceptance and trust*. For CPS 2.0 to be successfully adopted, it is essential to foster trust among users and stakeholders. CPS 2.0 systems must ensure openness, explanatory ability, and accountability to win over the public's acceptance and trust. They must also address

concerns regarding job loss, ethical ramifications, and the societal impact of autonomous systems.
- *Legal and regulatory frameworks.* Current legal and governmental frameworks have been tested by the rapid development of CPS 2.0. Problems that must be overcome include the ethical and legal ramifications of autonomous systems, liability concerns in the event of malfunctions or accidents, and the requirement for updated legislation to consider new technologies.
- *Skills and workforce development.* To implement CPS 2.0 successfully, a competent workforce that can create and manage these intricate systems is needed. A challenge that requires attention is closing the skills gap and offering instruction and training in artificial intelligence, cybersecurity, data analytics, and system integration.
- *Implications for society and ethics.* CPS 2.0 poses significant social and ethical issues. It is important to consider and try to minimize the issues associated with ensuring justice and transparency and resolving potential biases in algorithmic decision-making, as well as the influence on employment and societal institutions [1–6, 8, 18, 19, 22, 25, 32, 38–40, 43].

1.4.2 Future trends

The future trends of the CPS 2.0 are as follows:

- *Enhancements to cybersecurity.* With the increased use of CPS 2.0, cybersecurity will become more important. The development of strong security measures, such as secure communication protocols, cutting-edge encryption algorithms, and threat intelligence systems, will be a forthcoming pattern. Strategies such as anomaly detection, behavior analysis, and blockchain technology can be used to improve system security further.
- *Privacy-preserving solutions.* Privacy protection will become increasingly essential as CPS 2.0 systems collect enormous volumes of sensitive and personal data. Future developments will create privacy-preserving methods that protect data privacy while providing useful system functionality, such as differential privacy, federated learning, and secure multi-party computation [2, 10].
- *Cross-domain integration.* With the release of CPS 2.0, there will be a greater emphasis on cross-domain integration, promoting synergistic effects and improving system performance. An example of how dynamic energy management based on current traffic circumstances can be achieved is by linking smart grids with transportation networks. Cross-domain integration results in better coordination, more effective resource management, and improved system performance.

- *Green and sustainable CPS.* Sustainability and environmental concerns precede CPS 2.0 trends. Greener CPSs can be created by integrating them with renewable energy sources, optimization algorithms, and intelligent resource management. As a result, carbon emissions will decrease, energy use will be optimized, and sustainability objectives will be met.
- *Human-centered design.* This concept is emphasized in CPS 2.0. User experience, user-friendly interfaces, and seamless technology integration into people's lives will be prioritized in future trends. CPS 2.0 systems will be created to increase safety, expand human potential, and offer individualized and flexible services.
- *6G vision.* To meet the increasing demands of data-intensive applications, 6G is conceptualized to offer even higher data transfer speeds—possibly up to terabits per second—and increased capacity [12]. 6G has the potential to transform fields like holographic communication, sophisticated artificial intelligence, and immersive extended reality experiences, in addition to providing higher download speeds [22]. To be more intelligent, 6G will use AI and machine learning algorithms to optimize network resources dynamically [6]. In line with international initiatives toward sustainable technology development, there is a focus on improving the energy efficiency of 6G networks [32].

  6G is anticipated to provide synergy for innovative applications by integrating smoothly with cutting-edge technologies, including edge computing, blockchain, and sophisticated artificial intelligence [14]. Collaboration between different fields, including telecommunications, computing, and engineering, is encouraged by the multidisciplinary character of 6G development [21].

1.5 CONCLUSION

In conclusion, the chapter titled "Cyber-Physical Systems 2.0: Evolution, Innovations, and Challenges" provides a comprehensive overview of the evolving landscape of cyber-physical systems (CPS) 2.0. As technological advancements continuously reshape the integration of physical processes with computing systems, CPS 2.0 emerges as a paradigm that transcends its predecessor, incorporating more intricate interactions, enhanced intelligence, and unprecedented connectivity. This concluding reflection encapsulates the key themes discussed in the chapter, shedding light on the evolutionary trajectory, foundational innovations, and challenges that lie ahead.

The evolution of CPS into its 2.0 phase is marked by a shift toward more complex and interconnected systems. The section elucidates how CPS 2.0 surpasses traditional boundaries, emphasizing a deeper integration of digital and physical components. This evolution is propelled by the relentless pace of technological advancement, with progress in communication networks, sensor technologies, and artificial intelligence playing pivotal roles. The integration of 5G and the upcoming 6G vision is underscored as crucial catalysts, providing the groundwork for enhanced data transfer speeds, reduced latency, and the ability to support a high density of connected devices. Technological aspects such as the IoT, edge computing, and AI are intricately woven into the fabric of CPS 2.0. The chapter delves into the role of these technologies in shaping the intelligence and capabilities of CPS 2.0 applications. The concept of edge computing, in particular, is emphasized as a vital enabler for real-time processing and navigation, ensuring that CPS 2.0 applications can respond promptly to dynamic and unpredictable conditions.

While the prospects of CPS 2.0 are promising, the chapter also sheds light on the challenges accompanying this evolution. The complexities of managing vast amounts of data, ensuring the security and privacy of interconnected systems, and addressing the ethical considerations in deploying intelligent CPS applications are all highlighted as critical challenges. Additionally, the chapter underscores the importance of standardization and interoperability to facilitate the seamless integration of diverse CPS components. In summary, the chapter provides a holistic analysis of CPS 2.0, elucidating its evolution, the technological underpinnings, and the challenges that stakeholders must grapple with. As CPS 2.0 continues to evolve, the insights presented in this introductory chapter lay the groundwork for deeper exploration and understanding of this transformative paradigm. The integration of digital and physical elements in CPS 2.0 paves the way for a future where intelligent, adaptive, and interconnected systems play a central role in shaping various domains, from smart cities to health care and beyond.

REFERENCES

[1] Lee, J., Bagheri, B., & Jin, C. (2016). Introduction to cyber manufacturing. *Manufacturing Letters*, *8*, 11–15.

[2] Lun, Y. Z., D'Innocenzo, A., Smarra, F., Malavolta, I., & Di Benedetto, M. D. (2019). State of the art of cyber-physical systems security: An automatic control perspective. *Journal of Systems and Software*, *149*, 174–216.

[3] Marwedel, P., Mitra, T., Grimheden, M. E., & Andrade, H. A. (2020). Survey on education for cyber-physical systems. *IEEE Design & Test*, *37*(6), 56–70.

[4] Pishdad-Bozorgi, P., Gao, X., & Shelden, D. R. (2020). Introduction to cyber-physical systems in the built environment. In *Construction 4.0* (pp. 23–41). Routledge.

[5] Plakhotnikov, D. P., & Kotova, E. E. (2021). Design and analysis of cyber-physical systems. In *2021 IEEE Conference of Russian Young Researchers in Electrical and Electronic Engineering (ElConRus)* (pp. 589–593). IEEE.

[6] Raimi, L., Kah, J. M., & Tariq, M. U. (2022). The discourse of blue economy definitions, measurements, and theories: Implications for strengthening academic research and industry practice. In L. Raimi & J. Kah (Eds.), *Implications for Entrepreneurship and Enterprise Development in the Blue Economy* (pp. 1–17). IGI Global. https://doi.org/10.4018/978-1-6684-3393-5.ch001

[7] Chakraborty, S., Al Faruque, M. A., Chang, W., Goswami, D., Wolf, M., & Zhu, Q. (2016). Automotive cyber–physical systems: A tutorial introduction. *IEEE Design & Test*, *33*(4), 92–108.

[8] Raimi, L., Tariq, M. U., & Kah, J. M. (2022). Diversity, equity, and inclusion as the future workplace ethics: Theoretical review. In L. Raimi & J. Kah (Eds.), *Mainstreaming Diversity, Equity, and Inclusion as Future Workplace Ethics* (pp. 1–27). IGI Global. https://doi.org/10.4018/978-1-6684-3657-8.ch001

[9] Repetto, M., Colapinto, C., & Tariq, M. U. (2024). Artificial intelligence driven demand forecasting: An application to the electricity market. *Annals of Operations Research*. https://doi.org/10.1007/s10479-024-05965-y

[10] Tariq, M. U., & Rommel, P. S. (2024). *Emerging Innovation: Business Transformation in the New Normal, 111 Compact Case Studies*. Notion Press.

[11] Tariq, M. U. (2024). The transformation of healthcare through AI-driven diagnostics. In A. Sharma, N. Chanderwal, S. Tyagi, P. Upadhyay, & A. Tyagi (Eds.), *Enhancing Medical Imaging with Emerging Technologies* (pp. 250–264). IGI Global. https://doi.org/10.4018/979-8-3693-5261-8.ch015

[12] Ashfaq, M., Khan, I., Alzahrani, A., Tariq, M. U., Khan, H., & Ghani, A. (2024). Accurate wheat yield prediction using machine learning and climate-NDVI data fusion. *IEEE Access*, *12*, 40947–40961.

[13] Tariq, M. U. (2024). The role of emerging technologies in shaping the global digital government landscape. In Y. Guo (Ed.), *Emerging Developments and Technologies in Digital Government* (pp. 160–180). IGI Global. https://doi.org/10.4018/979-8-3693-2363-2.ch009

[14] Tariq, M. U. (2024). Equity and inclusion in learning ecosystems. In F. Al Husseiny & A. Munna (Eds.), *Preparing Students for the Future Educational Paradigm* (pp. 155–176). IGI Global. https://doi.org/10.4018/979-8-3693-1536-1.ch007

[15] Tariq, M. U. (2024). Revolutionizing health data management with blockchain technology: Enhancing security and efficiency in a digital era. In M. Garcia & R. de Almeida (Eds.), *Emerging Technologies for Health Literacy and Medical Practice* (pp. 153–175). IGI Global. https://doi.org/10.4018/979-8-3693-1214-8.ch008

[16] Tariq, M. U. (2024). Emerging trends and innovations in blockchain-digital twin integration for green investments: A case study perspective. In S. Jafar, R. Rodriguez, H. Kannan, S. Akhtar, & P. Plugmann (Eds.), *Harnessing Blockchain-Digital Twin Fusion for Sustainable Investments* (pp. 148–175). IGI Global. https://doi.org/10.4018/979-8-3693-1878-2.ch007

[17] Esterle, L., & Grosu, R. (2016). Cyber-physical systems: Challenge of the 21st century. *e & i Elektrotechnik und Informationstechnik*, *133*, 299–303. https://doi.org/10.1007/s00502-016-0426-6

[18] Kaleem, S., Sohail, A., Babar, M., Ahmad, A., & Tariq, M. U. (2024). A hybrid model for energy-efficient green Internet of Things enabled intelligent transportation systems using federated learning. *Internet of Things*, *25*, 101038.

[19] Kunze, H., La Torre, D., Riccoboni, A., & Galán, M. R. (Eds.). (2023). *Engineering Mathematics and Artificial Intelligence: Foundations, Methods, and Applications*. CRC Press.

[20] Tariq, M. U. (2024). Emotional intelligence in understanding and influencing consumer behavior. In T. Musiolik, R. Rodriguez, & H. Kannan (Eds.), *AI Impacts in Digital Consumer Behavior* (pp. 56–81). IGI Global. https://doi.org/10.4018/979-8-3693-1918-5.ch003

[21] Tariq, M. U. (2024). Fintech startups and cryptocurrency in business: Revolutionizing entrepreneurship. In K. Kankaew, P. Nakpathom, A. Chnitphattana, K. Pitchayadejanant, & S. Kunnapapdeelert (Eds.), *Applying Business Intelligence and Innovation to Entrepreneurship* (pp. 106–124). IGI Global. https://doi.org/10.4018/979-8-3693-1846-1.ch006

[22] Humayed, A., Lin, J., Li, F., & Luo, B. (2017). Cyber-physical systems security—A survey. *IEEE Internet of Things Journal*, *4*(6), 1802–1831.

[23] Tariq, M. U. (2024). Multidisciplinary service learning in higher education: Concepts, implementation, and impact. In S. Watson (Ed.), *Applications of Service Learning in Higher Education* (pp. 1–19). IGI Global. https://doi.org/10.4018/979-8-3693-2133-1.ch001

[24] Tariq, M. U. (2024). Enhancing cybersecurity protocols in modern healthcare systems: Strategies and best practices. In M. Garcia & R. de Almeida (Eds.), *Transformative Approaches to Patient Literacy and Healthcare Innovation* (pp. 223–241). IGI Global. https://doi.org/10.4018/979-8-3693-3661-8.ch011

[25] Jantunen, E., Zurutuza, U., Albano, M., di Orio, G., Maló, P., & Hegedus, C. (2017). The way cyber physical systems will revolutionise maintenance. In *30th Conference on Condition Monitoring and Diagnostic Engineering Management*.

[26] Tariq, M. U. (2024). Advanced wearable medical devices and their role in transformative remote health monitoring. In M. Garcia & R. de Almeida (Eds.), *Transformative Approaches to Patient Literacy and Healthcare Innovation* (pp. 308–326). IGI Global. https://doi.org/10.4018/979-8-3693-3661-8.ch015

[27] Tariq, M. U. (2024). Leveraging artificial intelligence for a sustainable and climate-neutral economy in Asia. In P. Ordóñez de Pablos, M. Almunawar, & M. Anshari (Eds.), *Strengthening Sustainable Digitalization of Asian Economy and Society* (pp. 1–21). IGI Global. https://doi.org/10.4018/979-8-3693-1942-0.ch001

[28] Tariq, M. U. (2024). Metaverse in business and commerce. In J. Kumar, M. Arora, & G. Erkol Bayram (Eds.), *Exploring the Use of Metaverse in Business and Education* (pp. 47–72). IGI Global. https://doi.org/10.4018/979-8-3693-5868-9.ch004

[29] Tariq, M. U., Abonamah, A., & Poulin, M. (2023). Artificial intelligence technologies and platforms. In *Engineering Mathematics and Artificial Intelligence* (pp. 211–226). CRC Press.

[30] Emmanouilidis, C., Pistofidis, P., Bertoncelj, L., Katsouros, V., Fournaris, A., Koulamas, C., & Ruiz-Carcel, C. (2019). Enabling the human in the loop: Linked data and knowledge in industrial cyber-physical systems. *Annual Reviews in Control*, *47*, 249–265.

[31] Tariq, M. U. (2023). Future health care and medical entrepreneurship in the age of pandemic. In *Medical Entrepreneurship: Trends and Prospects in the Digital Age* (pp. 133–149). Springer Nature.

[32] Huang, K., Zhou, C., Tian, Y. C., Yang, S., & Qin, Y. (2018). Assessing the physical impact of cyberattacks on industrial cyber-physical systems. *IEEE Transactions on Industrial Electronics*, *65*(10), 8153–8162.
[33] Tariq, M. U. (2023). Healthcare innovation & entrepreneurship, digital health entrepreneurship. In *Medical Entrepreneurship: Trends and Prospects in the Digital Age* (pp. 243–258). Springer Nature.
[34] Wolf, M., & Serpanos, D. (2017). Safety and security in cyber-physical systems and internet-of-things systems. *Proceedings of the IEEE*, *106*(1), 9–20.
[35] Wurm, J., Jin, Y., Liu, Y., Hu, S., Heffner, K., Rahman, F., & Tehranipoor, M. (2016). Introduction to cyber-physical system security: A cross-layer perspective. *IEEE Transactions on Multi-Scale Computing Systems*, *3*(3), 215–227.
[36] Zanero, S. (2017). Cyber-physical systems. *Computer*, *50*(4), 14–16.
[37] Zhang, C., Xu, X., & Chen, H. (2020). Theoretical foundations and applications of cyber-physical systems: A literature review. *Library Hi Tech*, *38*(1), 95–104.
[38] Monostori, L., Kádár, B., Bauernhansl, T., Kondoh, S., Kumara, S., Reinhart, G., . . . Ueda, K. (2016). Cyber-physical systems in manufacturing. *Cirp Annals*, *65*(2), 621–641.
[39] Möller, D. P., & Möller, D. P. (2016). Introduction to cyber-physical systems. In *Guide to Computing Fundamentals in Cyber-Physical Systems: Concepts, Design Methods, and Applications* (pp. 81–139). https://doi.org/10.1007/978-3-319-25178-3
[40] Kim, S., Park, K. J., & Lu, C. (2022). A survey on network security for cyber–physical systems: From threats to resilient design. *IEEE Communications Surveys & Tutorials*, *24*(3), 1534–1573.
[41] Tariq, M. U. (2024). Empowering educators in the learning ecosystem. In F. Al Husseiny & A. Munna (Eds.), *Preparing Students for the Future Educational Paradigm* (pp. 232–255). IGI Global. https://doi.org/10.4018/979-8-3693-1536-1.ch010
[42] Zhou, J., Zhou, Y., Wang, B., & Zang, J. (2019). Human–cyber–physical systems (HCPSs) in the context of new-generation intelligent manufacturing. *Engineering*, *5*(4), 624–636.
[43] Letichevsky, A. A., Letychevskyi, O. O., Skobelev, V. G., & Volkov, V. A. (2017). Cyber-physical systems. *Cybernetics and Systems Analysis*, *53*, 821–834.

Chapter 2

Communication and computing co-design under imperfect channel conditions for control and scheduling in CPS 2.0

Yifei Qiu, Shaohua Wu, Ying Wang, and Qinyu Zhang

LIST OF ABBREVIATIONS

AoI	age of information
AoII	age of incorrect information
CPS	cyber-physical systems
DDPG	deep deterministic policy gradient
DDQN	double deep Q-network
DQN	deep Q-network
D3QN	dueling double deep Q-network
DRL	deep reinforcement learning
ISA	International Society of Automation
IIoT	industrial Internet of Things
LTI	linear time-invariant
MDP	Markov decision process
mMTC	massive machine-type communication
MSE	mean squared error
PPO	proximal policy optimization
RL	reinforcement learning
SAC	soft actor-critic
SAGIN	space-air-ground integrated network
SMDP	semi-Markov decision process
TD3	twin delayed deep deterministic policy gradient
URLLC	ultra-reliable low-latency communication
VoI	value of information
WAIC	Wireless Avionics Intra-Communications

2.1 INTRODUCTION

In today's era, traditional sensors are rapidly evolving in the direction of wireless connectivity, versatility, and miniaturization, giving rise to the widely used intelligent sensors of the present. The market size of smart sensors is gradually expanding, and it is expected to reach 48.72 billion US dollars for the entire year of 2023. They find extensive applications in both industrial and consumer markets to collect and process required physical parameters.

DOI: 10.1201/9781003559993-2

With the improvement of communication infrastructure, equipment such as base stations, satellites, and drones can now serve communication, greatly expanding the range of communication. The concept of 5G was introduced in 2018. It was extensively deployed after confirming protocol details, making large-scale use of 5G networks possible. Thanks to the development of these two technologies, the Internet of Things (IoT) continues to evolve, and numerous applications are gradually appearing around us. In smart homes, people can instantly check the status of their homes and take corresponding control actions using their smartphones. In the field of smart driving, drivers can have a more comprehensive understanding of road conditions, leading to safer driving. In smart cities, people can gather information about various aspects of the city, facilitating allocating city resources and making urban operations smoother.

Beyond everyday applications, the combination of sensors and wireless communication has also made significant contributions in the industrial sector, giving rise to the concept of the industrial Internet of Things (IIoT) [1]. IIoT has attracted considerable academic attention due to its involvement in industrial production. Broadly, IIoT can be divided into four main aspects: perception, transmission, computation, and control. *Perception* involves the acquisition of physical parameters related to the monitored processes through sensors; *transmission* pertains to communication between various components in IIoT; *computation* summarizes all data processing; and *control* represents the ultimate goal of IIoT, which involves adjusting processes based on available information. Considering the coupling relationships among these four aspects, studying them as a whole at the system level is necessary, rather than focusing on individual aspects separately.

Around 2010, the concept of cyber-physical systems (CPS), which integrate computing and communication technologies with physical components to enhance human–network interaction, was introduced. CPS begins from the physical layer, emphasizing the collaboration of physical components within the network to achieve intelligent and efficient monitoring. Over the years, much research has been conducted on CPS. The notion of CPS 2.0 is progressively gaining traction within the academic community. This evolved version of the original CPS concept integrates cutting-edge technologies, including artificial intelligence, 5G, IoT, and wireless sensor networks (WSN). These technologies are propelling key functionalities—autonomous driving, traffic system management, and energy conservation—toward a future of improved decision-making capabilities and self-sufficient operations. This section primarily explores the design of network control systems in CPS from a communication perspective. The system diagram is depicted in Figure 2.1, where various devices in IIoT are classified as sensors, controllers, actuators, and controlled processes. Controllers, actuators, and the controlled processes are geographically close, while controllers

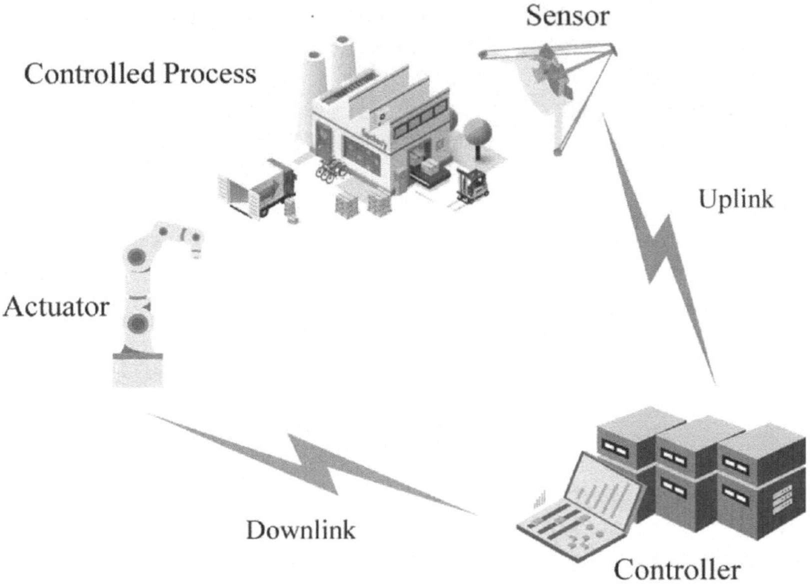

Figure 2.1 The system diagram of a remote-controlled CPS.

are at a distance. Sensors collect the state of the processes and send it to the controllers, which, after processing, generate control signals sent to the actuators to intervene in the processes, completing the entire closed-loop control. Communication mainly occurs in the links from sensors to controllers and controllers to actuators.

Unlike everyday life applications, the primary purpose of IIoT is to improve manufacturing processes and enhance industry. It places high demands on transmission stability and end-to-end latency. For example, haptic Internet and industrial automation require end-to-end latency within 1 ms and a packet loss rate lower than 10^{-5} [2]. Smart grids, on the other hand, require end-to-end latency within 1 s and a packet loss rate of 10^{-9} [3]. These requirements represent the shortcomings of wireless communication compared to wired communication. Constrained by the requirements of these indicators, wired communication dominated the industrial sector in the early years. However, with the widespread adoption of 5G communication and the introduction of concepts such as massive machine-type communication (mMTC) and ultra-reliable low-latency communication (URLLC), explicit specifications for the number of devices connected in the network, transmission delay, and transmission reliability have been established, ensuring the minimum requirements for communication in industrial production. Wireless communication offers advantages, such as lower cost, ease of deployment, and ease of maintenance, compared to wired communication. With the advancement of communication technology, wireless communication has gradually started to meet the performance requirements of industrial production, partially replacing wired communication.

With the use of wireless communication in CPS shown in Figure 2.1, wireless remote control of CPS has become more widely used and mentioned in the Industry 4.0 plan [4]. International organizations such as the Wireless Avionics Intra-Communications (WAIC) Alliance, the ZigBee Alliance, the Z-wave Alliance, and the International Society of Automation (ISA) [5] have also provided support for wireless remote-controlled CPS.

In the early stages of remote-controlled CPS research, there was a distinct focus on communication and computation as separate aspects. In communication research, emphasis was placed on communication metrics, such as transmission throughput and transmission rate, without considering the state of the controlled processes. This approach may lead to instability in a single process. On the other hand, research in computation primarily concentrated on the state of the controlled processes, employing techniques like filtering and fuzzy control to achieve more precise control. However, uplink and downlink transmission latencies were still subject to the limitations of wireless communication, hindering further performance improvements. In most previous studies, the design process started from the control system, defining the minimum communication requirements based on specified control performance, such as communication rate, latency, and reliability. Subsequently, the communication system was designed based on these criteria to obtain a fully functional wireless control system. With the widespread application of numerous new technologies in CPS 2.0, there is a consensus in the academic community regarding the joint consideration of communication and control.

Considering both communication and computation is a feasible approach to enhance system performance with the improvement of computational capabilities. When communication parameters are also considered in computation, although it makes the computation more complex, it also raises the upper limit of system performance. Under the influence of the collaborative control concept, communication and computation systems are treated as an integrated whole in designing scheduling schemes for CPS 2.0, leading to better system performance compared to independent design, albeit at the cost of increased design complexity. In collaborative design schemes, the focus is primarily on how key system parameters affect communication and computation, and appropriate parameters are selected after balancing the performance of both aspects [6]. This concept has led to more research directions in the CPS 2.0 field. The scheduling mechanism introduced in this section is one of the important directions, as it can enhance system performance without altering the system's structure.

2.2 KEY PARAMETERS OF A WIRELESS REMOTE CPS

This section primarily delves into the roles of various system structures, the key metrics considered in current research, and the evaluation criteria within joint optimization problems. This section initially introduces the system structures in the context of collaborative CPS design. Following this introduction, the main metrics considered in joint design will be discussed. Exploration of how these metrics influence the system's design and ultimate performance will also be conducted. Lastly, the section introduces some metrics for measuring remote-controlled CPS.

Generally speaking, in CPS, under joint design, there are four main components, namely, sensors, controllers, actuators, and controlled processes. The *controlled processes* refer to the physical processes of interest, which could be the assembly line in a factory, the temperature in a boiler, the humidity in a grain silo, and so on. In discrete-time systems, time is divided into equally spaced slots. The selection of slot length typically depends on industrial production requirements. In today's industrial applications, the control frequency mostly falls between 20 Hz and 50 Hz, so the slot length typically ranges from 20 ms to 50 ms. The state, control actions, and random noise at the current slot collectively determine the system's state at the next moment. The mathematical model for state transition can be described as either linear or nonlinear. In most scenarios, the system can be represented as a linear time-invariant (LTI) system. The system's objective is primarily related to the state of the physical process. When the state deviates from the desired state, corresponding control actions are required to reduce the error between the current and the desired states.

Sensors are mainly responsible for periodically or aperiodically collecting the state of the controlled process and then packaging it for transmission through the uplink channel to the controller. In *uplink transmission*, it is essential to consider the delays and packet loss that may occur in wireless communication, as both can significantly impact the final results. Once the controller receives the state data packet, it calculates the corresponding control instructions and sends the results to the actuators through the downlink channel. When the actuators successfully receive the control instructions, they execute the corresponding actions, enabling control of the process and completing the closed-loop control of the system.

In remote-controlled CPS, the settings of many parameters can significantly impact the ultimate control performance. Current scheduling research primarily focuses on adjusting parameters through resource allocation to ultimately improve system performance [7]. This section will introduce the important parameters in remote-controlled CPS in the order of closed-loop control discussed earlier. Diagrams illustrating these parameters are also provided in Figure 2.2 for readers' better comprehension.

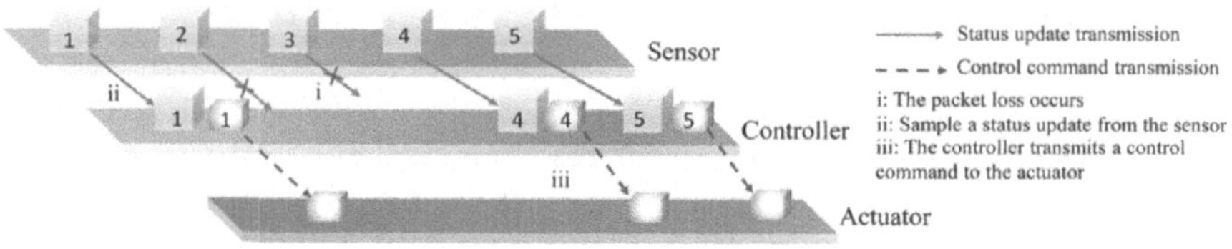

Figure 2.2 Communication processes in remote-controlled CPS.

As depicted in Figure 2.2, the wireless transmission for remote control of CPS is primarily divided into uplink and downlink transmission. In the uplink transmission, sensors transmit status data packets to the controller, represented by the larger cube in Figure 2.2. Subsequently, the controller generates control commands based on the received status and transmits them to the actuator, depicted as the smaller cube. Both uplink and downlink transmissions are subject to transmission delay and transmission failures. These specific parameters are discussed as follows:

- *Sampling interval.* In real-time control applications, accurate and timely state data packets aid the controller in generating precise control signals, forming the basis for enhancing system control effectiveness. Therefore, sensors must continuously sample the controlled process to allow the controller to estimate the system's state as accurately as possible. A sampling interval that is too lengthy prevents the controller from accurately understanding the state of the controlled process, resulting in suboptimal control instructions. However, resources like channel bandwidth and sensor transmission power are limited in communication. Therefore, indiscriminately increasing the sensor's sampling frequency is unrealistic and may lead to network congestion, data packet collisions, and a decline in system performance. Hence, for periodic sampling, the selection of the sampling interval is a trade-off between wireless transmission stability and timely state updates. Regarding event-triggered sampling strategies, the sampling interval depends on the state of the controlled process. When the controlled process significantly deviates from the expected trigger conditions, the sensor samples and sends state data packets to the controller. This is one of the sampling strategies explored in current research and can help conserve sensor energy. However, it still faces the issue of frequent sampling during sudden events.
- *Uplink and downlink transmission delay.* The numerical values of transmission latency significantly impact the generation of control instructions and have an important influence on the computation process at the controller's end. Sensors send state data packets to the

controller through the uplink channel to generate corresponding control instructions. After generating control instructions, they are transmitted through the downlink channel to be executed by the actuators, ultimately achieving control. When the uplink latency is minimal, the observed state values received by the controller are closer to the true values, which is more valuable for reference. Similarly, a shorter downlink latency results in the actuators' control being closer to the controller's predictions. From a communication perspective, *latency* includes channel access delay, transmission delay, propagation delay, queuing delay, and processing delay, influenced by factors such as state data packet length, bit rate, and scheduling policies [8].

- *Uplink packet loss*. Packet loss has a significant impact on system performance. When uplink transmission fails, the sensor needs to retransmit or send new state data, which respectively increases the uplink latency by an integer multiple of the sampling period. In practical applications, the uplink packet loss rate is usually influenced by packet length, bit rate, sensor power, channel fading, multipath effects, and transmission distance [9].
- *Downlink packet loss*. The remote control of CPS operates within a closed loop, where the transmission of status is coupled with the generation of control commands. The consequences of packet loss in the downlink are more severe than in the uplink. If the actuator fails to receive the control command successfully within the expected time frame, the control command becomes ineffective. Considering the computational delay of the controller, in the event of a downlink transmission failure, the closed-loop time will be increased by at least the sum of the downlink delay and the computation delay.

It is evident that various parameters within the system are often interconnected. Changes in communication parameters affect the computational processes. For instance, altering the uplink and downlink delays can impact predictive algorithms during control command generation. Likewise, the computational processes affect communication; for instance, computation time influences scheduling design and can alter the transmission timing of control commands or state data packets. This necessitates the consideration of system objectives while designing scheduling policy and achieving tradeoffs among various parameters.

Different application scenarios have different goals, with some focusing on control accuracy and others emphasizing control timeliness. Therefore, when establishing optimization problems for the researched systems, appropriate metrics must be selected to describe the degree of deviation between the system and its objectives. In other words, goal-oriented metrics must be created for the system design.

For specific scenarios, the timeliness of information plays a crucial role. For time-sensitive tasks, fresher data packets imply greater significance for the system, while older data packets have less significance. Take smart transportation, for instance, where data packets contain information like vehicle positions and traffic flow, which exhibit high rates of change. The time window for these data packets to be useful is limited. This means that data packets generated in such scenarios must be transmitted to the receiving end as quickly as possible. The longer the time since data packet generation, the less valuable the data packet becomes.

From this perspective, in real-time control systems, there is a need for a metric to describe the freshness of information. In this regard, Kaul and others, in 2012, measured the system's timeliness and introduced the age of information (AoI) metric [10]. Initially used in point-to-point systems, *AoI* is defined as the time elapsed at the receiving end since the last successful update. Although it appears to be a simple definition, it provides a precise quantitative measure of the freshness of information within the system. Figure 2.3 illustrates the specific concept of AoI.

Considering the uplink transmission of CPS in conjunction with Figure 2.3, S represents the time at which the controlled process generates a status packet to be sent, and R represents the time at which the controller receives the state data packet. The AoI experiences a decrease when the controller receives the state data packet with a value of $R_i - S_i$. W denotes the waiting time, and reducing this can increase the information transmission frequency. However, it also introduces challenges, such as increased power consumption and a higher probability of packet collisions within the system.

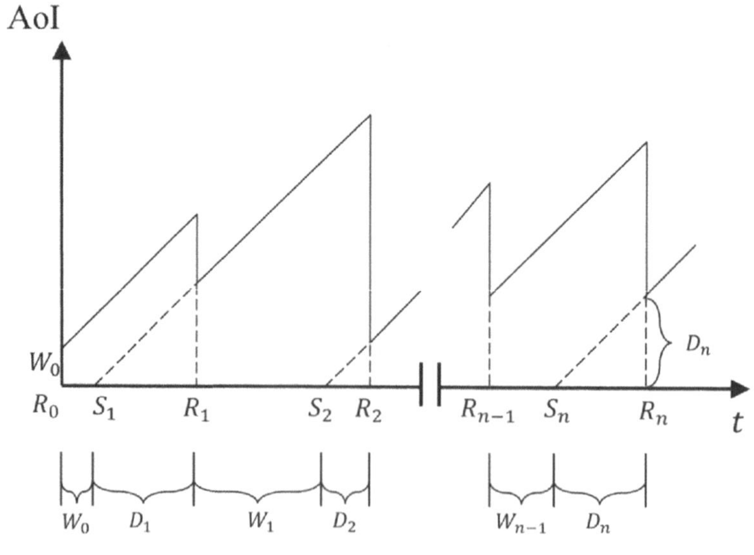

Figure 2.3 Schematic of the evolution.

Due to its concise yet effective characteristics, AoI has attracted the attention of numerous researchers. The concept of AoI is continuously expanding, and its applicability is becoming broader. Numerous studies have applied the concept of AoI to the research of remote-controlled CPS. In contrast to point-to-point observation systems, the goal of remote-controlled CPS is to bring the controlled process as close as possible to the expected state, adding downlink transmission and forming a closed-loop system [11]. The upper limit of control accuracy depends on the accuracy of the collected state. Even with predictive algorithms, estimated values gradually deviate from the true state as noise accumulates. Therefore, the starting time of AoI for remote-controlled CPS should be defined as the moment when the sensor collects the state [12, 13].

When the sensor collects the state and generates a data packet, it includes a timestamp of the sampling time. This timestamp remains with the state data packet until it reaches the controller and generates the corresponding control command. Finally, when the control command is successfully executed by the actuator, the system obtains the timestamp of the most recent successful control. Based on this, the AoI for remote-controlled CPS should be defined as the difference between the current time and the timestamp of the most recent successful control.

For the control loop as shown in Figure 2.1, the AoI of the controlled process at time t can be written as $t-\tau(t)$, where $\tau(t)$ is the timestamp of the most recent state collection, as illustrated in Figure 2.4.

AoI does not seem to differ significantly from the uplink and downlink delays in CPS. However, AoI considers factors beyond the delays, such as

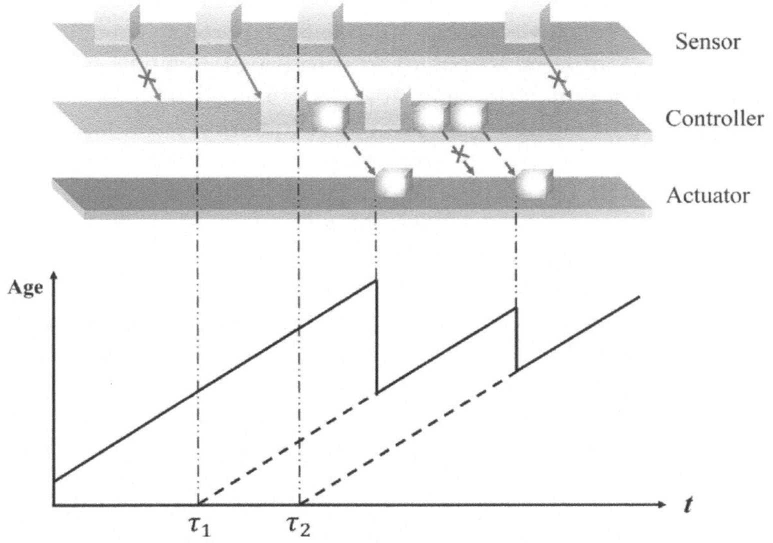

Figure 2.4 AoI evolution for remote-controlled CPS.

transmission losses and out-of-order packet arrivals. Its significance extends far beyond the uplink and downlink delays. At the same time, the uplink and downlink delays also determine the lower limit of AoI. When there are no losses in the uplink and downlink and successful control is achieved, the AoI is simply the sum of the uplink and downlink delays.

The simplicity of AoI is a significant reason for its widespread application, as researchers can conveniently apply it to their respective fields. However, AoI may appear too simplistic in specific scenarios and fail to perfectly describe the errors in these contexts. Consequently, many researchers have conducted studies or modifications based on AoI and proposed other excellent metrics.

Value of information (VoI) is one of the earliest AoI-based metrics, which considers the system's tolerance for different ages [14]. In VoI, an age penalty function is defined, which can be a linear, exponential, or logarithmic function. The VoI value is the product of AoI and the age penalty function. In practical applications, VoI has a broader scope because it allows the design of age penalty functions based on different contexts, not limited to the linear penalty of AoI.

AoI and VoI represent the freshness of wireless remote control in CPS, indicating whether timely control is achievable. However, as a content-agnostic metric, AoI cannot precisely describe the effectiveness of control. Only a concept that larger AoI leads to less accurate control is conveyed without a numerical description. Therefore, many current studies use metrics such as mean squared error (MSE) and age of incorrect information (AoII) [15] to measure the precision of control.

MSE reflects the precision of system control by calculating the deviation between the true value of the controlled process state and the expected target. When the mathematical model of the controlled process in the remote-controlled system is an LTI system, the system's MSE can be determined by the value of AoI [16]. Calculating system MSE using AoI significantly reduces simulation complexity. In Monte Carlo simulations, calculating AoI at each time step is simpler than computing errors at every moment and averaging them later.

AoII, in contrast to AoI, considers whether estimation errors have occurred. Its value equals the duration in which the controlled process remains in an incorrect state. AoII starts counting when the controlled process deviates from the target by more than a threshold, and it remains at 0 when the process state does not exceed the threshold. In practical applications, AoII can be combined with event-triggered control. It saves communication resources by not updating when the controlled process is in a normal state, and the longer the system stays in an abnormal state, the more significant the impact. In the field of remote control, AoII is mainly used in Markov processes and has a more limited application compared to MSE.

AoI is a crucial metric interconnecting the scheduling design of remote-controlled CPS. It comprehensively considers various indicators in an imperfect channel, including uplink and downlink transmission delays, packet loss, and sampling intervals, all while maintaining simplicity. It can be easily accessed by various parts of the system, and its value is reset when updates occur. When no updates are available, AoI linearly increases over time, providing decision-making information for existing scheduling schemes. Furthermore, AoI possesses excellent scalability, enabling redesign for specific scenarios. In cases where the controller has global knowledge, it greatly aids in scheduling design optimization through optimizing MSE or AoII. Even in situations where the controller has partial knowledge, it can be designed according to objectives through VoI.

2.3 SCHEDULING POLICY DESIGN METHODOLOGY

The system model shown in Figure 2.1 illustrates the basic process of single-process remote control, including the characteristics of wireless remote control. However, in practical applications, a system often consists of multiple controlled processes, making the analysis more complex than that of a single-process system [13]. Therefore, the considered system, as shown in Figure 2.5, corresponds to a controller and actuator for each controlled process. Each controlled process represents the same or different physical quantities, with differences in data size and rate of change depending on the monitored physical processes. For example, the data size for ship location

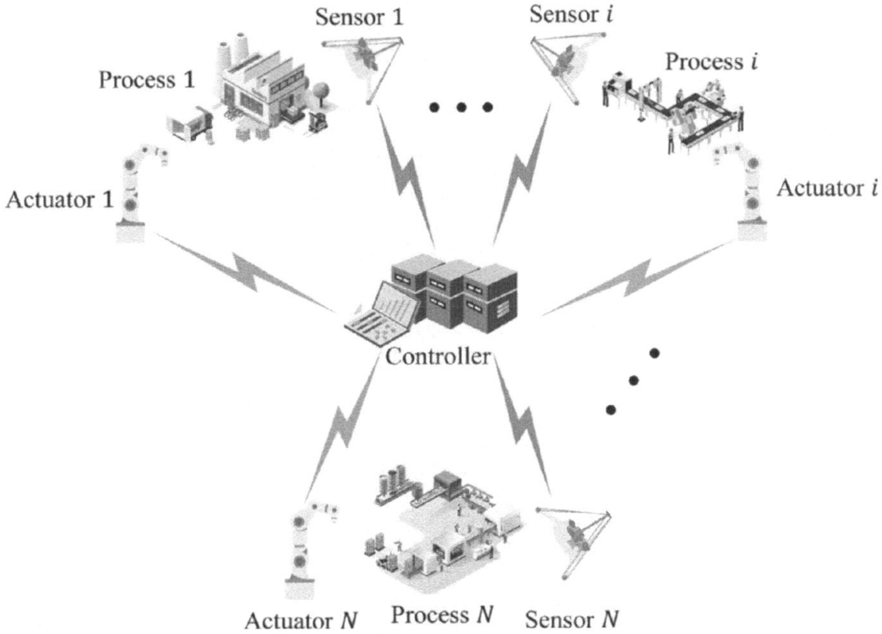

Figure 2.5 More common multi-loop remote-controlled CPS.

information (two-dimensional) and drone location information (three-dimensional) varies.

During wireless channel data transmission, the system requires available scheduling schemes. For example, there is a limited number of orthogonal channels, and too many sensors transmitting simultaneously can result in collisions of state data packets, leading to transmission failures. The controller side is a full-duplex device, capable of receiving or transmitting in the same time slot. Therefore, only partial data can be transmitted in discrete-time systems in each time slot. Thus, a scheduling strategy must be designed to ensure that each state in the system does not deviate from the expected target as much as possible.

As mentioned earlier, the objectives in different scenarios are not the same. Similarly, the wireless channel models in different scenarios are also different. Therefore, when designing scheduling policies, it is necessary to determine the channel model for the application scope and information such as wireless transmission distance, as these parameters greatly affect the probability of successful wireless channel transmission [17]. Common scenarios of wireless channel fading are summarized in Table 2.1.

Table 2.1 Common scenarios of wireless channel fading

Channel model	*Features*	*Typical scenario*
Rayleigh channel	Used to describe fast-fading channels, considering signal propagation through multiple paths.	Mobile communication, urban environments, and wireless local area networks
Rician channel	Takes into account the scenario in signal propagation, where there is a dominant path and multiple multipath scattering paths.	Satellite communication, suburban wireless communication, and elevated road communication
Clarke model	Represents the relationship between path loss, distance, and frequency. It is used to estimate the extent to which signal strength decreases as the distance increases.	Microwave communication and cellular mobile communication
Log-normal fading model	Used to describe the probability distribution of channel fading. It is commonly used to describe the variation in channel amplitude.	Sensor networks and urban communication
Rician log-normal fading model	Combining the characteristics of the Rician distribution and the logarithmic normal distribution.	Indoor communication and wireless local area networks (Wi-Fi)

In the joint design that comprehensively considers communication and computation, computation is also an essential component. The primary computational processes in the controller of remote-controlled CPS mainly involve two processes: state estimation and command generation.

The controlled processes are continually changing and influenced by noise, and the introduction of noise in sensor sampling affects the estimates in the controller. Furthermore, due to the introduction of wireless communication in remote-controlled CPS, data transmission is inevitably affected by noise and data packet collisions. Significant deviations may occur when there is a long duration without receiving state updates. Therefore, the first priority in computation is to achieve precision control, which is a prerequisite for generating accurate control commands.

In the remote control of CPS, Kalman filtering is commonly used for process estimation. This algorithm continuously updates estimates based on observed values and performs well in linear processes with perturbation Gaussian noise [18].

The control commands ultimately applied to the process directly impact the final system's performance. Currently, many control algorithms designed based on classical and modern control theories have been widely researched and adopted. Common control algorithms are summarized in Table 2.2 to cover most remote-controlled CPS scenarios.

Table 2.2 Common control algorithms

Approach	*Advantages*	*Limits*
PID	A model-free and easy tuning approach composed of proportion term, integral item, and derivative item.	Difficult to adapt to complex systems and not self-adaptive.
DP	Ensure the optimality of the solution by solving the Bellman equation.	State transfer probabilities are difficult to obtain and require a large amount of computation.
LQR	Good performance for linear processes under the perturbation of additive Gaussian noise.	System global information is needed to know.
MPC	Predicting future responses through realistic processes has low computational complexity.	It is a model-based algorithm, and accurate models are difficult to obtain.
RL	Large amounts of data can be used to solve complex systems by constantly interacting with the environment to approximate the optimal policy.	Difficult to converge to the optimal policy; no guarantee for system stability.

For the sake of analysis, in most research, remote-controlled CPS is considered as a discrete-time system. The state of the controlled process depends on the state of the previous time slot, the executed control commands, and noise. It is memoryless and conforms to the definition of a Markov process, where the current state is only related to the previous state. In the case of a fully observable system state, the observation of states corresponds to a Markov chain, while the controlled system corresponds to a Markov decision process (MDP).

In contrast to the Markov chain commonly used for state analysis, the Markov decision process introduces the concept of actions. This means that the state at the next time slot is related to the current state and the actions taken. Different actions, when taken in a given state, result in different state transition probabilities. In the design of scheduling mechanisms, there is a choice and comparison among different actions, which aligns well with the Markov decision process model.

A typical MDP is represented as shown in Figure 2.6, where the large circle represents the state, the small black circle in the solid core represents the action taken, and p represents the transfer probability. For simplicity of representation, only the case with two actions is drawn.

In discrete-state systems, this tool provides a framework to characterize CPS. By treating aspects like AoI and transmission results as states, and actions like transmission scheduling as the independent variable, it enables a detailed analysis of system performance, paving the way for further optimization.

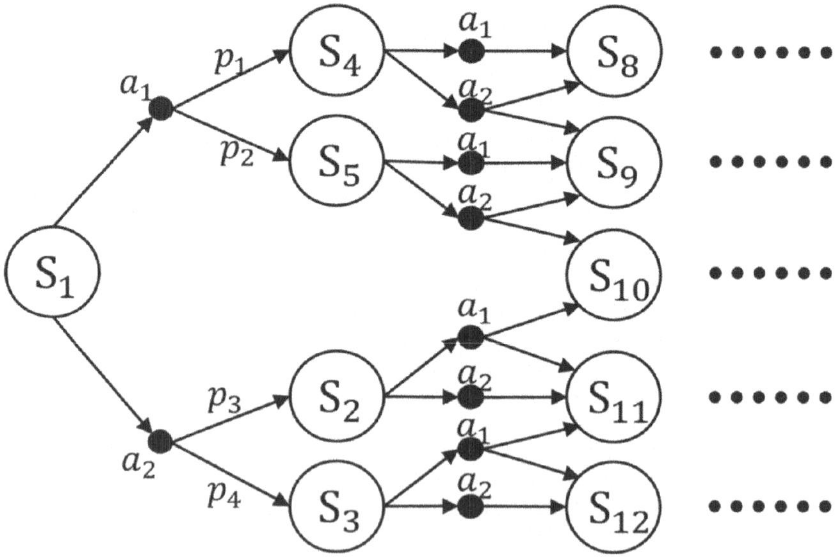

Figure 2.6 Schematic diagram of the markov decision-making process.

The Markov decision process is described by four variables (if described with five variables, it includes an additional discount factor γ) and can be represented as (S, A, P, R).

Where S represents the state space, which is defined the same as in a Markov chain (it is the set of all possible states s), and P represents the state transition matrix, with each element denoting the probability of state transitions. It is a matrix of size $|A| \times |S| \times |S|$, where $|A|$ and $|S|$ represent the number of elements in A and S, respectively. A stands for the action space, which is the set of all executable actions a. R represents the reward, jointly determined by the given state s and action a, denoted as $R(s,a)$.

In addition, there is a hyperparameter, the discount factor $\gamma \in [0,1)$, which measures the importance of current rewards versus future rewards, indicating the degree of emphasis on future rewards. When the value is closer to 1, it suggests a greater emphasis on future rewards, while values closer to 0 imply a greater emphasis on immediate rewards.

The aim of the remote-controlled CPS is to achieve timely and precise control through scheduling. In other words, when the controller observes a state s, it needs to make a scheduling decision a to maximize the long-term average reward. The scheduling policy is a mapping from the state space S to the action space A, denoted as $\pi : S \rightarrow A$. The design goal is to find the optimal scheduling policy π, which can maximize the long-term average reward of the system. After abstracting the studied scenario as an MDP, using various mature algorithms to approach the desired result becomes easier. Therefore, when global information is known, methods such as value iteration become an ultimate choice for scheduling.

The value iteration method is an iterative solution based on the value function, primarily using the Bellman equation for computation during the iterations. The value of each state and action is gradually stabilized by making the value function $V(s)$ and the reward function $R(s,a)$ iterative with each other. In turn, the optimal action for each state can be derived.

At the start of the algorithm, the values for each state of the process are initialized, with the value function for each state set to 0. In each iteration, the value function for each state is calculated using the Bellman equation, but the calculated value functions are not directly updated until a single cycle has computed the value functions for all states. Then, the value functions for all states are updated.

Through continuous iteration of the value functions, until the numerical values of the value functions for each state stabilize, the iteration is considered complete, and all value functions are obtained. Finally, using the Bellman equation, the optimal action for each state can be determined, leading to the optimal policy. In addition to the value iteration method, policy iteration is frequently used for solving problems. Like the value iteration method, policy iteration computes the optimal action for each state in every iteration, progressively converging toward the optimal policy.

The process begins with policy initialization, where random actions are assigned to each state. Then, it enters a loop, each cycle comprising two main steps. First, the Bellman equation is used to update the value function for each state. The second step is to update the optimal action for each state using the equation. By continuously looping through these two steps, the policy converges.

2.4 DESIGNING AND SOLVING SCHEDULING POLICIES IN REMOTE-CONTROLLED CPS

In the establishment of general scheduling policy optimization problems, both communication and computation impose limitations on the CPS. Limitations in the communication system include but are not limited to delay constraints, bandwidth constraints, and energy constraints. Constraints in computation, for example, include restrictions on the execution range. These factors determine the upper limit of system performance and significantly impact the design of system scheduling problems.

At the same time, different optimization goals lead to various design processes and scheduling strategy designs. For instance, the design objectives differ when pursuing timeliness and precision in control. Setting different goals also results in different scheduling policy solution processes under traditional methods, meaning, that different systems cannot share a single set of scheduling policy design methods, greatly increasing the complexity of the work.

Combining AoI and MDP space will provide an effective and universal method for designing CPS scheduling policies. Using a multi-process wireless control CPS as an example, the system model is described using MDP with appropriate extensions.

As shown in Figure 2.5, the system model contains multiple controlled processes. To avoid collisions during the transmission of data packets, data packets can only be transmitted through a limited set of orthogonal channels in each time slot. Optimizing system performance involves determining the transmission devices for each time slot.

First, the system is abstracted as an MDP, creating a state space S that contains all possible system states. For the design process of scheduling problems, an accurate and concise description of the state space is essential. Inaccurate abstractions reduce the precision of the description, resulting in suboptimal practical outcomes. Complex abstractions make finding the optimal solution more complex in terms of time and space, lacking practicality. In the description of multi-processes, if each process is simply discretized as the state environment in MDP, it results in an excessively large state space, demanding significant computing power and memory resources, and may even become unsolvable. The subsequent part will analyze the scheduling problem in remote-controlled CPS in the framework of MDP, extending from special cases to a general approach, comprehensively explaining the design methods for scheduling policy.

The AoI introduced earlier comprehensively considers various indicators in the CPS design, widely used in the study to describe the system's state. It only requires a single number to describe the state of each controlled process. Each state of a controlled process contains Δ to represent the AoI of each controlled process. Each state has an upper limit to ensure a finite state space, truncating the respective AoI values.

In the system, the uplink channel and the downlink channel handle state data packets and control commands, respectively, which are different types of data. Considering that the consequences of packet losses in the uplink and the downlink are different, it is necessary to distinguish the current data in its corresponding state. From a data perspective, each process's state can be divided into three states: uplink transmission, control command generation, and downlink transmission. In the case of control command generation and immediate transmission, the time for the controller to generate the command is equivalent to the downlink delay (as mentioned in the context of downlink packet loss). Thus, by considering the arrival of state data packets at the controller as a boundary, the data state of each process can be categorized into "currently in uplink transmission" and "downlink transmission."

When both uplink and downlink delays are set to 1 (for simplifying the system, in some studies, the time slots are set to be equal to the uplink and downlink delays), each process only requires one parameter, denoted as ϕ, to indicate whether the controller has generated the corresponding control command. When it is not generated, $\phi = 0$ means that the process is in uplink transmission, and when it is generated, $\phi = 1$ indicates that the state is in the process of executing downlink transmission. Considering the generation and computation delay of control commands, when the uplink and downlink delays are not both 1 and are not equal, for each controlled process, the transition time between the states of uplink transmission and downlink transmission is not the same. It follows a semi-Markov process. By setting the remaining uplink and downlink transmission time to standardize the transition time for each state, the original semi-Markov decision process (SMDP) is transformed into an MDP. To facilitate representation, this chapter defines the remaining uplink transmission time and the generalized remaining downlink transmission time as $\tau^{\uparrow}$ and $\tau^{\downarrow}$, respectively, with upper limits of T_{up} and $T_{down} + T_{com}$. In more general cases, especially when control commands are generated and awaiting transmission, an additional waiting time, denoted as $\tau^{\leftarrow}$, can be included in the downlink transmission time. This is used to describe the time it will take for the control command to be sent. When a state data packet arrives at the controller, $\tau^{\leftarrow}$ is at least T_{com}. This situation is commonly observed in half-duplex controllers [16].

In summary, when uplink and downlink transmission delays are the same, the state of each process can be represented as (Δ, ϕ). When uplink and downlink delays are unequal, each process's state can be represented as $\left(\Delta, \tau^{\uparrow}, \tau^{\downarrow}\right)$. In the more general case, each process's state can be represented as $\left(\Delta, \tau^{\uparrow}, \tau^{\leftarrow}, \tau^{\downarrow}\right)$. The overall system state is the collection of states for all processes, and the size of the state space S is the product of the possible states for each process.

The system state corresponds to various values within the analyzed system, used to cover all possible situations that may occur in the system. The action space, denoted as A, is meant to encompass all possible actions that can be executed, with each element used to describe the system's scheduling actions as independent variables. The design of actions depends on the specific research problem. For instance, in the case of multi-process scheduling design, one needs to decide in each time slot which portion of the controlled processes sends status information. For each process, a binary variable a_s can be defined, representing whether to send a status data packet or not in a given time slot. In the overall system, the number of data packets to be sent in each time slot cannot exceed a certain limit. For a half-duplex controller, one must also consider that sending and receiving cannot occur simultaneously. Since control commands require precise timing and can only be executed by actuators at specific time points, it is necessary to allocate waiting time on the actuator side, defined as a_c.

However, the design of the action space varies according to different research perspectives. For example, when optimizing system performance by allocating channels in each time slot, the elements in the action space represent the bandwidth allocated for each transmission. When studying the trade-off between transmitter power adjustment and energy consumption, the action space elements can represent each sensor's transmission power [19]. When researching different problems, designing the action space in the MDP according to the optimization variables is only necessary. Multiple perspectives can be considered in the analysis of problems, and when jointly considering optimization from various angles, it is not necessary to completely rethink the problem, reducing the steps and research complexity in the research process.

State transition is an essential step that links discrete states in wireless remote-controlled CPS at each moment, transforming it into a continuously running system. When studying the long-term average performance of the system, it is an indispensable factor.

State transition is uncertain under imperfect channels; during the transmission of uplink and downlink channels, different transmission outcomes lead to different states. When the uplink transmission is successful, the status information is successfully received at the controller, allowing the controller to begin computing control commands. When the downlink transmission is successful, it indicates that the control action has been received by the actuator, and Δ in the controlled process's state can be updated. The evolution of AoI throughout the process is shown in Figure 2.4.

In the special case of uplink and downlink delays, both being 1, if the uplink transmission is successful, the parameter ϕ is updated, indicating that downlink transmission will occur in the next time slot. Conversely, the parameter ϕ is not updated. When the downlink transmission is successful, the process's AoI and the parameter ϕ are reset; otherwise, only the parameter ϕ is reset, and AoI continues to increase linearly. In cases where the uplink and downlink delays are not the same, when deciding to perform uplink transmission, the remaining uplink transmission time $\tau^{\uparrow}$ is reset. When $\tau^{\uparrow}$ reaches zero and uplink transmission is successful, the downlink transmission time $\tau^{\downarrow}$ is reset. When $\tau^{\downarrow}$ reaches zero and downlink transmission is successful, the process's AoI is reset. Similarly, in the most general case, the readers only need to add the remaining calculation time for the controller.

The preceding analysis provides the possible next states when taking the corresponding action a under state s. However, for a more precise description of this process, it is necessary to clarify all state transition probabilities, which means obtaining the state transition matrix P in the MDP. Because remote-controlled CPS involves short packet transmission, errors result in the entire packet being discarded, making P a sparse matrix with only a few non-zero elements. The specific state transition probabilities depend on the communication parameters described earlier, and also on retransmission schemes, such as HARQ-CC and HARQ-IR. The values can be estimated based on the fading channel model provided earlier for the transmission success probability.

As seen from the previous discussion of AoI, it is a content-agnostic metric that only reflects the freshness of the controlled process and does not capture some inherent features of the system itself. If the scheduling scheme of the system only relies on AoI, it is not a communication–computation joint design, and the final results are far from reaching the upper limits of joint design. In the design, AoI plays a role in considering various communication metrics and providing optimization objectives for computing control commands. Therefore, in remote-controlled CPS, the reward R is related to the estimation and control instructions in the computation process, which can be derived from state parameters like AoI and action a.

When aiming for control precision, the relationship between AoI and corresponding long-term performance metrics varies under different control algorithms. This dependence specifically relies on the control algorithm employed in the controller. For instance, under the LQR and MPC control algorithms, given the known noise variance, the MSE of the controlled process can be calculated from AoI. When the goal is to reduce system energy consumption, the main consideration is the relationship between the reward R and action a. Sending fewer packets in a single time slot results in a larger reward for each state. Under a given objective, the reward function can be derived from state s and action a. This completes the transformation from AoI or other parameters to specific control objectives, achieving a joint design of communication and computation in the system.

After obtaining the complete MDP quadruple, it is possible to design the system scheduling policy π using methods like value iteration or policy iteration. This provides a mapping from the system state to scheduling actions, offering guidance for the scheduling actions in each time slot.

In summary, using MDP to describe remote-controlled CPS is analogous to an operating system on top of hardware. Just as an operating system allows for the convenient use of various software, describing remote-controlled CPS enables the more convenient optimization of various mathematical tools and learning algorithms. In cases with different structures and assumptions, it is necessary to design abstractions from the system model to the state space S, where the optimization problem's independent variables determine the action space A. The state transition matrix P plays a role in connecting the state space S and action space A, and these three elements are strongly interrelated in the design. The reward R depends on the system's objectives, considering the computational component beyond communication, making it an indispensable part of joint design.

This layered approach significantly simplifies the joint design of wireless remote-controlled CPS. Different objectives only require adjustments in the goal-oriented metrics, while different assumptions influence the choice of algorithms. For traditional methods like value iteration and policy iteration, their advantage lies in the ability to obtain high-performing scheduling strategies, but their algorithm complexity is high, requiring global knowledge of system communication parameters, which poses high demands on the system's controllers. When the state and action spaces are extensive, methods like value iteration face exponential growth in computational requirements. In situations where system parameters like channel mathematical models are unknown, traditional methods like value iteration become impractical. Hence, more widely applicable learning algorithms are used in scheduling algorithm generation.

Methods based on reinforcement learning (RL) or deep reinforcement learning (DRL) are extensively employed to overcome the demanding requirements of traditional methods [20, 21]. Similar to the design goals of remote-controlled scheduling strategies, RL and DRL focus on finding strategies that maximize cumulative rewards in the environment. As shown in Figure 2.7, in RL or DRL, there are two components: the environment and the agent. The agent continuously interacts with the environment, exploring

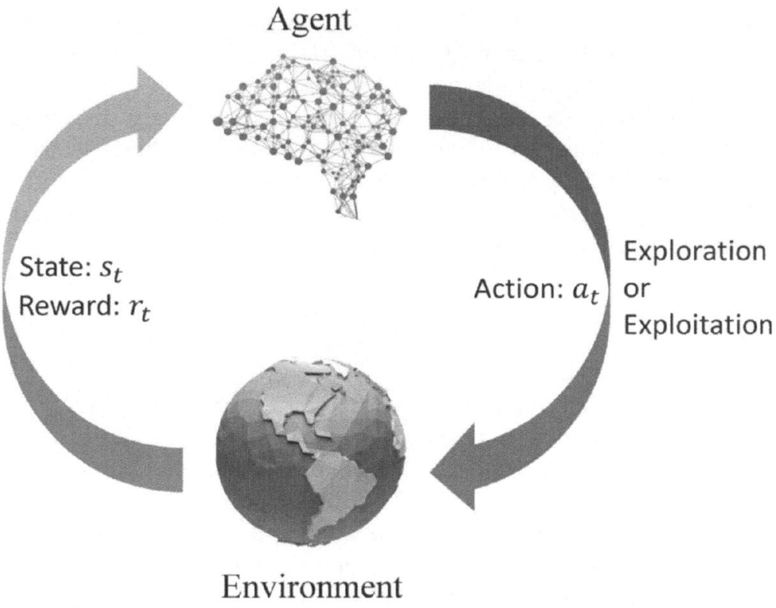

Figure 2.7 The framework of the RL algorithm.

potential scenarios. The environment is an unknown black box to the agent, with the primary goal not to obtain a mathematical model of the environment but to enable the agent to gradually accumulate more rewards through trial-and-error learning.

Compared to the MDP described earlier, using RL or DRL can reduce the requirements for the mathematical model of the system, allowing the state space S and action space A in the system to be continuous spaces, and the corresponding state transition probability matrix becomes a mapping from the state space S and action space A to the state space. This allows the scheduling method design process described earlier to be more applicable in remote-controlled CPS, such as the design of sensor transmission power or transmission rate.

Similar to the ideas of value iteration and policy iteration, RL or DRL can mainly be divided into three categories: value-based, policy-based, and hybrid methods, called actor-critic. An appropriate reinforcement learning algorithm can be selected after the system is abstracted into an MDP, ultimately leading to the scheduling policy. Taking a large number of DRL applications in remote-controlled CPS as an example, the choice of a specific algorithm depends on whether the action–state space is continuous or discrete. Q-learning algorithms are often used when both the action and state spaces are continuous. In cases where the action space is discrete and the state space is continuous, common algorithms include deep Q-network-class algorithms, such as deep Q-network (DQN), double deep Q-network (DDQN), and dueling double deep Q-network (D3QN). For cases where both the action and state spaces are continuous, common algorithms include deep deterministic policy gradient (DDPG), twin delayed deep deterministic policy gradient (TD3), proximal policy optimization (PPO), and soft actor-critic (SAC). In general, algorithms suitable for continuous spaces can also be used for discrete spaces, meaning, that discrete spaces are a special case of continuous spaces. However, when using DRL in practice, it is not recommended to discretize a continuous system directly to use the corresponding DRL algorithm, as this would lead to sparse rewards and make it difficult for the final policy to converge.

The DRL algorithms mentioned earlier have summarized their respective characteristics in a large amount of research on remote-controlled CPS. D3QN has been widely adopted in systems with discrete actions and has a relatively stable convergence process during training. The performance of TD3 is greatly influenced by hyperparameters and has high requirements for hyperparameter settings. PPO is more tolerant in terms of reward design and can provide good results even with relatively sparse rewards. SAC, on the other hand, emphasizes exploration of the environment and has a faster convergence speed. Readers can choose the appropriate DRL algorithm based on these characteristics.

It is worth noting that the RL and DRL algorithms mentioned in this chapter are just classical algorithms widely used in remote-controlled CPS at present, and many new algorithms have not been mentioned. Interested readers can explore more DRL algorithms to achieve better results in their own research on CPS joint design.

2.5 CHALLENGES AND OPPORTUNITIES

Despite the significant advancements in remote-controlled CPS research, continuous research is essential to adapt to more complex and dynamic communication environments and meet increasingly diverse requirements. Within the scope of research on joint design and scheduling strategies, several key aspects are highlighted:

Low-latency communication under space-air-ground integrated network (SAGIN). In the post-5G era, the development of integrated networks spanning land, sea, and sky has greatly expanded the scenarios for communication. However, the large scale and high dynamics of SAGIN present significant challenges for real-time state perception, transmission, computation, processing, and applications. Ideal assumptions for traditional mathematical models may become invalid, rendering existing methods ineffective. Moreover, high-precision remote-controlled systems and real-time decision-making applications demand timeliness in information. This challenges current remote-controlled CPSs, necessitating the development of more rational scheduling strategies between control centers and remote devices or actuators from a timeliness perspective.

Distributed control in large-scale CPS. In practical applications of remote-controlled CPS, remote devices are often sparsely distributed, and the types of information vary between different processes. Increasing the number of controllers in distributed control accelerates information processing in multi-process systems. However, the heterogeneity among multiple controllers, each with varying available information, storage, and computational resources, significantly complicates establishing system mathematical models. Therefore, for large-scale remote-controlled CPS, it is necessary to study technologies related to consistent control and cooperative control through scheduling policies.

Semantic communication. Remote-controlled CPS has received broad attention in both academic and industrial circles, and the significance of information in specific application scenarios has attracted substantial research interest. The ability to sustainably transmit the most critical information under limited system performance is becoming increasingly vital, emphasizing the importance of considering semantics. The realization of this concept must be based on a universal framework. However, from a measurement perspective, there is no unified framework for the study of semantic communication, making semantic communication one of the current hot research topics.

2.6 CONCLUSION

This chapter primarily explores remote-controlled CPS from a communication perspective and outlines the process of designing scheduling methods for communication and computation in a joint design context. Firstly, it describes the structure and system model of remote-controlled CPS, introducing common application scenarios. Subsequently, it introduces parameters in CPSs that exhibit strong coupling relationships, influencing both the communication and computation aspects of the system. In addition, taking inspiration from AoI and MDP, this chapter provides the process for designing scheduling policy under joint design and analyzes the work required for generating scheduling policy. Finally, it discusses the opportunities and challenges in remote-controlled CPS, elucidating the directions that require further research.

REFERENCES

[1] Boyes, Hugh, et al. "The Industrial Internet of Things (IIoT): An analysis framework." Computers in Industry 101 (2018): 1–12.
[2] Simsek, Meryem, et al. "5G-enabled tactile internet." IEEE Journal on Selected Areas in Communications 34.3 (2016): 460–473.
[3] Luvisotto, Michele, Zhibo Pang, and Dacfey Dzung. "Ultra high performance wireless control for critical applications: Challenges and directions." IEEE Transactions on Industrial Informatics 13.3 (2016): 1448–1459.
[4] Ma, He, and Shidong Zhou. "Noisy sensor scheduling in wireless networked control systems: Freshness or precision." IEEE Wireless Communications Letters 11.5 (2022): 1107–1111.
[5] Soleymani, Touraj, et al. "Feedback control over noisy channels: Characterization of a general equilibrium." IEEE Transactions on Automatic Control 67.7 (2021): 3396–3409.
[6] Klügel, Markus, et al. "Joint cross-layer optimization in real-time networked control systems." IEEE Transactions on Control of Network Systems 7.4 (2020): 1903–1915.
[7] Park, Pangun, et al. "Wireless network design for control systems: A survey." IEEE Communications Surveys & Tutorials 20.2 (2017): 978–1013.
[8] Zhou, Bo, and Walid Saad. "Minimum age of information in the Internet of Things with non-uniform status packet sizes." IEEE Transactions on Wireless Communications 19.3 (2019): 1933–1947.
[9] Liu, Wanchun, et al. "On the latency, rate, and reliability tradeoff in wireless networked control systems for IIoT." IEEE Internet of Things Journal 8.2 (2020): 723–733.
[10] Kaul, Sanjit, Roy Yates, and Marco Gruteser. "Real-time status: How often should one update?" 2012 Proceedings IEEE INFOCOM. IEEE (2012).
[11] Abd-Elmagid, Mohamed A., Nikolaos Pappas, and Harpreet S. Dhillon. "On the role of age of information in the Internet of Things." IEEE Communications Magazine 57.12 (2019): 72–77.
[12] Huang, Kang, et al. "Optimal downlink–uplink scheduling of wireless networked control for Industrial IoT." IEEE Internet of Things Journal 7.3 (2019): 1756–1772.
[13] Qiu, Yifei, et al. "On scheduling policy for multiprocess cyber–physical system with edge computing." IEEE Internet of Things Journal 9.19 (2022): 18559–18572.
[14] Sun, Yin, et al. "Update or wait: How to keep your data fresh." IEEE Transactions on Information Theory 63.11 (2017): 7492–7508.
[15] Maatouk, Ali, et al. "The age of incorrect information: A new performance metric for status updates." IEEE/ACM Transactions on Networking 28.5 (2020): 2215–2228.
[16] Huang, Kang, et al. "To sense or to control: Wireless networked control using a half-duplex controller for IIoT." 2019 IEEE Global Communications Conference (GLOBECOM). IEEE (2019).
[17] Zhang, Jianhua, et al. "A survey of massive MIMO channel measurements and models." Zte Communications 15.1 (2017): 14–22.

[18] Schenato, Luca. "Optimal estimation in networked control systems subject to random delay and packet drop." IEEE Transactions on Automatic Control 53.5 (2008): 1311–1317.

[19] Feng, Songtao, and Jing Yang. "Minimizing age of information for an energy harvesting source with updating failures." 2018 IEEE International Symposium on Information Theory (ISIT). IEEE (2018).

[20] Qiu, Yifei, et al. "Model-free control in wireless cyber–physical system with communication latency: A DRL method with improved experience replay." IEEE Transactions on Cybernetics 53.7 (2023, July): 4704–4717. https://doi.org/10.1109/TCYB.2023.3275150

[21] Bouteiller, Yann, et al. "Reinforcement learning with random delays." International Conference on Learning Representations (2021, May 4), Vienna, Austria. https://openreview.net/group?id=ICLR.cc/2021/Conference

Chapter 3

Software-defined networking in cyber-physical systems

Benefits, challenges, and opportunities

Chekwube Ezechi, Mobayode O. Akinsolu, Abimbola O. Sangodoyin, Folahanmi T. Akinsolu, and Wilson Sakpere

LIST OF ABBREVIATIONS

AI	artificial intelligence
API	application programming interface
BAS	building automation system
CoAP	constrained application protocol
CNN	convolutional neural network
CPS	cyber-physical system
CPU	central processing unit
DA-DIS	delay-based attack detection and isolation
DDoS	distributed denial-of-service
DRL	deep reinforcement learning
HTTP	hypertext transfer protocol
HVAC	heating, ventilation, and air-conditioning
ICS	industrial control systems
IDS	intrusion detection systems
IETF	Internet Engineering Task Force
IEEE	Institute of Electrical and Electronic Engineers
IoT	Internet of Things
IIoT	industrial Internet of Things
IBN	intent-based networking
IP	Internet Protocol
ITS	intelligent transportation system
IPS	intrusion prevention systems
KPI	key performance indexes
LLDP	link layer discovery protocol
MFI	mixed flow installation
ML	machine learning
MQTT	message queuing telemetry transport
NFV	network function virtualization
NIDS	network-based intrusion detection system
NOS	network operating system
NSF	National Science Foundation
ODL	OpenDaylight

DOI: 10.1201/9781003559993-3

ONF	Open Networking Foundation
OPC-UA	open platform communications united architecture
PC	personal computer
PROFINET	Process Field Network
PFIR	proactive flow installation re-routing
QoS	quality of service
RFI	reactive flow installation
SecaaS	security as a service
SCADA	supervisory control and data acquisition
SDDC	software-defined data centers
SDN-RM	software-defined network resilience manager
SDN	software-defined networking
SPOF	single point of failure
SRP	stream reservation protocol
TDRL-RP	trust-based deep reinforcement learning framework
TSN	time-sensitive networking
VANET	vehicular ad hoc network
VM	virtual machine
WAN	wireless access network
Wi-Fi	wireless fidelity
ZSM	zero-touch service network management

3.1 INTRODUCTION

Technological development in recent years has brought forth innovative paradigms that have transformed various industries. Among these, two prominent technologies, cyber-physical systems (CPSs) and software-defined networking (SDN), have emerged as game changers in networking and systems integration. One way to apply SDN in industrial automation settings is through the SDN Process Field Network (PROFINET), as proposed by [1]. By incorporating Internet of Things (IoT) developments, the study utilized smart factories and Industry 4.0 initiatives to automate industrial processes. The study suggested a demonstration displaying network properties related to SDN for remote supervision of and servicing dispersed industrial facilities. The study demonstrated the use of behavioral descriptions to identify possible relationships between components in industrial settings [2]. Integrating SDN into CPSs can significantly improve scalability, efficiency, and security. This is mainly because SDN enables network programmability and flexibility, while CPSs combine physical and computational elements to create intelligent systems. SDN is a networking method that separates the data plane from the control plane, enabling network managers to govern network traffic programmatically. CPS exemplifies the seamless fusion of physical systems with computational and communication capacities, resulting in intelligent systems that can communicate with the real world. Because of the robust levels of control, efficiency, and adaptability that the convergence of SDN and CPS offers, SDN-CPS can completely transform several industries, including manufacturing, energy, transportation, and health care [2].

A practical use of SDN-CPS is the proactive flow installation re-routing (PFIR) technique mentioned in [3]. It enables rapid reconfiguration of traffic types based on requests from the application layer. The PFIR's benefits include instantly modifying the data stream in response to alterations without pausing for new flow instructions from the SDN controller and accounting for the various quality of service (QoS) demands put forward by industrial applications. A concept for a mixed flow installation (MFI) which also considers the diverse QoS needs and considerably reduces the additional latency brought on by reactive flow installation (RFI) approaches was also proposed in [3]. The performance metrics addressed during the evaluations include packet violation ratio, packet loss rate, and end-to-end delay, all under the control of the OpenDaylight controller. The results show that the suggested methods outperformed both conventional and hybrid flow production techniques to reach the desired result. The concepts of SDN and CPS, their integration (that is, SDN-CPS), and the motivations guiding this fusion are explored in this chapter to complement existing similar works. Specifically, the benefits, possible issues that should be carefully considered, and future paths that might result in important breakthroughs are discussed to identify gaps in the current research landscape and recommend new directions for research and development.

3.2 OVERVIEW OF SOFTWARE-DEFINED NETWORKING (SDN)

SDN is an architectural method used in network architecture that divides the data and control planes. Because the network's management logic is isolated from the underlying hardware, network managers can regulate and control network behavior via a centralized software-based controller [4]. This separation enables dynamic network programmability, centralized management, and the ability to introduce new services and applications without requiring changes to the underlying network infrastructure. An organization that advocates SDN, the Open Networking Foundation (ONF), states that SDN is a dynamic, manageable, cost-effective, and flexible developing architecture that is appropriate considering the high bandwidth and dynamic nature of contemporary applications [5]. The SDN controller can specify how network traffic moves through the network by using the OpenFlow protocol, frequently used as a standardized interface between the controller and the forwarding devices in the data plane. [6] showcased how the OpenFlow protocol can implement fine-grained and dynamic control over network traffic in campus environments. Specifically, the work in [6] described designing and deploying an OpenFlow-based network architecture in a university campus network, which allowed for advanced traffic engineering, load balancing, and security measures. They also highlighted how OpenFlow can control network flows, making it easier to manage and configure network devices uniformly. In traditional networks, network devices like routers and switches have tightly integrated control and data planes, making network management complex and rigid. Network policies and routes are defined and managed centrally by a software controller using SDN, offering users a detailed view of the network.

3.2.1 SDN architecture

SDN is the networking industry's largest transformation to date [4]. The division of control logic and network devices is the fundamental idea behind the typical SDN architecture. Compared to a standard Internet Protocol (IP) network, this split facilitates easier administration of applications and increased accessibility to management and development of a variety of network equipment and devices. The application, control, and data layers are the three distinct levels that make up a typical SDN architecture (see Figure 3.1).

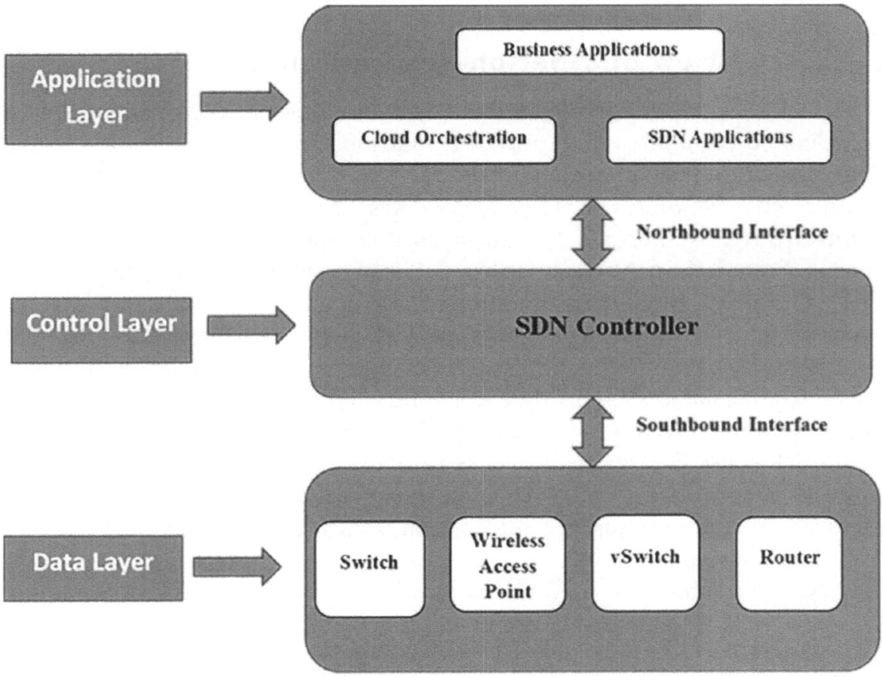

Figure 3.1 Typical SDN architecture.

- *Application layer*. This layer oversees managing all business and security apps. This layer controls several critical software services, including firewall implementation, mobility management, intrusion detection and intrusion prevention systems (IDS and IPS), metering, routing, QoS, and load balancing. This layer uses northbound application interfaces to communicate with a lower layer [7].
- *Control layer*. The network operating system (NOS), also referred to as the network controller, is housed under the application layer and is responsible for overseeing the overall operations of the network. Using programming, a logically centralized controller manages the network as a whole and decides on packet dropping, flow forwarding, and routing [8]. Logically centralized and physically dispersed, west- and eastbound interfaces allow the controller, which is an environment, to connect with other levels as well as with itself. This layer uses southbound application programming interfaces (APIs) like OpenFlow to connect with the layer beneath it or the data layer.
- *Data layer*. This layer forwards packets in compliance with the rules and guidelines that the controller has established and assigned. It comprises physical network devices (such as switches, routers, access points, etc.) and virtual switches (such as Open vSwitch, Indigo, Pica8, Nettle, OpenFlow, etc.) [9].

3.2.2 Overview of cyber-physical system (CPS)

A typical CPS can be broadly defined as combining physical processes with coordinated communication and computational capabilities, leading to the convergence of the physical and digital worlds. CPSs can also be viewed as representative of advanced embedded systems, where computational elements are tightly integrated with physical components [10]. The National Science Foundation (NSF) describes CPSs as designed systems that rely on the fusion of physical components with computational algorithms [11]. In other words, CPS merges the physical and digital worlds, combining sensors, actuators, communication networks, and computational capabilities to create sophisticated systems that can communicate with their surroundings [12]. To enable the monitoring, control, and optimization of physical processes, CPSs use data gathered via sensors in real time and engage with the physical environment. There are numerous fields in which CPS applications are used, including industrial automation, smart grids, self-driving vehicles, and health-care monitoring.

A typical example of a CPS is the collaborative human–robot assembly presented in [13] for potential use in manufacturing facilities. Conventionally, a human–robot assembly may be used and guided to undertake hazardous mining procedures with very modest modifications, resulting in safer working environments and lowering the risks of miners' exposure to hazardous substances. Since the operations of such robots are often automated, there would be little or no physical or human intervention or labor required. Drones, or unmanned aerial vehicles, are another common class of CPSs that can be used for low-weight distributions over short distances to reduce traffic congestion for short-distance deliveries [14]. The environmental effects of regular deliveries are greatly reduced when drones are used for deliveries. Companies such as Amazon are already leveraging this technology, with plans to use drones for the deliveries of items weighing 5 lb (2.268 kg) or less, with an expected delivery time of around 30 min. Smartwatches having sensors and computational capabilities also constitute another form of CPS [15].

Based on the examples presented, a typical CPS comprises three primary components: a physical system, a networking and communication element, and a distributed cyber system. To put it another way, CPSs are built from a variety of distributed hardware, software, and network elements that are incorporated into real-world settings and systems. The most important component is the software, which comprises all software applications for information processing, filtering, and storage. CPSs interact with the physical system via networks [16]. The main CPS components are described in the next subsections.

3.2.3 Physical system

The physical entities in CPSs are the processes being monitored, controlled, or optimized. These components represent the tangible, real-world elements of the CPS. The physical system interacts with its environment, often with sensors and actuators, to gather data and perform actions. Sensors gather information from the physical environment about things like motion, pressure, temperature, and chemical composition. Actuators, on the other hand, are responsible for effecting changes, such as controlling the movement of robotic arms, adjusting valves in a chemical plant, or steering a self-driving car in the physical world. Unlike traditional embedded systems, CPSs communicate with the physical world directly. CPSs are a group of hybrid systems that operate together and communicate through networks, especially wireless networks, much like hybrid systems that also manage physical processes. The cyber-physical system's core three "C" ideas are depicted in Figure 3.2. The physical platforms that support CPSs have five functionalities available. These are computation, communication, autonomy, precise control, and remote collaboration, according to [17].

3.2.4 Networking and communication element

The digital elements are responsible for data processing, decision-making, and control. These include algorithms, software, communication networks, and computing devices. The physical and cyber systems can communicate

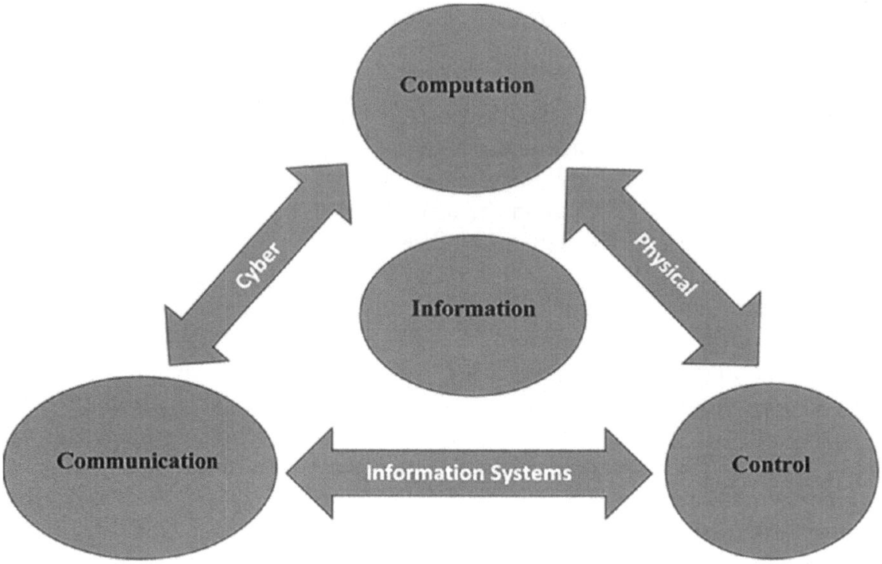

Figure 3.2 The 3Cs of a typical CPS.

more easily because of these CPS components. They include various communication protocols, network infrastructure, and data transmission technologies. CPS relies on a network infrastructure to enable data exchange between the physical components and the cyber system. This infrastructure can include wired and wireless networks, the Internet, and protocols such as Wi-Fi (wireless fidelity), Ethernet, or cellular networks. Generated data by sensors and actuators need to be transmitted efficiently and reliably to the cyber system for processing and control. Communication protocols and technologies, including hypertext transfer protocol (HTTP), message queuing telemetry transport (MQTT), constrained application protocol (CoAP), and custom protocols, are used to enable data transfer.

3.2.5 Distributed cyber systems

Many CPS applications involve human interaction, where human operators monitor and configure the system and intervene, as may be required. The distributed cyber system is the digital counterpart of the CPS. It comprises the computing resources that process and evaluate the information obtained from the physical system. These include microcontrollers, embedded systems, servers, cloud computing platforms, and edge devices. Transmitting control signals back to the physical systems, distributed cyber systems process and interpret data received from the physical systems. Monitoring, analyzing, and making decisions in real time are frequently part of this process. According to [17], CPSs are sophisticated systems that enable the integration of 3C technology (as depicted in Figure 3.2). Together, these three components create a system that seamlessly blends the digital and physical worlds. Put differently, they combine cyber skills (computation and communication) with physical capabilities (sensors and actuators). Due to their numerous uses in industries like manufacturing, building systems, transportation, electric power grids, and health-care systems, CPSs can be said to be ubiquitous. In terms of dependability, scalability, security, and real-time responsiveness, the combination of physical and cyber components in CPS presents both unique difficulties and potential [18].

3.2.6 Motivation for integrating SDN into CPS

Integrating SDN into CPS can be said to be motivated by the need to address the evolving requirements and challenges of CPS environments. CPSs often involve many interconnected devices and sensors, such as IoT devices and sensors in smart grids or industrial automation. These systems need to accommodate increasing numbers of devices efficiently while maintaining performance and reliability. CPS applications require adaptable and programmable network infrastructure to accommodate changing communication patterns and application-specific requirements. Traditional network infrastructures are often rigid and difficult to modify in response to changing needs [6]. SDN allows for dynamic reconfiguration of the network to adapt to changing CPS requirements. SDN's programmability also enables fine-grained control over network behavior, allowing CPS applications to adjust network configurations on the fly. This flexibility is crucial in CPS scenarios, where real-time adjustments are necessary based on the physical environment and system behavior [19]. Centralized administration and control of network resources are made possible by SDN's scalable architecture, which divides the control plane from the data plane [19]. This makes it well-suited for CPS environments with dynamic and growing requirements. By combining the network programmability offered by SDN with the real-time data analysis capabilities of CPSs, SDN can help optimize resource usage by dynamically allocating and reallocating network resources based on application needs. CPS environments also demand robust network management and security mechanisms to guarantee the dependability and security of vital operations. The centralized control of SDNs facilitates better network management and security in CPSs [16].

3.3 BENEFITS OF SOFTWARE-DEFINED NETWORKING (SDN) IN CYBER-PHYSICAL SYSTEMS

As CPSs are becoming increasingly pervasive in several sectors, such as industry and critical infrastructure, ensuring efficient, secure, and flexible network management becomes paramount. Integrating various devices

Table 3.1 Future research direction and opportunity

Research direction	*Brief description*	*Key references*
Interoperability and standardization	Ensuring interoperability and standardization of SDN solutions for CPS is crucial for seamless integration and widespread adoption. Future research should focus on developing common protocols, interfaces, and APIs that enable different SDN-enabled CPSs to work harmoniously.	[2]
Resilient SDN for CPS	Enhancing the resilience of SDN-enabled CPS applications against various types of failures and disruptions is critical. This will encompass the design and development of fault-tolerant algorithms, redundancy mechanisms, and adaptive control strategies to ensure continuity when failures are present.	[20]
ML and AI in SDN-CPS	Leveraging ML and AI techniques in SDN can allow more intelligent and autonomous network management. These algorithms can optimize network resource allocation, detect anomalies, and predict future demands in CPS applications.	[21]
Edge and fog computing in SDN-CPS	As CPS applications demand low latency and real-time processing, integrating edge and fog computing with SDN can bring computation closer to the data source. This can lower communication overhead and improve the efficiency of CPSs.	[22]

and applications in CPSs creates complex networks that demand efficient management and control. The diverse and dynamic nature of CPS presents difficulties for traditional network architectures. However, SDN offers promising solutions in CPS by enabling centralized control, programmability, and adaptability. This section explores the benefits of SDN in CPSs, with a focus on improved network management, enhanced flexibility, resource utilization, scalability, and security.

3.3.1 Improved network management and control

Improved network management and control is one of SDN-CPS's main advantages. In conventional network topologies, it can be difficult and time-consuming to manage many devices and services. SDN eliminates this bottleneck by isolating the network control plane from the data plane and offering centralized network management via a logically distinct controller. The centralized controller's dynamic network management permits administrators to control and set up network elements from a unified point. This centralized approach simplifies the management of complex CPS networks, leading to reduced operational overheads and enhanced network control [23]. With centralized control, administrators can efficiently monitor and manage the entire network from a single interface. This simplifies network configuration, troubleshooting, and policy enforcement. As a result, CPS administrators can respond more quickly to network events and optimize network performance, enhancing overall system efficiency [24]. SDN enables automated and rapid provisioning of network resources in CPS. The centralized controller enables dynamic resource allocation, depending on application needs and network conditions, resulting in effective network resource use and reduced provisioning time.

3.3.2 Enhanced flexibility and programmability

CPS environments often require frequent changes in network behavior due to dynamic conditions and varying application requirements [25]. In traditional networks, implementing these changes can be arduous and time-consuming. SDN offers enhanced flexibility and programmability, allowing administrators to modify network policies and behavior on the fly [6]. Through SDN's programmable nature, network administrators can deploy, modify, and manage network services and protocols through software-based applications (open APIs), reducing dependence on proprietary solutions [7]. This programmability encourages creativity and makes it easier to incorporate new services and applications into the CPS ecosystem. In CPS environments, where the network requirements can change rapidly due to dynamic physical processes, the flexibility of SDN is invaluable. It enables administrators to adapt the network to changing conditions, accommodate new devices or components, and deploy specialized network functions as may be needed [24]. This agility ensures the network can respond to varying demands, making CPS applications more adaptable and resilient. For instance, in a smart grid, SDN can allocate more bandwidth to critical energy distribution systems during peak hours, optimizing the overall power distribution process [25].

3.3.3 Efficient resource utilization

Resource utilization is critical in CPSs, where the energy, memory, and computing capacity of devices are frequently constrained. In traditional networks, resource allocation is usually static, leading to underutilization or overutilization. SDN provides the ability to monitor network resources and make intelligent decisions for resource allocation. SDN controllers can collect data on network conditions in real time, device capabilities, and traffic load [19]. When SDN is used in CPSs, the controller can logically allocate resources by taking into account traffic patterns, application requirements, and network circumstances. As a result, CPS applications operate more smoothly and make effective use of network resources. Important CPS applications are given the resources they require thanks to this dynamic resource management, preventing bottlenecks and congestion while improving the overall system performance [25]. SDN's network virtualization and slicing capabilities permit the coexistence of numerous virtual networks on the same physical infrastructure. This adaptability is vital in CPS environments, where diverse applications often coexist.

3.3.4 Scalability and adaptability

Scalability is a crucial factor in CPSs, as CPS networks often need to accommodate many devices and services [25]. With SDN, administrators can easily add new devices and services to the CPS network without the need for complex reconfigurations. Traditional networks might face scalability challenges due to their distributed control planes. SDN's centralized control plane eliminates this limitation, making it highly scalable [4]. SDN's centralized control enables horizontal scalability, where additional resources can be seamlessly added to the network without affecting existing infrastructure. This is crucial for rapidly expanding CPS deployments. The entire network may be effectively managed by the SDN controller even as the size of the CPS increases mainly via the number of connected devices. SDN's programmable nature and real-time control will also allow CPS networks to adapt to changing conditions and application requirements [4]. For instance, in an industrial CPS such as vehicular ad hoc network (VANET), SDN can allocate network resources on the fly to prioritize mission-critical traffic during periods of high demand, ensuring seamless and reliable operations [26]. Additionally, SDN's adaptability allows for the ease of integration of new devices and services into the CPS network. This ability to scale and adapt ensures that CPSs can accommodate future technological advancements without major overhauls [26].

3.3.5 Security and resilience

Security and resilience are paramount in CPSs, as any disruption or compromise can have severe consequences. For instance, in the context of smart grids, a cyberattack can disrupt the distribution of electricity, leading to widespread power outages [27]. SDN enables network segmentation, where CPS components are logically isolated to limit the points of access that an unauthorized person could use to enter the system. This enhances security and helps contain potential breaches within the isolated segments. SDN makes it possible to see the network globally by centralizing the control plane, enabling comprehensive security policy enforcement. SDN's centralized control architecture allows for rapid identification and reduction of security risks. In this way, the SDN controller can enforce security policies and take immediate action to prevent attacks from propagating throughout the CPS network [4]. SDN also allows for the enforcement of access control, implementation of firewall rules, and dynamic re-routing of traffic to avoid potential attack paths. SDN's programmability can also enable the deployment of customized security measures tailored to meet the unique requirements of CPS applications. In terms of resilience, SDN facilitates quick network recovery time (up to 50% faster than traditional networks) in the event of link failures [4]. The centralized controller can reconfigure the network and adapt traffic paths to bypass faulty components, minimizing downtime and ensuring continuous operation.

3.4 CHALLENGES OF SDN IN CYBER-PHYSICAL SYSTEMS

CPSs have emerged as an integral part of today's society, connecting physical devices and infrastructures with computational systems. The seamless operations of a typical CPS heavily rely on the underlying networking infrastructure to support real-time communication, data exchange, and control. SDN has become a viable approach for managing the complexity of CPS networks. Flexibility, manageability, scalability, and efficiency are some of the important benefits of SDN adoption in CPS. The use of SDN in CPSs (that is, SDN-CPSs) promises increased flexibility and effectiveness in the operations of CPSs. However, this integration also poses some challenges that demand careful consideration. A summary of some of these challenges and the proposed solutions (including the identified limitations of the solutions) in the existing literature is given in Table 3.2. In this section of the chapter, the key challenges of SDN-CPSs are further highlighted by focusing on security concerns, network latency, scalability, and integration with legacy systems.

3.4.1 Network availability and reliability

Network availability and reliability are crucial for the proper functioning of CPSs, as any downtime or failure can have severe consequences on their safety and operational efficiency [28]. SDN introduces new points of potential failure, such as the central controller, which becomes an SPOF (single point of failure) [4]. The reliance on centralized control introduces new vulnerabilities and increases the potential for targeted attacks on the controller [29]. In addition to this, communication failures and link disruptions can also lead to delayed or incorrect actions, impacting real-time processes and

Table 3.2 SDN-CPS challenges, proposed solutions, and identified limitations

SDN-CPS challenges	*Proposed solutions*	*Identified limitations*	*Key references*
Network latency	Optimization of data flow management and leveraging edge computing to support data processing closer to where it is generated within SDN-CPSs	Latency reduction might compromise data integrity and data security with SDN-CPSs.	[22]
Integration with legacy systems	Development of hybrid networks that allow for the integration of SDN-CPS with existing legacy systems to ensure interoperability and gradual transition without disrupting existing operations	Integrating SDN-CPSs with existing legacy systems could be technically and financially challenging in terms of the resources required.	[22]
Scalability	Dynamic resource allocation and predictive modeling of network traffic and resource utilization patterns using ML algorithms within SDN-CPSs to facilitate the optimization of network resources and improve efficiency and scalability	High costs of implementation, biases in ML algorithms, and slow convergence speeds of applicable optimization frameworks.	[26]
Security concerns	Implementation of advanced encryption techniques, authentication mechanisms, and intrusion detection/prevention systems within SDN-CPS architectures to mitigate security threats	Sophisticated security and resilience features in SDN-CPS architectures could introduce overhead and complexity, potentially impacting overall system performance and management.	[27]

causing inefficiencies [29]. In response to these challenges, a great deal of study has been done on the development of robust fault tolerance mechanisms. These mechanisms have become essential in ensuring the dependable operation of SDN-based systems. One prominent strategy entails the strategic deployment of hot redundancy within the control plane. This usually involves having backup controllers ready to seamlessly assume control in case the main controller fails, guaranteeing redundancy [30].

To ensure network reliability and availability, distributed SDN controllers also provide a promising solution [31]. By distributing the control logic across multiple entities, SDN-based CPSs can significantly mitigate the potential disruptions caused by a particular controller malfunction [31]. Specifically, this method leverages a network of controllers that work collaboratively by sharing the network load and responsibilities. This not only reduces the likelihood of a failure but also enhances the scalability and overall resilience of the network. When these techniques are thoughtfully integrated into the SDN-CPS infrastructure, the result is a network that exhibits remarkable improvements in reliability and availability, effectively ensuring that critical applications and services always remain operational in the face of adverse events or component failures.

3.4.2 Complexity and scalability

Managing, and growing, the underlying SDN infrastructure becomes more complex as CPSs get bigger and more complicated. The dynamic nature of SDN and the requirement for regular reconfiguration can also significantly raise the complexity of network management. Pushing all control functionality to a central controller is one potential SDN approach. An SDN controller can manage 30,000 requests, according to early benchmarks [32]. Although this might be adequate for a sizable company network, it might pose a serious issue in scenarios such as data centers with high-flow initiation rates. Additionally, the growing number of connected devices in a typical CPS may exceed the controller's capacity to handle the increasing flow of table entries. In [33], an early SDN security solution, a controller is responsible for setting forwarding status on switches on a per-flow basis. This reactive method of flow management provides a great deal of flexibility (for instance, simple fine-grained, high-level network-wide policy enforcement, as reported in [33]), but it also delays flow setup and, relative to how it is implemented, may limit scalability.

Many research projects tackling the scaling issues with SDN fall into one of three categories: control plane, data plane, or hybrid. DevoFlow (a variant of the OpenFlow model) and software-defined counters are approaches that focus on the data plane; they assign some responsibilities to the forwarding devices, thereby reducing the control plane's overhead [34]. Instead of asking the controller for a decision on each flow, switches might, for instance, choose the flows (elephant flows, for example) that might require the control plane applications to make more advanced decisions. Adding more potent general-purpose central processing units (CPUs) to the forwarding devices to enable software-defined data centers (SDDCs) is another illustration. A general objective of enabling software-based implementations of operations for data aggregation and compression (for example, CPU and SDDCs) is to open new opportunities for lowering the control plane overhead.

3.4.3 Privacy and security concerns

SDN-CPS creates new privacy and security issues. For example, CPS applications are appealing targets for cyberattacks because they frequently deal with sensitive data and crucial processes. A malicious entity can take over the SDN controller and gain almost total control over the CPS network. Highly successful attacks could be conducted by a breached SDN controller in an industrial control system setting, disrupting physical operations and the real-time nature of industrial protocols [35]. Potential attackers see the centralized control plane of SDN as a high-value target, and a successful attack on the controller could have disastrous effects on the entire system. Additionally, an attacker may exploit network components, including servers, switches, and personal computers (PCs) or workstations, to initiate attacks like distributed denial-of-service (DDoS). Regarding the available literature, it can be noted that researchers and network security administrators have prioritized preventing DDoS attacks [35]. Distributed and decentralized control techniques that restrict the accessibility of sensitive data to a single point have been suggested to alleviate privacy and security issues [34]. To further safeguard data integrity and confidentiality in SDN-CPS, encryption methods and secure communication protocols have been proposed in [36].

3.4.4 Interoperability and standardization

Interoperability is a crucial concern in CPSs because heterogeneous devices and protocols must seamlessly communicate and coordinate with each other in a typical CPS. The lack of standardization and interoperability in SDN solutions poses significant challenges to the integration of SDN in diverse CPS environments. Different vendors may implement SDN protocols differently, leading to compatibility issues between network elements and controllers. This can then result in limited flexibility and vendor lock-in, hindering innovation and the deployment of heterogeneous CPS architectures. Interoperability and latency become complex and open challenges when more heterogeneous devices and communication protocols are employed, which makes data collection challenging. For low-latency data gathering, an open and flat industrial IoT (IIoT) architecture based on SDN and the open platform communications united architecture (OPC-UA) are proposed in [37] to address some of these issues. The OPC foundation created the OPC-UA machine-to-machine communication standard, which is best characterized as an industrial automation tool [37].

Standardization organizations like the Internet Engineering Task Force (IETF) and the Institute of Electrical and Electronic Engineers (IEEE) have also shown some interest in the application of SDN in CPSs [38]. To improve interoperability, IEEE and IETF have been actively creating protocols and frameworks for SDN-CPS [38]. The IEEE 802.1 time-sensitive networking

(TSN) task group, which is concerned with the first (that is, protocols for SDN-CPS), promotes interoperability across networked devices, with a concentration on layers 1 and 2 [38]. The stream reservation protocol (SRP, 802.1Qcc) is one of the Ethernet upgrades that the TSN work group suggested for enabling admission control, resource reservation, and parallel redundancy mechanisms [38]. Given that IEEE 802.1 TSN systems enable the computation and establishment of necessary network resources by a central control element, SDN and TSN complement each other well and can be adopted conjunctively. The IEEE 802.1CF project also claims that there are issues with service control, security, and provisioning in contemporary heterogeneous networks like smart grids, home automation, and IoT [38]. As a result, it adheres to SDN principles to remove obstacles to innovative technologies and network operators, unifying shared network control interfaces [38].

3.5 SDN-CPS APPLICATIONS

CPSs play a crucial role in smart applications. So the convergence of SDN and CPSs is a transformative force in shaping several smart applications. However, the complexity and size of CPSs are growing, and this poses substantial problems for real-time control, network administration, and security in smart applications. SDN has become a promising strategy to overcome these obstacles and improve the performance, security, and flexibility of CPSs. With a focus on smart infrastructure, an overview of SDN-CPS for smart applications is presented in this section by exploring the current challenges and prospects as reported in recent studies.

3.5.1 SDN-enabled CPS for smart infrastructure

Smart infrastructure demands intelligent and dynamic networking solutions to efficiently manage and control diverse interconnected systems. Interconnected systems such as CPSs combine computation, communication, and control to govern physical processes. CPSs play an important role in various critical smart infrastructures, including smart grids, intelligent transportation systems (ITSs), telemedicine, and industrial automation [39]. Modern information and communication technologies are incorporated into these smart systems to improve their efficiency, dependability, and sustainability. As an example, smart grids require intelligent management and control to optimize energy distribution and consumption. SDN can offer a reliable solution by enabling dynamic network reconfiguration and resource allocation in this context. SDN can be used in smart grids to improve fault detection and recovery mechanisms. Different data and control flow types with varying degrees of criticality exist in smart grids. SDN allows for the dynamic prioritization of packet flows in situations where the smart grid network is busy or targeted by a denial-of-service (DoS) attack. How SDN might strengthen the security of typical smart grids against such malicious attacks has been investigated in [40].

The centralization of SDN enhances the ability to manage complex industrial networks, including supervisory control and data acquisition (SCADA) systems that are at the core of present-day industrial automation [41]. For example, an SDN-based strategy to update SCADA systems is presented in [41]. The work carried out in [41] also provided an adequate means to stop eavesdroppers from recording communication flows amongst SCADA modules. This led to the proposal of an intrusion detection system (network-based intrusion detection system [NIDS]) that makes use of SDN and SCADA characteristics for traffic classification [41]. NIDS primarily works by employing SDN to collect network data and information to allow it to keep track of communication between the components of the ICS.

SDN is also a pivotal technology in the development of smarter and more energy-efficient buildings, which are the core of smart infrastructure. As smart buildings increasingly incorporate CPSs to enhance energy efficiency, comfort, and security in buildings, SDN naturally becomes the network architecture in this context. A very good example of the application of SDN in smart buildings is Cisco's digital building solutions platform [42]. This platform employs SDN principles in building intelligent networks that manage and optimize building services, such as lighting, heating, and cooling [42]. In this way, building management services can be adjusted dynamically by utilizing data and information in real time to optimize energy use and improve occupants' comfort with what is often referred to as building automation systems (BASs). BASs are directly linked to the Internet in smart buildings [43]. Using sensors and actuators, BASs typically enable the control and management of several building components, including heating, ventilation, and air-conditioning (HVAC) components and subsystems, security, safety, lighting, shading, and entertainment accessories and/or equipment, amongst others. BASs also have the potential to improve building maintenance and energy usage [43].

3.6 OPPORTUNITIES FOR FUTURE RESEARCH AND DEVELOPMENT IN SDN-ENABLED CPSs

The convergence of SDN and CPS, that is, SDN-CPS, or SDN-enabled CPS, offers numerous opportunities for future research and development. This section of the chapter discusses five key areas of potential advancement in SDN-enabled CPSs. These areas are quality of service (QoS), fault tolerance and resilience, machine learning (ML) and artificial intelligence (AI), edge computing, and the fifth-generation mobile network (5G network).

- *Quality of service (QoS)*. The efficient delivery of services in SDN-enabled CPSs relies on QoS. QoS ensures that the system meets specific performance requirements and provides reliable communication between physical devices and the underlying network. Ensuring optimum QoS in SDN-enabled CPSs is crucial because it directly impacts the performance and reliability of CPS applications. Typically, QoS management involves maintaining low-latency, high-bandwidth, and reliable connectivity to measure up to the stringent constraints of real-world use cases. By ensuring high-bandwidth availability and minimizing network congestion through SDN's dynamic resource allocation capabilities, QoS mechanisms aid in the optimization of throughput for various CPS applications, including those with high-data demands, like autonomous vehicles, telemedicine, and smart manufacturing [4]. Several studies have focused on QoS in SDN-enabled CPSs by addressing various aspects, such as traffic engineering, resource allocation, and congestion control. Since ensuring security alongside QoS is vital to protecting CPSs from cyberattacks and unauthorized access, the combination of SDN with techniques like network function virtualization (NFV) and implementation of advanced security mechanisms can significantly enhance the robustness of SDN-enabled CPS networks [44]. To support this, a QoS-aware routing system in SDN for IoT applications was proposed in [44].
- *Fault tolerance and resilience*. Network failures, hardware issues, and cyberattacks are just a few of the potential problems that can affect SDN-enabled CPSs. Therefore, ensuring fault tolerance and resilience in such systems is vital to maintaining continuous and reliable operation, where system failures can have severe consequences in real-world applications. SDN's ability to manage and reconfigure networks in real time provides opportunities to enhance fault tolerance and resilience in CPS networks. Fault-tolerant and resilient SDN controllers have been proposed in recent times—an example is the delay-based attack detection and isolation technique (DA-DIS), a fusion of ML with a route-handoff mechanism, proposed to prevent hostile switches from accessing the routes [45]. DA-DIS improves the resilience of industrial networks by predicting and labelling malicious switches [45]. Another example is Ravana, which works by using the full performance of multi-stage action and previous state restoration [46]. To achieve a good degree of fault tolerance, Ravana employs a range of generally acknowledged distributed application design best practices, specifically, a two-phase replication method that expressly guarantees event messages are received and processed properly while storing all the data in a shared in-memory log [46]. Ravana parallelizes event logging and transaction processing, ensuring that its fault tolerance contributes little to the throughput overhead [46]. A real-time delay attack detection and isolation method for fault-tolerant SDN in industrial settings

is also presented in [47]. The goal is to minimize the resilience in the SDN resilience manager (SDN-RM) system that had been previously proposed [47], where the link layer discovery protocol (LLDP) packets are delayed when the intruder breaches the OpenFlow switch and commences an intrusion.

- *Artificial intelligence and machine learning (AI/ML).* Integrating AI and ML with SDN in CPSs opens new possibilities for intelligent decision-making and resource optimization. AI/ML algorithms can analyze network data and adaptively adjust SDN configurations to improve performance and resource utilization. As an example, VANET issues have been addressed by researchers using a variety of ML-based approaches [48]. In [48], to choose the best routing strategy for an intelligent transport system (ITS) application, a deep reinforcement learning (DRL) approach was proposed. The proposed DRL method makes use of a deep Q-network that has a new artificial agent that learns its rules from high-dimensional inputs, and it outperforms more well-known ML methods [48]. A software-defined trust-based deep reinforcement learning framework (TDRL-RP) was presented in [21]. TDRL-RP comprises a deep Q-learning algorithm integrated into an illogically centralized SDN controller. Using typical VANETs as a case in point, a convolution neural network (CNN) is used by the SDN controller to find the highest routing path trust value based on a trust model that evaluates neighbors' packet-forwarding behavior, demonstrating the use of AI/ML for traffic prediction, anomaly detection, and network optimization in SDN environments [21]. TDRL-RP and similar AI/ML frameworks can be extended to SDN-CPSs to optimize network traffic flow, reduce latency, and enhance overall system performance. For example, a study that examines the state of the art to identify potential applications for ML-based self-healing in CPSs has been carried out [21]. Identification of anomalies, alerting of faults, and fault auto-remediation are the three main aspects of systems' self-healing functionality that are presented in [21]. There are many reasons to believe that the development of self-healing capabilities in CPSs will have a significant impact on future digital technologies. This is because self-healing capabilities can seamlessly integrate self-organizing and self-restoration functions into CPSs, enhancing user experience and system security.
- *Edge computing.* The integration of edge computing with SDN in CPS can greatly improve the system's effectiveness and reduce latency. Edge computing enhances the proximity of CPS devices and available computational resources to enable faster data processing, leading to speedy decision-making. In [49], edge computing frameworks have been designed and used via SDN for various applications. The evaluation focused on the network layer of the framework (that is, the access

network and wireless access network [WAN]). By leveraging SDN in edge computing environments, researchers can design efficient data processing and resource management strategies. It has also been suggested that SDN can control and manage a broad class of IoT devices via applications such as EdgeIoT [50]. In EdgeIoT, users' privacy remains a pressing issue to be addressed. In [50], all the IoT devices are reported to be linked to a few wireless access points close to edge servers that provide services to them. To safeguard their anonymity, every user employs a proxy virtual machine (VM). The IoT gadgets might, however, wander off to different access sites. To reduce traffic to the core network, proxy VM migration is therefore required. As an enhancement to the work in [50], an SDN/NFV-enabled edge node was presented in [22] for IoT services using integrated cloud/fog and network resources through end-to-end (E2E) SDN. Provisioning E2E network services in the architecture is done via a multidomain SDN orchestrator. To manage the OpenFlow-enabled switches, edge SDN controllers were revealed, and the IoT world testbed was used for the evaluations [51].

- *5G and 6G networks*. The 5G and beyond 5G (that is, 6G) networks offer reduced latency, higher data rates, and improved connectivity, making the 5G network a game changer for CPS applications. Integrating SDN with 5G networks can further enhance the capabilities of SDN-CPS. A few recent studies have explored the use of SDN in 5G networks to carry out operations like network slicing, network resource allocation, and network traffic management [52]. In a way, these studies have provided a premise for the integration or co-use of SDN- and 5G-based CPSs [52]. The widespread adoption and implementation of 5G-enabled SDN-CPS can be said to be relative to how security and privacy issues are addressed, among several other issues and challenges. Hence, CPS implementation in the 5G network is still very much embryonic. As technology develops further, the prospects and wider applications beyond 5G, specifically 6G technologies, that constitute the next generation of cellular networks are being discussed [53]. Even though 6G can be considered to still be in its infancy, it is anticipated to provide even faster speeds, lower latency, and more sophisticated features in comparison to 5G. The incorporation of 6G technologies into SDN-enabled CPSs presents novel prospects for innovation and progress across multiple industries. For example, 6G-enabled CPSs in smart cities can improve resource management, urban mobility, and environmental sustainability by optimizing key infrastructure in real time [53]. Also, 6G SDN-enabled CPSs can also be anticipated to revolutionize the way health-care services are accessed and provided, by enabling telemedicine, tailored health-care delivery, and remote patient monitoring [53].

- *Intent-based networking (IBN).* With its emphasis on business outcomes and intent, IBN marks a paradigm shift in networking. It provides a greater level of abstraction and intelligence in network administration by automating the translation of business intent into network policies and configurations [54]. Various technologies, including AI models, network function virtualization (NFV), and SDN, automate network administration processes. Zero-touch service and network management (ZSM), a next-generation management system, can only be realized with the aid of IBN [55]. IBN enables users to independently construct the underlying network infrastructure based on their intentions and to articulate broad, high-level interests or concerns (such as service, performance, data storage, etc.) in a declarative manner using human natural language. It is more than a game changer; with the help of sophisticated data analytics techniques, intentions are converted into executable scripts, or network rules, and then put into action via the network's IT automation capabilities to improve the IT infrastructure's agility in every way conceivable. Specifically, IBN develops a closed-loop automation system that substitutes intelligent and sophisticated software procedures for manual and error-prone network settings. This system includes the steps of intent expression, intent refining, intent activation, and intent assurance [54]. The field of IBN holds significant promise for revolutionizing network management, yet many of its foundational components remain in early developmental stages. Notably, major international standards organizations such as the IETF and leading industry players like Cisco and Huawei have been instrumental in shaping a comprehensive reference model for IBN [54]. This concerted effort underscores the significance of IBN. When considering the intersection of IBN with SDN-CPSs, the implications for various applications and advantages such as heightened levels of automation, adaptability, and intelligence in network management become evident.
- *Network function virtualization (NFV).* To allow network functions to operate as software on general-purpose hardware, NFV decouples network functions from proprietary hardware appliances [56]. The structure and functionality of CPSs have been completely transformed in recent years by the combination of NFV and SDN. NFV primarily entails establishing network functions as software-based virtual instances by abstracting them from specialized hardware appliances, whereas SDN offers centralized control and programmability over resource management. Considerable progress in the development of CPSs has resulted from this convergence, allowing network services to be dynamically provisioned, managed, and orchestrated to adapt to changing application needs. By separating network operations from the underlying hardware, NFV, according to [56], improves the

flexibility and scalability of CPSs and permits the quick deployment of new services and effective use of available resources. Security as a service (SecaaS) is provided by an ML-based cybersecurity architecture that was established in [57]. SecaaS provides an extensible, flexible, affordable, and reliable security solution by dynamically deploying security devices throughout the network using NFV and SDN. An autonomous system that restores an NFV version to a prior version when a failing version is identified has also been proposed using an ML model [58].

3.7 CONCLUSION

SDN and CPS represent groundbreaking paradigms that can augment and transform various industries. The amalgamation of SDN and CPS (that is, SDN-CPS) facilitates dynamic and programmable network configurations, merging physical processes with computing capabilities. This convergence yields numerous benefits, such as enhanced network management, real-time responsiveness, improved security, resource optimization, scalability, flexibility, and efficient service orchestration. These benefits collectively contribute to the successful deployment of CPS applications by ensuring their efficiency, resilience, and adaptability in dynamic environments. While SDN presents promising opportunities for enhancing the capabilities of CPSs, some challenges remain in terms of reliable and secure operations. Issues such as network reliability and availability, real-time constraints, complexity and scalability, privacy and security concerns, and interoperability and standardization demand further research and innovation. Hence, collaborations between researchers and practitioners are crucial to devising innovative SDN-CPS solutions that will ultimately yield safer, more efficient, and resilient CPS applications across various domains. Looking ahead, numerous prospects for future work and development arise from the integration of SDN and CPS. Areas such as QoS, fault tolerance, machine learning (ML) and artificial intelligence (AI), edge computing, and 6G networks present avenues for exploration. Researchers and developers can also leverage SDN-enabled CPS to transform industries like health care, transportation, and industrial automation by addressing contemporary challenges.

REFERENCES

[1] Ahmed, A. A., Nazzal, M. A., and Darras, B. M. (2021). Cyber-physical systems as an enabler of circular economy to achieve sustainable development goals: A comprehensive review. *International Journal of Precision Engineering and Manufacturing-Green Technology*, vol. 8, pp. 1–21.

[2] Urrea, C., and Benítez Machado, D. (2021). Software-defined networking solutions, architecture and controllers for the industrial Internet of Things: A review. *Sensors*, vol. 21, pp. 1–20.

[3] Josbert, N. N. *et al.* (2021). A framework for managing dynamic routing in industrial networks driven by software-defined networking technology. *IEEE Access*, vol. 9, pp. 74343–74359.

[4] Kreutz, D. *et al.* (2015). Software-defined networking: A comprehensive survey. *Proceedings of the IEEE*, vol. 103, no. 1, pp. 14–76.

[5] Open Networking Foundation. (2012). Software-defined networking: The new norm for networks. *ONF White Paper*, vol. 2, pp. 2–6.

[6] McKeown, N. *et al.* (2008). OpenFlow: Enabling innovation in campus networks. *Computer Communication Review*, vol. 38, no. 2, pp. 69–74.

[7] Voellmy, A. *et al.* (2012). Procera: A language for high-level reactive network control. In *Proceedings of the First Workshop on Hot Topics in SDN*, pp. 43–48.

[8] Gude, N. *et al.* (2008). NOX: Towards an operating system for networks. *ACM SIGCOMM Computer Communication Review*, vol. 38, no. 3, pp. 105–110.

[9] Hu, F., Hao, Q., and Bao, K. (2014). A survey on software-defined network and OpenFlow: From concept to implementation. *IEEE Communications Surveys & Tutorials*, vol. 16, no. 4, pp. 2181–2206.

[10] Tsiatsis, V. *et al.* (2019). Autonomous vehicles and systems of cyber-physical systems. In *Internet of Things* (2nd ed., pp. 299–305). Academic Press.

[11] The National Science Foundation (NSF). (2015). Cyber-physical systems. www.nsf.gov/funding/pgm_summ.jsp?pims_id=50328671 (Accessed on 26 August 2023).

[12] Amato, A., Quarto, A., and Di Lecce, V. (2020). An application of cyber-physical systems and multi-agent technology to demand-side management systems. *Pattern Recognition Letters*, vol. 141, pp. 23–31.

[13] Wang, X. V. *et al.* (2017). Human-robot collaborative assembly in cyber-physical production: Classification framework and implementation. *CIRP Annals–Manufacturing Technology*, vol. 66, pp. 5–8.

[14] Goodchild, A., and Toy, J. (2018). Delivery by drone: An evaluation of unmanned aerial vehicle technology in reducing CO_2 emissions in the delivery service industry. *Transportation Research Part D*, vol. 61, pp. 58–67.

[15] Zhao, J. *et al.* (2019). A fully integrated and self-powered smartwatch for continuous sweat glucose monitoring. *ACS Sensors*, vol. 4, pp. 1925–1933.

[16] Harkat, H. *et al.* (2024). Cyber-physical systems security: A systematic review. *Computers & Industrial Engineering*, vol. 188, no. 109891.

[17] Wan, K. *et al.* (2011). Investigation on composition mechanisms for cyber-physical systems. *International Journal of Design, Analysis and Tools for Circuits and Systems*, vol. 2, no. 1, pp. 30–40.

[18] Tripathi, D. *et al.* (2022). An integrated approach to designing functionality with security for distributed cyber-physical systems. *The Journal of Supercomputing*, vol. 78, pp. 14813–14845.

[19] Han, B. *et al.* (2015). SDN-based cyber-physical systems: A case study in smart grid. *IEEE Network*, vol. 29, no. 4, pp. 32–38.

[20] Smith, J., Johnson, M., and Brown, R. (2018). Enhancing smart grid resilience with software-defined networking. *IEEE Transactions on Power Systems*, vol. 33, no. 4, pp. 4191–4200.

[21] Johnphill, O. *et al.* (2023). Self-healing in cyber-physical systems using machine learning: A critical analysis of theories and tools. *Future Internet*, vol. 15, no. 7, pp. 244.
[22] Vilalta, R. *et al.* (2016). End-to-end SDN orchestration of IoT services using an SDN/NFV-enabled edge node. In *Proceedings of the Optical Fiber Communications Conference and Exhibition (OFC)*, pp. 1–3.
[23] Smith, J. *et al.* (2020). Advancements in software-defined networking: A comprehensive review. *Journal of Networking Technology*, vol. 42, no. 3, pp. 123–136.
[24] Tyagi, A. K., and Sreenath, N. (2021). Cyber physical systems: Analyses, challenges and possible solutions. *Internet of Things and Cyber-Physical Systems*, vol. 1, pp. 22–33.
[25] Johnson, C. R., and White, E. M. (2019). Software-defined networking for cyber-physical systems: A review of challenges, solutions, and opportunities. *IEEE Transactions on Industrial Informatics*, vol. 15, no. 4, pp. 2160–2167.
[26] Zhang, D., Yu, F., and Yang, R. (2018). A machine learning approach for software-defined vehicular Ad Hoc networks with trust management. In *2018 IEEE Global Communications Conference (GLOBECOM)*, pp. 1–6.
[27] Gupta, S., and Mohan, S. (2017). Security and resilience in smart grid cyber-physical systems: A review. *IEEE Transactions on Industrial Informatics*, vol. 13, no. 6, pp. 2999–3006.
[28] Liu, Y. *et al.* (2017). Dynamic quality of service provisioning for SDN-based industrial cyber-physical systems. *IEEE Transactions on Industrial Informatics*, vol. 13, no. 6, pp. 3169–3180.
[29] Porras, P. *et al.* (2015). A security enforcement kernel for OpenFlow networks. In *Proceedings of the 2015 ACM Conference on Special Interest Group on Data Communication*, pp. 27–28.
[30] Khan, A. (2017). SDN-based fault tolerance in cyber-physical systems. In *Proceedings of the IEEE International Conference on Cybernetics and Intelligent Systems*, pp. 268–273.
[31] Zhang, C. *et al.* (2016). Toward distributed software-defined networking for cyber-physical systems. *IEEE Transactions on Industrial Informatics*, vol. 12, no. 6, pp. 2158–2168.
[32] Tavakoli, A. *et al.* (2009). Applying NOX to the datacenter. *ACM Workshop on Hot Topics in Networks*. http://conferences.sigcomm.org/hotnets/2009/papers/hotnets2009-final103.pdf (Accessed on 11 November 2023).
[33] Casado, M. *et al.* (2007). Ethane: Taking control of the enterprise. *ACM SIGCOMM Computer Communication Review*, vol. 37, no. 4, pp. 1–12.
[34] Curtis, A. *et al.* (2011). DevoFlow: Scaling flow management for high-performance networks. In *Proceedings of the ACM SIGCOMM 2011 Conference, SIGCOMM'11*, pp. 254–265.
[35] Ohri, P., and Neogi, S. (2022). Software-defined networking security challenges and solutions: A comprehensive survey. *International Journal of Computing and Digital Systems*, vol. 12, pp. 383–400.
[36] Liao, Z. *et al.* (2018). A privacy-preserving decentralized control scheme for the Industrial Internet of Things. *IEEE Transactions on Industrial Informatics*, vol. 14, no. 1, pp. 428–437.
[37] Ladegourdie, M., and Kua, J. (2022). Performance analysis of OPC UA for industrial interoperability towards Industry 4.0. *IoT 2022*, vol. 3, pp. 507–525.

[38] Molina, E., and Jacob, E. (2017). Software-defined networking in cyber-physical systems: A survey. *Computers and Electrical Engineering*, vol. 66, pp. 407–419.
[39] Dong, X. *et al.* (2015). Software-defined networking for smart grid resilience: Opportunities and challenges. In *Proceedings of the 1st ACM Workshop on Cyber-Physical System Security (ACM)*, pp. 61–68.
[40] Ramirez, G. D. *et al.* (2016). Enhancing smart grid reliability through software-defined networking. *IEEE Transactions on Smart Grid*, vol. 7, no. 6, pp. 2785–2796.
[41] da Silva, E. *et al.* (2015). Capitalizing on SDN-based SCADA systems: An anti-eavesdropping case-study. In *Proceedings of the 2015 IFIP/IEEE International Symposium on Integrated Network Management (IM 2015)*, pp. 165–173.
[42] Cisco Digital Building Solutions. (2024). www.cisco.com/c/en/us/solutions/smart-building/what-is-a-smart-building.html (Accessed on 3 January 2024).
[43] O'Grady, T. *et al.* (2021). A systematic review and meta-analysis of building automation systems. *Building and Environment*, vol. 195, no. 3, p. 107770.
[44] Bera, S. *et al.* (2021). Q-Soft: QoS-aware traffic forwarding in software-defined cyber-physical systems. *IEEE Internet of Things Journal*, vol. 9, no. 12, pp. 9675–9682.
[45] Liatifis, A. *et al.* (2022). Fault-tolerant SDN solution for cybersecurity applications. In *Proceedings of the 17th International Conference on Availability, Reliability and Security (ARES '22). Association for Computing Machinery*, New York, no. 73, pp. 1–6.
[46] Katta, N. *et al.* (2015). Ravana: Controller fault-tolerance in software-defined networking. In *Proceedings of the 1st ACM SIGCOMM Symposium on Software Defined Networking Research (SOSR '15). Association for Computing Machinery*, no. 4, pp. 1–12.
[47] Ramani, S., and Jhaveri, R. H. (2022). ML-based delay attack detection and isolation for fault-tolerant software-defined industrial networks. *Sensors*, vol. 22, no. 18, pp. 6958.
[48] Yu, X. *et al.* (2018). Machine learning for software-defined networking: A survey. *IEEE Communications Surveys and Tutorials*, vol. 20, no. 4, pp. 3032–3071.
[49] Nain, A. *et al.* (2024). Resource optimization in edge and SDN-based edge computing: A comprehensive study. *Cluster Computing*, pp. 1–29.
[50] Sun, X., and Ansari, N. (2016). EdgeIoT: Mobile edge computing for the Internet of Things. *IEEE Communications Magazine*, vol. 54, no. 12, pp. 22–29.
[51] Serra, J. *et al.* (2014). Smart HVAC control in IoT: Energy consumption minimization with user comfort constraints. *Scientific World Journal*, vol. 2014, pp. 161874.
[52] Alotaibi, D., Thayananthan, V., and Yazdani, J. (2021). The 5G network slicing using SDN-based technology for managing network traffic. *Procedia Computer Science*, vol. 194, pp. 114–121.
[53] Murroni, M. *et al.* (2023). 6G—Enabling the new smart city: A survey. *Sensors*, vol. 23, no. 17, pp. 7528.
[54] Leivadeas, A., and Falkner, M. (2023). A survey on intent-based networking. *IEEE Communications Surveys & Tutorials*, vol. 25, no. 1, pp. 625–655.
[55] Rajab, M., Yang, L., & Shami, A. (2024). Zero-touch networks: Towards next-generation network automation. *Computer Networks*, vol. 243, no. 2024, pp. 110294.

[56] Alarifi, A. *et al.* (2020). SDN and NFV for security enhancement in CPS: State-of-the-art and research challenges. *IEEE Access*, vol. 8, pp. 58355–58373.
[57] Gardikis, G. *et al.* (2017). Shield: A novel NFV-based cybersecurity framework. In *2017 IEEE Conference on Network Softwarization (NetSoft) (IEEE)*, pp. 1–6.
[58] Ahrens, J. *et al.* (2018). An AI-driven malfunction detection concept for NFV instances in 5G. arXiv preprint arXiv:1804.05796. https://doi.org/10.48550/arXiv.1804.05796

Chapter 4

Toward future metasystems

From today's CPS to tomorrow's cyber-physical-social systems in the emerging metaverse

Minoo Soltanshahi, Nika Hosseini, and Martin Maier

ABBREVIATIONS

3GPP	third-generation partnership project
AI	artificial intelligence
AR	augmented reality
CC	circular causality
CI	collective intelligence
CPS	cyber-physical systems
CPSS	cyber-physical-social systems
DAC	decentralized autonomous corporation
DAOs	decentralized autonomous organizations
DAS	distributed autonomous system
DeSci	decentralized science
ETSI	European Telecommunications Standards Institute
F5G	fifth generation
GAI	generative AI
GANs	generative adversarial networks
GDM	generative diffusion models
GPT	generative pre-trained transformers
ICPS	industrial cyber-physical systems
IoM	Internet of Minds
IoT	Internet of Things
ITU-R	International Telecommunication Union Radiocommunication Sector
NFT	non-fungible token
NGMN	next-generation mobile networks
VR	virtual reality
XR	extended reality

4.1 INTRODUCTION

The rapid advancement of technology has ushered in an era marked by the convergence of physical and digital realities. Central to this transformative landscape is the concept of the metaverse, a virtual universe seamlessly blending the real and digital realms. It represents a profound shift

 DOI: 10.1201/9781003559993-4

in human interaction with technology, promising a unified, immersive, and interconnected experience. This concept of the metaverse is part of a broader technological and societal transformation. Key to this transformation is the evolution of communication networks, transitioning from conventional networks to the F5G initiative, laying the groundwork for digital innovations such as digital avatars and sensory Internet experiences [1, 2].

Simultaneously, the multiverse concept has emerged, proposing eight distinct types of reality that bridge the gap between virtual and augmented realities. These concepts reshape human–computer interactions, offering immersive experiences beyond traditional 2D interfaces. Technologies like AI, XR, and blockchain are essential for realizing the metaverse's vision. As technology advances, the multiverse becomes the foundation for extended XR experiences, integrating physical and non-physical dimensions [3]. The Fourth Industrial Revolution, driven by IoT and CPS, has given rise to CPSS, which seamlessly integrates technology and humanity [4]. Transitioning from Industry 4.0 to Industry 5.0 and Society 5.0 emphasizes a human-centric approach [3]. This evolution incorporates digital twins, AI, edge computing, and 5G/6G networks, forming an industrial metaverse. Ericsson's 6G vision emphasizes trustworthiness, sustainability, automation, and digitalization. The successful realization of these visions relies on coordinated efforts among organizations like ITU-R, 3GPP, ETSI, NGMN Alliance, and Next G Alliance, defining the framework for 6G technologies.

As we embark on this journey into the 6G era, the convergence of real and digital worlds embodied by the metaverse and multiverse promises to redefine how we interact, communicate, and create value. A human-centric approach is crucial, ensuring technology benefits to individuals and society. Within the realm of complex systems, the concept of CPSS has emerged, emphasizing the integration of social dynamics into system design. CPSS envisions seamless coordination across cyberspace, physical space, and human social spaces, introducing citizen sensing and actuation. These concepts enhance environmental intelligence and social applications within CPSS [4]. Blockchain technology plays a vital role in CPSS by creating an O2O society and simplifying complex coordination through smart contracts. Parallel intelligence bridges the cognitive gap in CPSS, aligning with the vision of decentralized autonomous organizations [5].

This evolution from CPS to CPSS leads to the concept of "metasystems," blurring the lines between the physical and virtual worlds. Further, stigmergy and collective intelligence shape the trajectory of a future *stigmergic Society 5.0*, enhancing decentralized coordination within a digital ecosystem [6]. In addition, we delve into generative AI, exemplified by lifelike digital organisms, and the emerging concept of *interbeing*, which underpins the intricate web of connections within the metaverse's forthcoming virtual society. Together, these concepts offer a visionary outlook on interconnected, intelligent societies [7].

The remainder of this book chapter is structured as follows. Section 4.2 explores the technical details of the 6G vision, delving into the conceptualization of amalgamating real-world and digital experiences. Section 4.3 will examine the emerging metaverse, providing insights into recent advancements within this dynamically evolving domain. Sections 4.4 and 4.5 examine CPSS and their impact on future metasystems. Specifically, Section 4.4 explores CPSS components, applications, and blockchain integration, while Section 4.5 discusses extended stigmergy in dynamic media, highlighting its role in decentralized coordination and collective intelligence in the emerging metaverse. Section 4.6 explores the profound impact of generative AI on the metaverse, emphasizing how this transformative technology contributes to the design of lifelike digital entities and immersive experiences within the virtual realm. Finally, Section 4.7 concludes this book chapter, summarizing the key insights and implications from our comprehensive exploration and outlining possible future research avenues.

4.2 6G VISION: FUSION OF REAL AND DIGITAL WORLDS

An intriguing concept, the *multiverse*, has been introduced to bridge the gap between the real and the digital domains [8]. The multiverse comprises eight distinct types of reality, including but not limited to VR and AR. Note that the recently emerging metaverse is a closely related term poised to become the subsequent evolution beyond today's mobile Internet. The metaverse represents the fusion of real and digital worlds, seamlessly integrating interoperable and immersive virtual ecosystems through user-controlled avatars. It blurs the boundaries between virtual and physical realms by enabling tangible interactions through VR, haptic feedback, and AR. Its inclusive nature ensures no single company owns the metaverse, promoting collaboration among entities. Immersive experiences transcend conventional 2D interactions, fostering real-life-like connections among users and allowing thousands to coexist in a single session while sharing interactions globally. The metaverse's ecosystem supports comprehensive services, bridging the gap between the physical and the virtual worlds.

Figure 4.1 depicts the multiverse as an architecture of advanced XR experiences. It consists of three dimensions, six variables, and eight realms. The metaverse resonates with emerging paradigms, like the multiverse, in the ever-evolving landscape, signifying a transformative shift in human–computer interactions [9]. The metaverse's transformative potential is immense, as it is envisioned to become a cyber-physical space for content creation, virtual economy, and social interactions, ultimately impacting the real world. To achieve this ambitious vision, integrating AI, XR, 6G networks, and blockchain technology is deemed crucial [10]. User engagement plays a pivotal

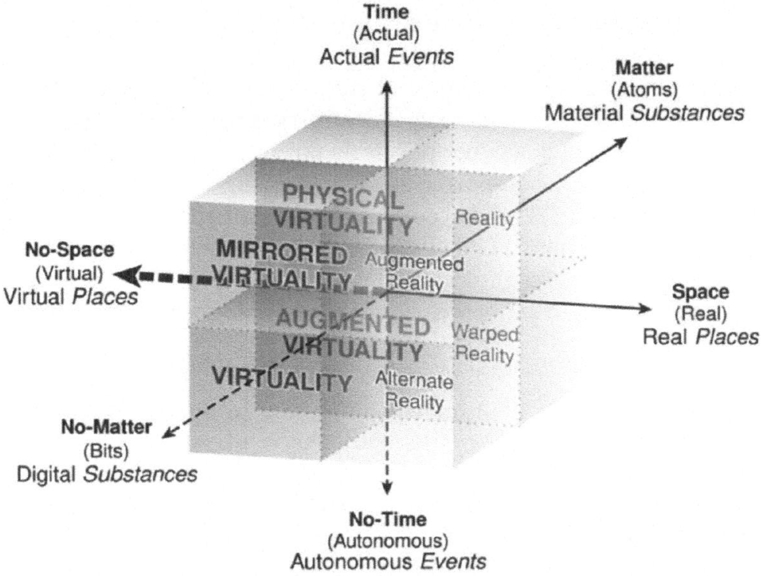

Figure 4.1 The multiverse: an architecture of advanced XR experiences.

Source: [3].

role in the success of the metaverse, as it relies heavily on massive online users to realize its social value [11]. The metaverse aims to blur the line between the virtual and the physical worlds, offering tangible forms for users to interact with physically, employing technologies such as VR, AR, and haptic feedback.

The multiverse concept introduces an interactive platform that seamlessly merges real and virtual elements in XR, resulting in cross-reality environments known as third spaces. This concept aligns with Mark Weiser's concept of embodied virtuality, more widely known as ubiquitous computing [12]. As XR technologies continue to advance, they promise to unveil novel and unforeseen types of reality, whereby *X* represents unexplored possibilities on the digital frontier. The multiverse is an architectural foundation for advanced XR experiences, incorporating three well-established physical dimensions (space, time, and matter) and three non-physical dimensions (no-space, no-time, and no-matter) collectively forming the virtual world. These digital dimensions are not constrained by physical limitations, enabling the creative design of innovative XR experiences using six variables. Within the multiverse's architectural framework, eight distinct realms of reality emerge from combining three opposing physical and digital dimensions.

These realms encompass diverse realities, ranging from conventional VR and AR to more sophisticated forms of reality, such as mirrored virtuality, warped reality, and alternate reality.

The vision for 6G revolves around the convergence of the real and the digital worlds, aiming to establish a seamless integration of physical and digital realms by creating a cyber-physical continuum. Ericsson's 6G research outlook underscores the importance of addressing societal challenges and adapting to technological advancements to meet the demands of the 6G era. The driving forces behind 6G development are centered on trustworthiness, sustainability, accelerated automation, and digitalization. To realize the 6G vision, several paradigm shifts are proposed to guide the transformation [13]:

- Moving from secure communication to building trustworthy platforms
- Transitioning from data management to granting data ownership
- Shifting focus from energy efficiency to achieving sustainable transformation
- Evolving from terrestrial 2D to global 3D connectivity, encompassing land, sea, and air areas
- Embracing learning networks over manually controlled ones, utilizing intelligence to achieve goals
- Adapting from predefined services to flexible user-centricity, enabling networks to cater to user needs
- Progressing from separate physical and digital worlds to a cyber-physical continuum that merges realities for seamless interaction and immersive experiences
- Expanding the role of networks from mere data links to encompassing services beyond communication

Figure 4.2 illustrates Ericsson's envisioned cyber-physical continuum for the 6G era to fully merge realities and allow immersive experiences; 6G technology facilitates the seamless movement between the physical and the digital worlds by means of network intelligence and synchronization. The physical world is equipped with sensors sending real-time data to update its digital representation, while actuators execute commands from the digital realm. This enables real-time tracing, analysis, and action, akin to the metaverse concept, where digital and physical objects coexist to enhance the real world with mixed reality.

The transition to 6G also involves addressing the challenge of value creation and capture, necessitating a shift from individual technology-driven innovations to fostering innovation within platforms and ecosystems and ultimately benefiting society as a whole. The successful implementation of 6G heavily relies on the widespread availability of sensing and computing capabilities supported by distributed neural networks, which will pave the

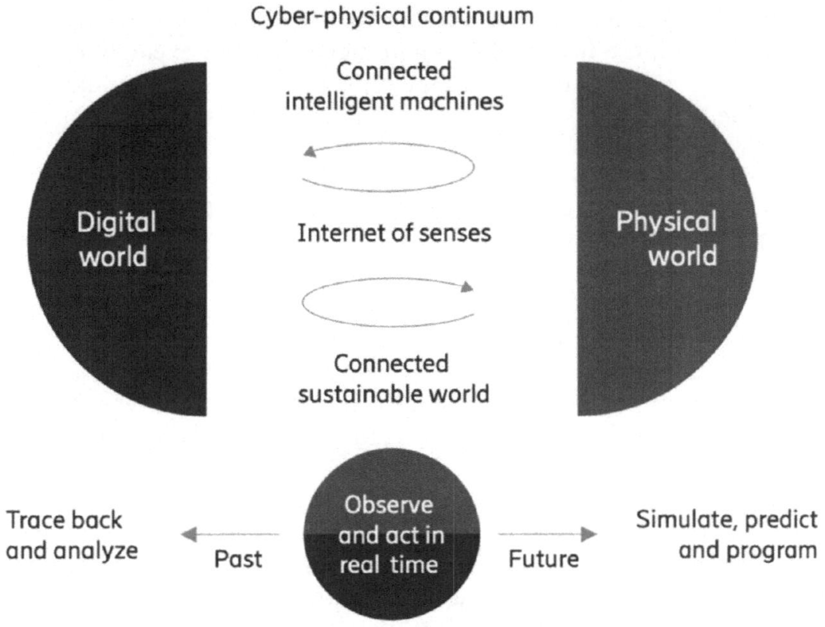

Figure 4.2 Ericsson envisioned a cyber-physical continuum for the 6G era.
Source: [13].

way for the emergence of perceptive mobile networks (PMNs). In the 6G era, wireless networks will go beyond their conventional role and function as large-scale sensor networks, enabling the deployment of innovative applications, such as human activity recognition and vehicle-to-everything (V2X) communication. Quantum-enabled 6G wireless networks are expected to harness the potential of quantum information technology, leading to significant enhancements in security, computing, and communication efficiency. Concurrently, integrating blockchain technology will be pivotal in instilling trust, ensuring security, and offering fault tolerance in 6G networks. Blockchain's application will extend to various domains, including network management and resource sharing. At the core of 6G's evolution lies the concept of AI-native networks, where edge AI will drive efficiency and low-latency AI services. This transformative shift in network design will emphasize task-oriented communication by seamlessly integrating communication, computation, and learning processes [14, 15].

To achieve the vision of 6G, a coordinated and cooperative effort among multiple organizations and alliances is essential. Key players such as ITU-R, 3GPP, ETSI, NGMN Alliance, and Next G Alliance are collaboratively working to establish the framework and specifications for 6G technologies via a unified approach toward turning the 6G vision into reality. As we enter the 6G era, the fusion of real and digital worlds, exemplified by the metaverse and multiverse, will reshape how we interact, communicate, and create value.

4.3 EMERGING METAVERSE: RECENT PROGRESS AND STATE OF THE ART

Recently, much progress has been made on many important aspects of the metaverse. Pertinent publications can be categorized into different areas. Specifically, the areas covered include architecture design, use cases, security, educational technology, and health care. The following comprehensive overview serves as a roadmap for navigating the diverse landscape of metaverse research.

Several seminal contributions have redefined how virtual environments are structured and managed in architectural design. Notably, decentralized science (DeSci) MetaMarkets were introduced as a revolutionary concept that enables virtual representations of decentralized science markets. These MetaMarkets serve as platforms for knowledge distribution, bridging the gap between humans, robots, and digital entities. Furthermore, DeMana, a management framework within MetaMarkets, optimizes decision-making processes by facilitating efficient resource allocation and improved knowledge management. Integrating virtual and real markets holds tremendous potential for strategic optimization and efficient resource utilization.

More specifically, the authors of [8] extensively explore collaborative sensing, edge-assisted rendering, and the allocation of resources within the context of developing virtual cities. Similarly, in [9], the authors introduce a metaverse framework empowered by digital twin technology that seamlessly integrates IoT, XR, and blockchain components to create a comprehensive and immersive experiential realm. The authors of [16] delve into optimizing user incentives within the Ethereum network, aligning contributions with incentivization strategies to elevate service quality. This intricate interplay between the discipline of architectural design and various other facets of metaverse investigation weaves together a multifaceted narrative that calls for in-depth examination.

In [17], the authors initiate a web evolution beyond Web 2.0, presenting an AIB-metaverse-based Web 3.0 architecture that seamlessly blends physical and virtual worlds. The presented design includes layers for interaction, space rendering, smart decision, and secure storage. Enabling decentralization, AI-driven decisions, and secure blockchain data custody helps advance decentralized intelligence. Further, the authors of [18] address challenges related to splitting XR traffic over 5G networks. They propose power-saving strategies for XR devices that optimize the latency of XR experiences.

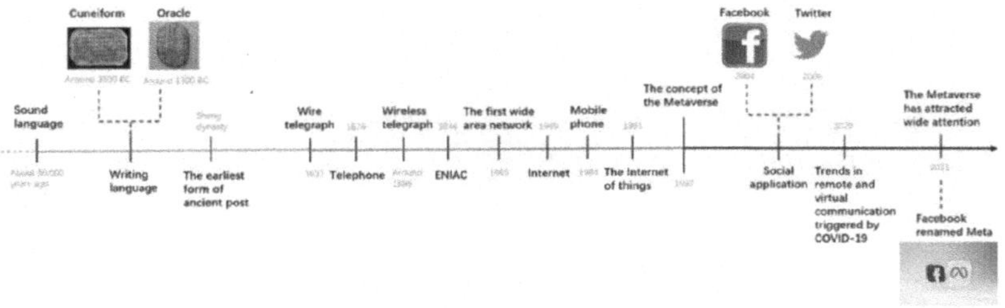

Figure 4.3 Timeline of the development of communication methods.

Source: [19].

As surveyed in [19], emerging metaverse communication technology and media are continuously advancing. For illustration, Figure 4.3 shows that electronic communication has overcome temporal and spatial limits, resulting in improved information exchange efficiency. The metaverse's emergence further magnifies this transformation by blending reality and virtuality seamlessly, thus reshaping our perception of time and space.

In [20], uncertainty modeling for control engineering tools in industrial cyber-physical metaverse smart manufacturing systems (ICPMSMSs) is described in more detail. The proposed approach involves decision matrices, estimation methods, and ranking tools for effective tool selection in complex systems. It provides a structured framework for evaluating tools. In [21], a framework is proposed to optimize accuracy and latency in edge intelligence for immersive multimedia applications. By integrating predictive models, deep reinforcement learning, and meta-learning, the framework enhances content delivery, user interaction, and communication latency, thus improving overall application performance.

By analyzing blockchain cryptocurrency networks, the authors of [22] introduce the SVRP method that captures network structural identity using random walks. This method advances representation learning for intricate BCNs. In [23], the focus is on metaverse xURLLC services in wireless networks. The authors introduce models and contract designs to optimize user experience and utility. This is achieved through strategic resource allocation that enhances both quality of experience and utility. Moreover, the authors of [24] address resource allocation in MEC-enabled metaverse environments using cooperative multi-agent game theory. The article introduces Dec-POMDP and reward functions to optimize allocation to improve user QoE and resource balance in the metaverse.

The application domains of the metaverse, often referred to as its use cases, have captured the attention of researchers seeking to harness its transformative potential. In this context, the authors of [25] investigate the mediatization–metaverse relationship, offering an intricate analysis of value

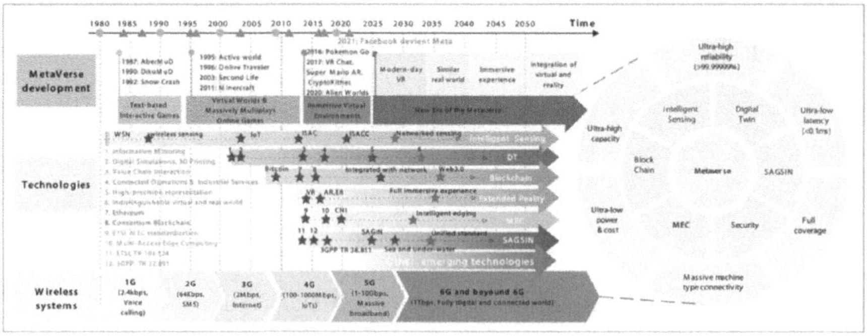

Figure 4.4 Metaverse roadmap: communication, networking, and enabling technologies.
Source: [27].

generation and structured interactions in this emergent landscape. Furthermore, in [26], the authors examine the integration of 6G networks into the metaverse, aiming to elevate vertical industries and immersive experiences. The work in [27] extensively explores communication and networking technologies for real-time interactions in the metaverse. For illustration, Figure 4.4 depicts the roadmap of how the metaverse evolves with these technologies. The roadmap defines the metaverse and emphasizes its strict communication and networking requirements. Fulfilling these demands is vital for the metaverse's realization.

In [28], the authors highlight the integration of pivotal technologies, including digital twins, AI, edge computing, and 5G/6G, within the context of Industry 5.0 advancements. This integration leads to the emergence of an industrial metaverse, a virtual counterpart running parallel to the physical industrial system. Within this virtual space, global experts collaborate on comprehensive product planning with a transformative impact on conventional manufacturing methods, enhancing productivity and promoting iterative advancement. Figure 4.5 visually outlines the technological progression of the industrial metaverse for advanced manufacturing. Interestingly, the role of non-fungible tokens (NFTs) in smart cities is further examined in [29], where the authors present innovative solutions for enhancing efficiency, security, and transparency.

Addressing security concerns remains important in the metaverse ecosystem, prompting researchers to pioneer solutions that protect its participants. The study in [30] meticulously scrutinizes security vulnerabilities in NFT trading platforms, shedding light on plausible attack vectors and proposing remedies to enable security. In a complementary manner, the study in [31] introduces an NFTPrivate protocol that leverages cryptographic commitments and zero-knowledge proofs to uphold the privacy and confidentiality of NFT transactions. In [32], the authors also tackle essential aspects

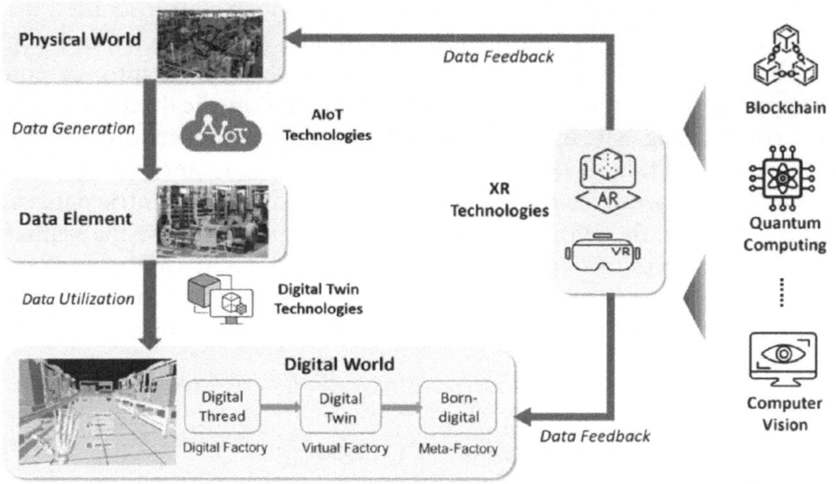

Figure 4.5 Technology roadmap of the industrial metaverse for advanced manufacturing.
Source: [28].

of continual authentication through federated learning using an adaptable security framework tailored to virtual reality. Similarly, the authors of [33] outline a two-factor authentication model tailored for metaverse avatars, which assures traceability and consistency of identity.

Other important aspects related to educational technology in the metaverse have been covered in [34–37], where the authors explore innovative approaches using technology to enhance learning. In [34], an edu-metaverse powered by AI, AR, and blockchain is introduced, focusing on embodied cognition and learner engagement. Similarly, the authors of [35] propose a metaverse for language learning using constructivist principles, exemplified by the Learningverse platform that emphasizes the impact of data collection. This platform is further investigated in [36] as a tool for interactive education that seamlessly integrates avatars, virtual spaces, and networks. The education metaverse framework, outlined in [37], prioritizes adaptability across five layers and underscores the role of blockchains in sustainability and user participation.

In health care, the metaverse holds the potential to revolutionize medical practices and patient experiences. Recent scholarly explorations are notably inclined toward harnessing the potential of the metaverse for addressing prevailing cognitive health challenges. The conceptualization of "meta-hospitals" and "meta-laboratories" ingeniously leverages digital twin technology, alongside augmented and virtual reality, to pave the way for remote consultations, automated testing, and analysis. The meticulously designed approach embraces diverse techniques, such as transcranial direct current stimulation, VR-based exercise systems, and AR-enhanced cognitive rehabilitation platforms [38]. In [39], the authors delve into the potential applications of the metaverse in telemedicine, medical education, and more, spotlighting XR, AI, distributed computing, and decentralized ledger technology (DLT). In [40], the authors exploit NFTs to address health-care challenges, proposing their use in supply chain and patient-centric data management. In [41], the authors also explore how the metaverse may enhance health-care services through innovative technologies and approaches.

Each of the aforementioned publications offers insights that collectively contribute to a comprehensive understanding of the evolving metaverse landscape.

4.4 CPSS: INTEGRATING INTELLIGENCE IN FUTURE METASYSTEMS

The concepts of open complex giant system and collaborative integration workshop were initially developed by Chinese scientists in the 1990s. They paved the way for creating the so-called *artificial societies*, *computational experiments*, and *parallel execution (ACP)* approach. This approach aims to manage and control complex systems via closed-loop feedback control between physical and artificial systems. Through ACP, events in a complex system can be computed and tested in a software-defined "laboratory," thereby providing decision and verification support [42].

The foundation of parallel systems in CPSS relies on knowledge automation technology, which combines knowledge representations and machine learning. Knowledge representations convert knowledge into qualitative descriptions and data structures. For example, parallel learning, proposed by Li and Lin, extracts relevant information from big data using knowledge from software-defined artificial systems. CPSS development is key to the "five-grid-in-one" theory, integrating transportation, energy, information, IoT, and the Internet of Minds (IoM). IoM aims to bridge human and machine intelligence gaps by creating a social ecosystem of interconnected intelligent agents, addressing challenges in knowledge acquisition, representation, transmission, association, and utilization, facilitating intelligent operations in CPSS [42].

The parallel society model comprises three main steps: (1) modeling and representing the real social system, (2) computing and experimenting with historical events to adjust parallel system behaviors, and (3) controlling and managing parallel execution between real and artificial systems. The model utilizes online data for real-time collection, fusion, and analysis of social signals. ACP-based social computing, data mining, natural language processing, and machine learning are integrated to enable comprehensive social analysis.

The results obtained guide, manage, and visually provide feedback to social personnel, facilitating remote multi-party collaboration and societal transformation [42].

In the ever-evolving landscape of CPS, the concept of CPSS has emerged as a groundbreaking paradigm that promises a new era of interconnected smart enterprises and industries. This concept, first introduced in 2010, emphasizes the integration of social and human dynamics as integral components in CPS design, ushering in the potential for seamless coordination between cyberspace, the physical world, and human resources. In recent years, researchers have witnessed a resurgence of interest in CPSS, paying particular attention to the role of individual human beings within these systems. The paradigm shift from traditional CPS to CPSS is often called *enhanced living environments (ELEs)*. The critical question at the heart of CPSS revolves around humans' roles and how they can be effectively integrated into these systems. Humans in CPSS can both consume services and contribute to system functionality, known as citizen sensing and actuation. *Citizen sensing* collects data using citizen sensors, while *citizen actuation* generates actionable items through citizen actuators [43].

Any given CPSS emphasizes human involvement in today's CPS. As shown in Figure 4.6, the CPSS ecosystem has three dimensions: cyberspace, physical space, and human-social space, also known as hyperspace. To meet given deployment challenges, novel architectural designs have been proposed, including wireless network virtualization and software-defined networking (SDN) for improved resource utilization and scalability. Other design

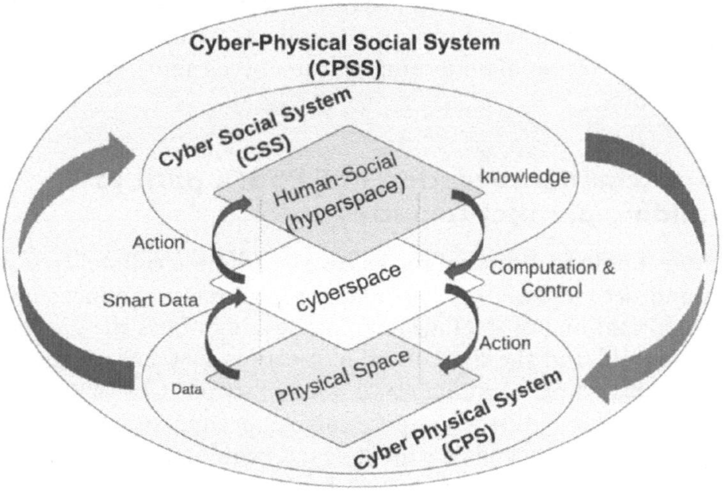

Figure 4.6 CPSS ecosystem.

Source: [43].

solutions involve integrated caching, computing, and networking to accelerate information retrieval and enhance data processing. While incorporating humans within CPSS has opened up new possibilities for enhanced environmental intelligence through innovative social applications like citizen sensing and actuation, it also presents challenges due to the inherent variability of human behavior.

In [43], a comprehensive CPSS taxonomy was outlined, comprising the following six distinct categories, alongside their associated fields and disciplines:

1. *Components.* CPSS consists of physical, cyber, and social systems.
2. *Assets.* These are valuable resources within CPSS, tangible (hardware, software) and intangible (information, human interactions).
3. *Applications and platforms.* CPSS applications cover domains like smart cities, health care, transportation, and energy. They provide services for monitoring and managing the CPSS environment.
4. *Technologies.* CPSS integrates technologies like IoT, cyber-physical systems, social networking, and computing to address challenges related to data fusion, computation, communication, privacy, and security.
5. *Design.* Focuses on effectively combining human and artificial intelligence using component-based, model-based, and contract-based approaches.
6. *Supporting theory.* ACP is introduced as a supporting theory for CPSS.

This taxonomy has been crafted with the explicit purpose of providing a structured framework that not only enables prospective research endeavors but also fosters the creation of prototype products within the CPSS domain. It also serves as an essential underpinning for a better understanding of their intricate dynamics.

4.4.1 Blockchain integration in CPSS: a path to building parallel societies

Blockchain technology is known for its trustworthiness, reliability, usability, efficiency, and decentralized nature, making it a vital component of CPSS-driven transformations of intelligent industries. It bridges the gap between the physical world and the digital realm, creating an O2O parallel society that incorporates complex human and social factors. CPSS-based parallel blockchain technology utilizes computational experiments and parallel optimizations to enhance blockchain's intelligence, enabling it to seamlessly integrate with online big data and IoT to form the Blockchain of Things (BoT) and ensure the secure management of physical devices. Smart contracts, an integral part of blockchains, are crucial in CPSS-based knowledge automation. Intelligent smart contracts go beyond traditional legal provisions, taking "IF–THEN" scenario–response rules to various unforeseen "WHAT–IF" scenario-deduction rules. Intelligent smart contracts simplify and reduce CPSS's complexity by addressing human cooperation and coordination uncertainties, thereby enhancing social efficiency.

The aforementioned parallel society concept introduces ACP-based parallel intelligence methods into social management systems using multi-agent modeling, social computing, virtual reality, machine learning, and social network analysis. In doing so, it creates an artificial social system mirroring real societal structures and allowing interactions, coevolution, and feedback with the actual social system. Importantly, this artificial system becomes a dynamic laboratory for conducting computational experiments to guide and control complex social processes [44].

CPSS-based parallel organizations and societies, characterized by substantial uncertainty, diversity, and complexity, rely on blockchain technology and smart contracts as foundational infrastructure. They offer robust decentralized data structures and interactive mechanisms essential for distributed social systems and AI. Nodes executing smart contracts act as software agents, understanding the environment and autonomously executing contracts through negotiation processes, leading to the emergence of decentralized autonomous organizations (DAO), decentralized autonomous corporations (DAC), and distributed autonomous systems (DAS).

ACP seamlessly aligns with blockchain technology to realize parallel organizational and societal governance through smart contracts. Within this framework, blockchain's attributes, programmable smart contracts, and the fusion of blockchain with IoT create a dynamic environment for optimal organizational and societal governance akin to parallel execution. In the age of advanced intelligence industries, relying solely on AI is inadequate. A profound shift in philosophy and thinking is essential to succeed in this new era. This transformation involves embracing Karl Popper's three-world model, expanding our approach from "two Bs" to "three Bs" (being, becoming, believing), and fostering a new norm for scientific activities [44].

4.4.2 Metasystems: the paradigm shift from CPSS to metaverses and beyond

The concept of metaverses has gained significant attention from technologists and the general public, representing a profound transformation in our approach to complex systems. This transformation is closely linked to the evolution from traditional CPS toward the broader domain of CPSS. Including metaverses within CPSS offers a novel perspective on the convergence of physical and virtual dimensions, creating a truly complex space. To have a clear understanding, it is essential to delve into the origins and definitions of metaverses. The term *metaverses* intrigues with its plural form, suggesting a multitude of interconnected virtual worlds. It traces its roots to Norbert

Wiener's concept of circular causality (CC) in cybernetics. CC embodies a feedback-controlled purpose, where observed outcomes guide further actions to maintain specific conditions. Initially, cybernetics explored teleological mechanisms for regulating living beings and machines. Metaverses emerge from various fields, including computer science/systems, communication science/systems, control science/systems, cognitive science/systems, and cyberspace/systems [44].

The excitement around metaverses prompts a deep dive into their potential impact. Technologies like parallel intelligence, digital twins, blockchain, smart contracts, Web3, DAOs, DeSci, and DeSoc are crucial for shaping automatic control and intelligent automation, forming the groundwork for intelligent industries and smart societies. Integrating physical and digital realms reflects the transformative shift from real to complex numbers, akin to the evolution of 6G technology.

The initial phase of the current metaverse development revolves around digital twins, representing physical entities and their digital counterparts in CPS. Digital twins use historical and real-time data to create virtual entities, aiding decision-making in control systems. However, predicting human and social behaviors in CPSS remains a challenge. Parallel intelligence, introduced in 2004, addresses this challenge by incorporating ACP. Unlike digital twins, parallel intelligence models human behaviors in artificial systems, emphasizing transparency and openness in CPSS management. This versatile approach is applicable across different domains, as depicted in Figure 4.7.

The term "DeMetaverses" signifies the creation of decentralized autonomous metaverses by merging blockchain and Web3 technologies. These platforms promote trustworthy, efficient, and effective societies by integrating decentralized intelligence and ensuring data security. Within DeMetaverses, users can create avatars, engage in DAO collaboration, and participate in community activities, fostering innovation and collaboration [45].

Recent technological advancements, particularly in AI and organizational coordination, have driven a significant transition. These innovations empower decision-making and reshape organizational dynamics, leading to the emergence of intelligent metasystems. These systems seamlessly integrate human intelligence and mechanized operations within the immersive domains of metaverses, offering transformative potential. Essentially, a *metasystem* is a complex entity that intertwines physical, artificial, and mental realms, as well as physical space and cyberspace. It blurs the traditional boundaries between physical and virtual worlds by leveraging advanced technologies like metaverses, parallel intelligence, blockchain, and smart contracts [45].

The concept of metasystems assumes a pivotal role in shaping the trajectory of control intelligence, automation, and human–machine interactions within the expansive realm of the emerging metaverse. This transformative endeavor takes inspiration from the concept of circular causality, which artfully bridges ancient Greek teleology with modern machinery while remaining in harmony

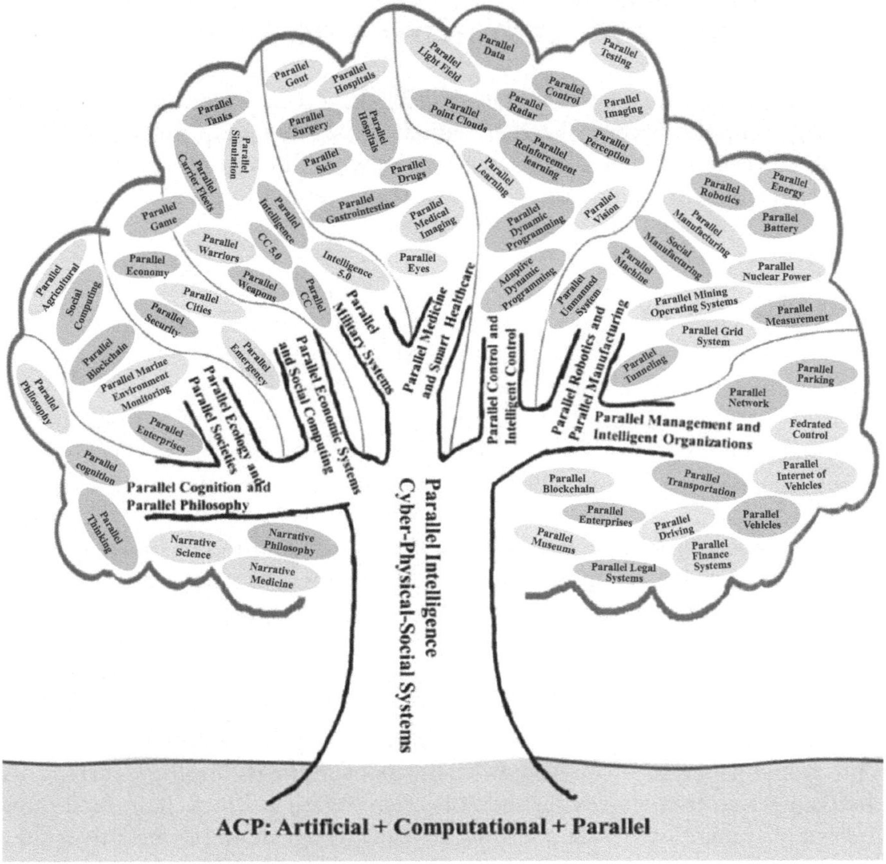

Figure 4.7 Applications of parallel intelligence and CPSS.

Source: [45].

with Karl Popper's three-world model. The forthcoming future envisions the realization of the aforementioned "three Bs," thereby paving the way for the emergence of "6S" intelligent societies characterized by safety, security, sustainability, sensitivity, service orientation, and smart functionality [45].

The transition from conventional CPS to complex CPSS, alongside the emergence of metaverses and DeMetaverses, signifies a profound paradigm shift in our engagement with intricate metasystems. These transformative advancements carry the potential to usher in a future characterized by innovation, thereby fundamentally reshaping societal structures and fundamentally transforming our dynamic interface with advanced meta-technologies, as explained next for the case of extended stigmergy in dynamic media and their role in realizing proactive collective intelligence in the emerging metaverse.

4.5 TOWARD PROACTIVE COLLECTIVE INTELLIGENCE: EXTENDED STIGMERGY IN DYNAMIC MEDIA

Stigmergy and collective intelligence (CI) drive our ever-changing world. Stigmergy enables decentralized coordination, while CI taps into group wisdom for problem-solving. In the following, we will explore how these concepts shape a future stigmergic Society 5.0 and revolutionize dynamic media such as the emerging metaverse.

4.5.1 Stigmergy

In an attempt to better understand CI, Alex Pentland's book *Social Physics* serves as a foundational guidepost by emphasizing the pivotal role of social interactions in evolution [46]. Pentland draws an intriguing parallel, likening human behavior to that of social insects, particularly bees. This analogy opens the door to a compelling proposition: the potential integration of ancient decision-making processes observed in bee colonies into the fabric of human society, a vision that hints at the transformation of future techno-social systems. This visionary concept gains further resonance through Max Borders's introduction of the concept of "social singularity" [47]. Borders delineates the social singularity as the threshold at which humanity transitions toward a hive mind, akin to the collective consciousness observed in social insects. According to Borders, humans will increasingly adopt behaviors reminiscent of bees, acting as neurons in an expansive human hive mind. This transformation is underpinned by blockchain technology, serving as the connective tissue weaving the collective together. Blockchain facilitates the establishment of virtual pheromone trails, programmable incentives that guide and incentivize coordinated actions on a grand scale.

The concept of stigmergy, deeply intertwined with these ideas, forms the core of our exploration. Stigmergy, often associated with the collective behaviors of social insects, such as bees and ants, represents a decentralized coordination mechanism driven by indirect communication through environmental modifications. Within the realm of stigmergy, we find the key to understanding how these visionary concepts can be practically applied in shaping the future of techno-social systems [46].

4.5.2 Stigmergic society 5.0

Expanding upon the foundational concept of stigmergy, we delve into its practical application within the framework of our envisioned concept of a future stigmergic Society 5.0 [6]. This paradigm represents a significant departure from conventional societal structures, drawing inspiration from the innate coordination mechanisms observed in the natural world. In the context of stigmergic Society 5.0, stigmergy's principles form the bedrock for decentralized coordination and collaboration. Many DAO members, encompassing humans, social robots, and embodied AI entities, function as offline agents within a meticulously designed online environment. This digital ecosystem harnesses an extensive array of Ethereum blockchain technologies, including on-chaining oracles, tokenized digital twins, and precisely engineered smart contracts designed to generate purpose-driven tokens. These tokens, operating as potent incentives, effectively stimulate offline agents to contribute toward predefined objectives actively.

Remarkably, the transformation of a given CPSS to stigmergic Society 5.0 vividly demonstrates the synergistic interactions among diverse members of the DAO. These interactions are choreographed through the principles of stigmergy, amplifying CI by leveraging blockchain-driven coordination. Our empirical investigation, as elaborated in [48], presents compelling evidence of the transformative potential inherent in blockchain transactions serving as stigmergic traces. These traces function as catalysts that steer collective human behavior toward significantly elevated levels of CI, as exemplified by human intelligence tasks showcased in Figure 4.8.

As we delve deeper into these concepts, our exploration extends to the domain of *extended stigmergy in dynamic media*, encompassing the burgeoning metaverse landscapes and the creation of highly intelligent digital organisms inspired by a diverse range of biological superorganisms that transcend the conventional boundaries observed in the behaviors of bees and ants [6].

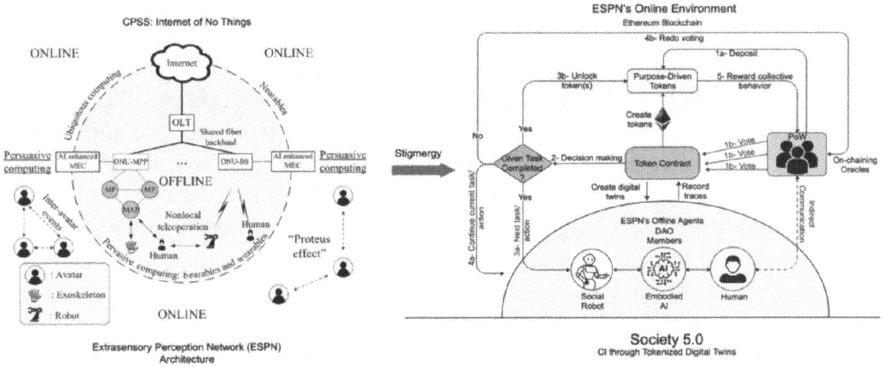

Figure 4.8 CPSS integrates human beings into today's cyber-physical systems at the physical, social, and cognitive levels. Society 5.0 targets the fusion of the digital and the physical worlds via various types of VR/AR/XR, embodied AI, humans, and social robots.

Source: [6].

4.5.3 Extended stigmergy and dynamic media

Within the rapidly evolving metaverse, a profound synergy is unfolding that bridges the realms of the Internet and human existence. This emergent phenomenon, which we have recently introduced as *interbeing* in [7], represents a transformative connection that weaves together the virtual realm of the Internet with the tangible world inhabited by humans, a concept yet to find its place in conventional dictionaries. Interbeing, in essence, encapsulates the core of this profound symbiotic relationship between the Inter(net) and (human) being within the metaverse. The foundational framework defines and elucidates the intricate web of connections, interactions, and coexistence that characterizes metaverse's forthcoming virtual society. Our exploration tries to shed light on the multifaceted dimensions of interbeing, offering insights into its implications for the design of future human-AI-cybernetic organisms, dynamic media, and extended stigmergy. In this dynamic context, the concept of interbeing assumes a pivotal role, enabling a deeper understanding of the transformative potential inherent in the evolving metaverse. Interbeing is a catalyst for profound change, ushering in an era where the boundaries between the real and the virtual worlds blur ever more distinctly. This paradigm shift promises to reshape the fabric of our societal, cognitive, and technological landscapes. Within this overarching framework, we embark on an extensive examination of the principles and manifestations of interbeing, unveiling its multifaceted facets and profound implications for the emerging metaverse.

Beyond conceptual exploration lies the practical realm of extended stigmergy, a mechanism that orchestrates active agents within a dynamic human-AI-cybernetic organism environment, where the medium actively alters stigmergic traces. This innovative approach offers substantial advantages. Rooted in evolutionary principles, it emphasizes the preference for self-organizing behaviors embodied by stigmergy. This adaptability allows for the emergence of diverse collective responses from simple individual behaviors of agents, particularly within ever-changing environments.

Recent strides in AI research signify a transformative shift. While past endeavors focused on AI algorithms mimicking human intelligence, the true breakthrough emerges from algorithms emulating the intricate processes of evolution. These algorithms empower AIs to generate their training environments, ushering in the era of self-generating AI, commonly known as *generative AI* (GAI). Generative AI encompasses a diverse array of techniques, including generative pre-trained transformers (GPT), generative adversarial networks (GANs), and generative diffusion models (GDM) [48].

Through the lens of meta-learning, where AI enhances its intelligence, generative AI promises to introduce a unique form of intelligence, one that transcends the boundaries of human cognition. Instead of instructing machines to mimic human thought, this approach empowers machines to introduce novel cognitive paradigms, challenging our existing cognitive frontiers. The convergence of digital evolution and biology emerges as a pivotal catalyst in crafting lifelike digital organisms. These entities represent genuine instances of natural evolution, far exceeding the confines of mere simulations. Remarkably, these digital organisms demonstrate the capacity to devise ingenious solutions previously unexplored, often surpassing human capabilities [49].

Figure 4.9 encapsulates our envisioned trajectory toward the convergence of digital evolution and biology, culminating in realizing the reality–virtuality continuum within metaverse's forthcoming virtual society. Notably, stigmergic communication, observed in nature both aboveground (e.g., bees) and belowground (e.g., ants), relies on informative chemical cues, termed infochemicals, to shape interactions among various agents and their environment. These agents encompass social insects, flora (e.g., trees), and fauna (e.g., giraffes). Our investigation delves into the merits of emulating extended stigmergy as an indirectly mediated mechanism for coordinating active agents within a dynamic human-AI-cybernetic organism-environment. In this context, stigmergic traces left in the medium are significantly influenced by these agents, thus enhancing the robustness and adaptability of the collective system. Evolutionary principles further underscore the resilience of stigmergic self-organizing principles, where the simplicity of individual agent behaviors can yield diverse collective responses in ever-shifting environments. Crucially, intelligent agents operating within and exploring

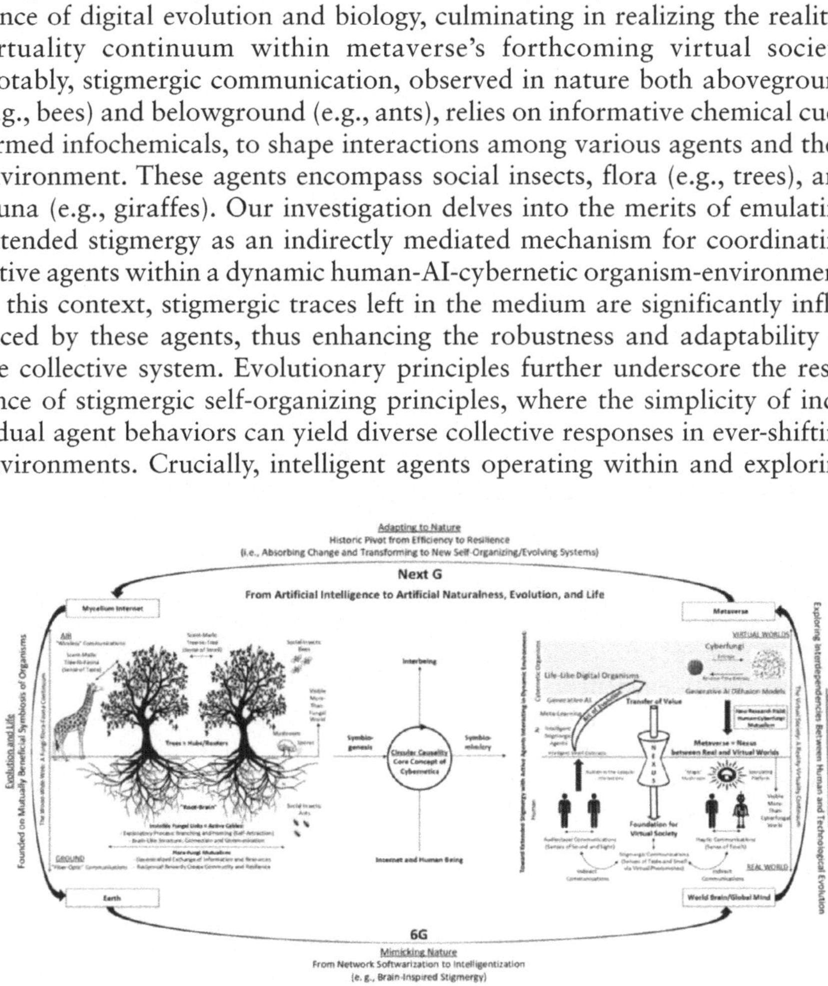

Figure 4.9 The virtual society's symbiosis of Inter(net) and (human) beings toward interbeing in the metaverse: extended stigmergy with active agents interacting in a dynamic human-AI-cybernetic organisms environment.

Source: [7].

diverse environments exhibit the capacity for independent action as they anticipate and adapt to the evolving configurations of their surroundings. Extended stigmergy presents a comprehensive framework for understanding the decentralized control of complex collective behavior [7].

Within this context, we design cybernetic organisms that are lifelike digital entities harnessing generative AI diffusion models, giving rise to *cyberfungi*, a distinctive breed of lifelike digital organisms. Fungi, occupying a unique ecological niche bridging fauna and flora, offer intriguing features. Their cell walls, constructed from chitin, distinguish them from plants, aligning them more closely with insects. In natural ecosystems, forests utilize intricate underground fungal networks for regulation, akin to the social networks found in nature. These fungal networks facilitate the exchange of information and resources, such as carbon and sugars, between trees and plants. Strikingly, these fungal networks bear a remarkable resemblance to neural networks found in human brains, contributing to what is termed the "root–brain" hypothesis. These recent discoveries fundamentally reshape our understanding of trees, highlighting their role as interconnected, communicative entities rather than isolated competitors. These profound insights hold great promise of exciting prospects for the development of novel, lifelike digital organisms [7].

4.6 USE CASE: REALIZING METAVERSE'S VIRTUAL SOCIETY VIA WEB3 BLOCKCHAIN TECHNOLOGIES

This section represents the earlier visionary concepts' practical realization, bridging the evolving metaverse's theoretical and practical dimensions. It extends from the foundational ideas presented in Section 4.5, which examined the concept of interbeing and the intricate interplay between the Internet and human existence.

In this section, we pivot from the abstract exploration of how technology, nature, and human society intersect and coalesce within the metaverse to a pragmatic examination of the methodologies and tools instrumental in translating these concepts into tangible reality. We elucidate how advanced technology, notably generative AI, such as OpenAI's denoising diffusion probabilistic model (DDPM), catalyzes the transformation of these visionary ideas into palpable experiences. Integrating intelligent stigmergic agents and smart contracts adds a dynamic and responsive dimension to the metaverse, fundamentally altering digital interactions. This transition deepens our understanding of the transformative potential of technology in shaping the future of the metaverse, where the convergence of technology, nature, and human society unlocks novel possibilities and reshapes the dynamics of engagement within this digital realm. Digital entities, exemplified by our proposed cyberfungi, emerge as dynamic contributors enriching the immersive fabric of the virtual society.

4.6.1 Generative AI's impact on the metaverse

Generative AI, a transformative force in content creation, has ushered in a paradigm shift that transcends traditional boundaries. It encompasses various algorithms and models meticulously crafted to autonomously generate high-quality content spanning text, images, videos, and more. At the forefront of this technological evolution stands DDPM, a pioneering creation by OpenAI. DDPM's revolutionary approach leverages a diffusion process to metamorphose data distributions into Gaussian noise and then subsequently orchestrates the reverse transformation to reconstruct the original data distribution faithfully. This innate denoising prowess assumes a pivotal role in generating realistic and coherent content, positioning DDPM as the linchpin in crafting lifelike digital entities within the immersive landscape of the metaverse.

In the context of the metaverse, DDPM emerges as a foundational force that is instrumental in the proliferation of lifelike digital organisms in our work on cyberfungi. These digital entities represent a distinct manifestation of natural evolution, diverging from conventional simulations. Through rigorous experimentation with DDPM, we have achieved the creation of cyberfungi that exhibit behaviors and attributes mirroring those of real-world organisms, effectively bridging the chasm between digital and biological evolution. Our developed cyberfungi operate as dynamic and autonomous agents, faithfully replicating the actions and characteristics of their real-world counterparts. Their interactions with humans and other digital entities engender intricate ecosystems, thereby deepening the sense of immersion within the virtual expanse of the metaverse.

DDPM's numerical techniques offer the means to fine-tune the behaviors and traits of cyberfungi, thereby yielding a diverse array of digital life-forms that enrich the fabric of the metaverse. Beyond content generation, DDPM extends its influence on creating intricate network topologies that faithfully mimic real-world systems. Through its unique hallucination capabilities, DDPM conjures scale-free networks imbued with small-world properties, an essential element in modeling the intricate web of interactions within the metaverse. Drawing inspiration from nature's *wood-wide web* found in forests, these networks stand as the architectural backbone connecting cyberfungi with humans, thus serving as conduits for social contagion and the propagation of new social norms and behaviors within the metaverse. In the forward progression, our deployed DDPM starts with a conventional 64-node ring lattice, a well-structured network configuration. Subsequently, DDPM systematically increases the entropy or randomness within this lattice, transforming it into a dynamic and random network. This transformation reflects the system's capacity to introduce disorder and unpredictability, a hallmark of small-world networks.

In the reverse-time entropy process, DDPM's remarkable capability comes into play. It hallucinates scale-free networks possessing small-world properties, a feat of paramount significance. These networks are adept at achieving equilibrium regarding social contagion among all nodes. This equilibrium is accomplished within a considerably reduced time frame compared to conventional networks. The key parameter here is denoted as r, representing the spreading probability. Various values of r are considered, and their impact on the network dynamics is examined. It is important to note that these scale-free networks were not present in our synthetic training data.

Notwithstanding, a series of formidable challenges looms on the path to establishing a symbiotic nexus that facilitates the seamless transfer of real-world information to the metaverse. As we explore the transformative role of smart contracts in conjunction with intelligent stigmergic agents in the next section, we shall witness the orchestration of processes, the enforcement of rules, and the enablement of trustless interactions, all while transcending predefined rules and functioning as intelligent agents responsive to the evolving environment. This innovation empowers individuals with greater control over their digital identities and assets, fostering a more adaptive and self-organizing digital ecosystem. This research represents a significant leap in the evolution of the metaverse, where technology, nature, and human society converge to unlock new possibilities and redefine the dynamics of engagement within this digital realm.

4.6.2 Intelligent stigmergic agents and smart contracts

The metaverse, a dynamic and transformative realm, draws its very essence from Web3, a foundational framework introducing groundbreaking components poised to reshape the digital landscape. Among these components are self-evolving smart contracts and adaptive digital agents, which stand as keystones in this era of technological metamorphosis. This paradigm shift is not limited to technology; it extends its empowering reach to individuals, affording them unprecedented control over their digital identities, assets, and interactions. Simultaneously, it engenders a profound redefinition of societal structures, governance models, and cooperative dynamics within the metaverse. Blockchain technology is central to this transformative landscape, the bedrock upon which trust, transparency, and immutability are established. Blockchain's pivotal role enables the creation of unique virtual assets known as NFTs, each representing a distinct and irreplaceable digital entity. In our work, the linchpin of this transformation is the concept of intelligent smart contracts, acting as the neural network of the digital realm. Unlike traditional static code, these contracts transform remarkably into dynamic agents, mirroring the adaptability observed in natural systems.

The role of intelligent smart contracts is multifaceted and transformative. These dynamic agents are the vital bridge between the digital landscape and human entities coexisting within the metaverse. Their adaptability and situational awareness drive the flourishing of metaverse's forthcoming virtual society within this complex and ever-evolving environment. This transformative perspective elevates smart contracts from static, predefined scripts to dynamic and responsive entities. One of the key innovations in our work is the integration of intelligent stigmergic agents alongside smart contracts, thereby introducing a groundbreaking dimension to the metaverse's digital ecosystem. These agents orchestrate processes, enforce rules, and enable trustless interactions, all while possessing the unique ability to transcend predefined rules. Their agility allows them to function as intelligent agents responsive to the evolving environment, effectively mirroring the adaptability inherent in natural systems.

With regard to the design and challenges associated with intelligent stigmergic agents and smart contracts, our ongoing research pursues several promising avenues:

- *Dynamic contract evolution.* We are exploring methods for smart contracts to adapt and evolve over time, learning from interactions and dynamically updating their rules and behavior to optimize outcomes.
- *Enhanced situational awareness.* Our focus includes enhancing the agents' ability to perceive and understand the evolving context within the metaverse, allowing for more informed decision-making and coordination.
- *Inter-agent communication.* Facilitating seamless communication and cooperation among intelligent stigmergic agents and smart contracts is critical to our research, promoting the emergence of sophisticated digital ecosystems.
- *Security and privacy.* Ensuring the security and privacy of interactions within the metaverse remains a paramount concern.

4.7 CONCLUSION

Incorporating CPSS principles enhances the intelligence of the metaverse, giving rise to parallel blockchain societies and simplifying complex interactions and coordination through smart contracts. Additionally, CPSS introduces concepts like citizen sensing and citizen actuation, reshaping the interactions of human entities within the metaverse. This transformative synergy between

CPSS and the metaverse redefines the landscape of technology and human interaction, promising a future where the boundaries between the virtual and the real worlds blur ever more distinctly. Of particular significance is the metaverse, signifying a fundamental shift in human–computer interactions, with its reliance on technologies like AI, XR, 6G, and blockchains. This paradigm shift has implications for communication, collaboration, and content creation. Maintaining a human-centric approach remains essential as we move into the 6G era, characterized by the convergence of the real and the digital worlds within the metaverse and multiverse. Our work revolves around the transformative landscape of advanced applications beyond 2030, driven by concepts like network 2030, the metaverse, and the multiverse. These concepts aim to seamlessly merge the digital and the real worlds, offering innovative experiences and applications.

Several open research challenges exist in the realm of human-in-the-loop AI, involving intelligent smart contracts and intelligent stigmergic agents that operate within dynamic environments shaped by human-AI-cybernetic entities and extended stigmergy. These challenges span various critical domains, such as the imperative to facilitate the dynamic adaptation and evolution of smart contracts over time, augmenting the situational awareness of intelligent agents navigating the intricacies of the metaverse, establishing mechanisms for seamless communication and collaborative endeavors among these agents and contracts, and ensuring security and confidentiality of interactions within this ever-evolving digital landscape. Our exploration of intelligent stigmergic agents and smart contracts signifies a substantial stride forward in the continuous progression of the metaverse. This advancement can potentially wield transformative influence over technological paradigms and human interactions. These developments provide exciting prospects for forthcoming research and innovation on the frontier of the ever-evolving digital domain.

REFERENCES

[1] Focus Group Technologies for Network 2030 (FG-NET-2030). Network 2030 – A Blueprint of Technology, Applications and Market Drivers Towards the Year 2030 and Beyond. pp. 1–19. ITU-T, 2019.

[2] ETSI. The Fifth Generation Fixed Network (F5G): Bringing Fibre to Everywhere and Everything. pp. 1–24. White Paper No. 41, 1st ed., 2020.

[3] M. Maier, "6G and Onward to Next G: The Road to the Multiverse", Wiley-IEEE Press, 2023.

[4] Y. Zhou, F. R. Yu, J. Chen, and Y. Kuo, "Cyber-Physical-Social Systems: A State-of-the-Art Survey, Challenges and Opportunities", IEEE Communications Surveys & Tutorials, vol. 22, no. 1, pp. 389–425, 2020.

[5] S. Wang, L. Ouyang, Y. Yuan, X. Ni, X. Han, and F.-Y. Wang, "Blockchain-Enabled Smart Contracts: Architecture, Applications, and Future Trends", IEEE

Transactions on Systems, Man, and Cybernetics: Systems, vol. 49, no. 11, pp. 2266–2277, 2019.

[6] A. Beniiche, S. Rostami, and M. Maier, "Society 5.0: Internet as if People Mattered", IEEE Wireless Communications, vol. 29, no. 6, pp. 160–168, 2022.

[7] M. Maier, N. Hosseini, and M. Soltanshahi, "INTERBEING: On the Symbiosis between INTERnet and Human BEING", IEEE Consumer Electronics Magazine, IEEE Xplore Early Access, pp. 1–8, 2023.

[8] B. J. Pine II and K. C. Korn, "Infinite Possibility: Creating Customer Value on the Digital Frontier", Berrett-Koehler Publishers, 2011.

[9] W. Y. B. Lim, Z. Xiong, D. Niyato, X. Cao, C. Miao, S. Sun, and Q. Yang, "Realizing the Metaverse with Edge Intelligence: A Match Made in Heaven", IEEE Wireless Communications, vol. 30, no. 4, pp. 64–71, 2023.

[10] M. Aloqaily, O. Bouachir, F. Karray, I. A. Ridhawi, and A. E. Saddik, "Integrating Digital Twin and Advanced Intelligent Technologies to Realize the Metaverse", IEEE Consumer Electronics Magazine, vol. 12, no. 6, pp. 47–55, 2023.

[11] G. Zhang, J. Wu, G. Jeon, Y. Chen, Y. Wang, and M. Tan, "Towards Understanding Metaverse Engagement via Social Patterns and Reward Mechanism: A Case Study of Nova Empire", IEEE Transactions on Computational Social Systems, vol. 10, no. 5, pp. 2165–2176, 2023.

[12] M. Weiser, "The Computer or the 21st Century", Scientific American, vol. 265, no. 3, pp. 94–104, 1991.

[13] Ericsson. 6G-Connecting a Cyber-Physical World: A Research Outlook Toward 2030. pp. 1–31. White Paper GFTL-20:001402, 2022.

[14] D. C. Nguyen, M. Ding, P. N. Pathirana, A. Seneviratne, J. Li, D. Niyato, O. A. Dobre, H. Vincent Poor, "6G Internet of Things: A Comprehensive Survey", IEEE Internet of Things Journal, vol. 9, no. 1, pp. 359–383, 2022.

[15] F. Liu, Y. Cui, C. Masouros, J. Xu, T. Han, Y. C. Eldar, and S. Buzzi, "Integrated Sensing and Communications: Toward Dual-Functional Wireless Networks for 6G and Beyond", IEEE Journal on Selected Areas in Communications, vol. 40, no. 6, pp. 1728–1767, 2022.

[16] D. M. Doe, J. Li, N. Dusit, Z. Gao, J. Li, and Z. Han, "Promoting the Sustainability of Blockchain in Web 3.0 and the Metaverse Through Diversified Incentive Mechanism Design", IEEE Open Journal of the Computer Society, vol. 4, pp. 171–184, 2023.

[17] X. Zhang, G. Min, T. Li, Z. Ma, X. Cao, and S. Wang, "AI and Blockchain Empowered Metaverse for Web 3.0: Vision, Architecture, and Future Directions", IEEE Communications Magazine, vol. 61, no. 8, pp. 60–66, 2023.

[18] P. Hande, P. Tinnakornsrisuphap, J. Damnjanovic, H. Xu, M. Mondet, H. Y. Lee, and I. Sakhnini, "Extended Reality Over 5G – Standards Evolution", IEEE Journal on Selected Areas in Communications, vol. 41, no. 6, pp. 1757–1771, 2023.

[19] H. Ning, H. Wang, Y. Lin, W. Wang, S. Dhelim, F. Farha, J. Ding, and M. Daneshmand, "A Survey on the Metaverse: The State-of-the-Art, Technologies, Applications, and Challenges", IEEE Internet of Things Journal, vol. 10, no. 16, pp. 14671–14688, 2023.

[20] A. A. Zaidan, H. A. Alsattar, S. Qahtan, M. Deveci, D. Pamucar, and M. Hajiaghaei-Keshteli, "Uncertainty Decision Modeling Approach for Control Engineering Tools to Support Industrial Cyber-Physical Metaverse Smart Manufacturing Systems", IEEE Systems Journal, vol. 17, no. 4, pp. 5303–5314, 2023.

[21] Z. Wang, J. Liu, and W. Zhu, "Edge Intelligence Empowered Immersive Media: Challenges and Approaches", IEEE MultiMedia, vol. 30, no. 2, pp. 8–17, 2023.
[22] B. Tao, H.-N. Dai, H. Xie, and F. L. Wang, "Structural Identity Representation Learning for Blockchain-Enabled Metaverse Based on Complex Network Analysis", IEEE Transactions on Computational Social Systems, vol. 10, no. 5, pp. 2214–2225, 2023.
[23] H. Du, J. Liu, D. Niyato, J. Kang, Z. Xiong, J. Zhang, and D. Kim, "Attention-Aware Resource Allocation and QoE Analysis for Metaverse xURLLC Services", IEEE Journal on Selected Areas in Communications, vol. 41, no. 7, pp. 2158–2175, 2023.
[24] Z. Long, H. Dong, and A. E. Saddik, "Human-Centric Resource Allocation for the Metaverse with Multiaccess Edge Computing", IEEE Internet of Things Journal, vol. 10, no. 22, pp. 19993–20005, 2023.
[25] C. Wang, C. Yu, and Y. Li, "Toward Understanding Attention Economy in Metaverse: A Case Study of NFT Value", IEEE Transactions on Computational Social Systems, vol. 10, no. 5, pp. 2177–2188, 2023.
[26] P. Schwenteck, G. T. Nguyen, H. Boche, W. Kellerer, and F. H. P. Fitzek, "6G Perspective of Mobile Network Operators, Manufacturers, and Verticals", IEEE Networking Letters, vol. 5, no. 3, pp. 169–172, 2023.
[27] F. Tang, X. Chen, M. Zhao, and N. Kato, "The Roadmap of Communication and Networking in 6G for the Metaverse", IEEE Wireless Communications, vol. 30, no. 4, pp. 72–81, 2023.
[28] W. Xiang, K. Yu, F. Han, L. Fang, D. He, and Q.-L. Han, "Advanced Manufacturing in Industry 5.0: A Survey of Key Enabling Technologies and Future Trends", IEEE Transactions on Industrial Informatics, vol. 20, no. 2, pp. 1055–1068, 2024.
[29] A. Musamih, A. Dirir, I. Yaqoob, K. Salah, R. Jayaraman, and D. Puthal, "NFTs in Smart Cities: Vision, Applications, and Challenges", IEEE Consumer Electronics Magazine, vol. 13, no. 2, pp. 9–23, 2024.
[30] Y. Gao, M. Saad, A. Oest, J. Zhang, B. Han, and S. Chen, "Can I Own Your NFTs? Understanding the New Attack Surface to NFTs", IEEE Communications Magazine, vol. 61, no. 9, pp. 64–70, 2023.
[31] Y. Xiao, L. Xu, C. Zhang, L. Zhu, and Y. Zhang, "Blockchain-Empowered Privacy-Preserving Digital Object Trading in the Metaverse", IEEE MultiMedia, vol. 30, no. 2, pp. 81–90, 2023.
[32] R. Cheng, S. Chen, and B. Han, "Toward Zero-Trust Security for the Metaverse", IEEE Communications Magazine, vol. 62, no. 2, pp. 156–162, 2024.
[33] K. Yang, Z. Zhang, T. Youliang, and J. Ma, "A Secure Authentication Framework to Guarantee the Traceability of Avatars in Metaverse", IEEE Transactions on Information Forensics and Security, vol. 18, pp. 3817–3832, 2023.
[34] Z. Han, Y. Tu, and C. Huang, "A Framework for Constructing a Technology-Enhanced Education Metaverse: Learner Engagement with Human–Machine Collaboration", IEEE Transactions on Learning Technologies, vol. 16, no. 6, pp. 1179–1189, 2023.
[35] S.-M. Lee, "Second Language Learning Through an Emergent Narrative in a Narrative-Rich Customizable Metaverse Platform", IEEE Transactions on Learning Technologies, vol. 16, no. 6, pp. 1071–1081, 2023.

[36] Y. Song, J. Cao, K. Wu, P. L. H. Yu, and J. C.-K. Lee, "Developing 'Learningverse'—A 3-D Metaverse Platform to Support Teaching, Social, and Cognitive Presences", IEEE Transactions on Learning Technologies, vol. 16, no. 6, pp. 1165–1178, 2023.
[37] X. Chen, Z. Zhong, and D. Wu, "Metaverse for Education: Technical Framework and Design Criteria", IEEE Transactions on Learning Technologies, vol. 16, no. 6, pp. 1034–1044, 2023.
[38] S. P. Ramu, G. Srivastava, R. Chengoden, N. Victor, P. K. R. Maddikunta, and T. R. Gadekallu, "The Metaverse for Cognitive Health: A Paradigm Shift", IEEE Consumer Electronics Magazine, vol. 13, no. 3, pp. 73–79, 2024.
[39] A. Musamih, I. Yaqoob, Kh. Salah, R. Jayaraman, Y. A. Hammadi, M. Omar, and S. Ellahham, "Metaverse in Healthcare: Applications, Challenges, and Future Directions", IEEE Consumer Electronics Magazine, vol. 12, no. 4, pp. 33–46, 2023.
[40] A. Musamih, I. Yaqoob, K. Salah, R. Jayaraman, Y. A. Hammadi, M. Omar, and S. Ellahham, "NFTs in Healthcare: Vision, Opportunities, and Challenges", IEEE Consumer Electronics Magazine, vol. 12, no. 4, pp. 21–32, 2023.
[41] H. Ullah, S. Manickam, M. Obaidat, S. U. A. Laghari, and M. Uddin, "Exploring the Potential of Metaverse Technology in Healthcare: Applications, Challenges, and Future Directions", IEEE Access, vol. 11, pp. 69686–69707, 2023.
[42] J. J. Zhang et al., "Cyber-Physical-Social Systems: The State of the Art and Perspectives", IEEE Transactions on Computational Social Systems, vol. 5, no. 3, pp. 829–840, 2018.
[43] S. Pasandideh, P. Pereira, and L. Gomes, "Cyber-Physical-Social Systems: Taxonomy, Challenges, and Opportunities", IEEE Access, vol. 10, pp. 42404–42419, 2022.
[44] F.-Y. Wang, "The DAO to MetaControl for MetaSystems in Metaverses: The System of Parallel Control Systems for Knowledge Automation and Control Intelligence in CPSS", IEEE/CAA Journal of Automatica Sinica, vol. 9, no. 11, pp. 1899–1908, 2022.
[45] X. Wang, J. Yang, J. Han, W. Wang, and F.-Y. Wang, "Metaverses and DeMetaverses: From Digital Twins in CPS to Parallel Intelligence in CPSS", IEEE Intelligent Systems, vol. 37, no. 4, pp. 97–102, 2022.
[46] A. Pentland, "Social Physics: How Good Ideas Spread—The Lessons from a New Science", Penguin Press, 2014.
[47] M. Borders, "The Social Singularity, Social Evolution", 2018. https://socialevolution.com/book/
[48] M. Jovanović and M. Campbell, "Generative Artificial Intelligence: Trends and Prospects", Computer, vol. 55, no. 10, pp. 107–112, 2022.
[49] J. J. Lehman et al., "The Surprising Creativity of Digital Evolution: A Collection of Anecdotes from the Evolutionary Computation and Artificial Life Research Communities", Artificial Life, vol. 26, no. 2, pp. 274–306, 2020.

Chapter 5

Health-care cyber-physical system for smart cities

Implementation, challenges, and solution

K. Hemant Kumar Reddy, Manjula Gururaj Rao, and Diptendu Sinha Roy

LIST OF ABBREVIATIONS

AI	artificial intelligence
AR	augmented reality
CM	citizen engagement
CPS	cyber-physical system
DDDK	data-driven decision-making
DI	dynamic interaction
DSTS	distributed smart transport service
EHR	electronic health record
IC	interconnected components
ICT	information and communications technology
IoT	Internet of Things
LMTS	level monitoring task scheduling
PME	power management equipment
QoS	quality of service
RTMC	real-time monitoring and control
SR	security and resilience
UAV	unmanned aerial vehicle
UHD	ultra-high definition
V2X	vehicular-to-everything
VR	virtual reality

5.1 INTRODUCTION

In the year 2050, there will be a dramatic increase in the number of people living in metropolitan areas worldwide. Smart cities have entered the scene because of population growth. The adaptable qualities of smart cities are numerous. Communication between machines and humans is possible in smart cities. Cyber-physical systems (CPSs) are key components of the smart city. These are needed for the execution of difficult tasks, data analysis, and the efficient management and administration of operations.

DOI: 10.1201/9781003559993-5

5.1.1 Cyber-physical systems

The components of the CPS are described in this section, containing the following features:

- *Interconnected components (IC).* This component includes both physical and cyber properties. Roads, buildings, transportation networks, utilities, and other tangible features comprise the physical infrastructure. Information and communication technology (ICT), sensors, actuators, communication networks, and computational systems are all examples of cyber components.
- *Real-time monitoring and control (RTMC).* Sensors, communication networks, and control systems are examples of real-time monitoring and control components. Actuators and sensors embedded in physical infrastructure capture real-time data on various aspects, including traffic movement, environmental conditions, energy usage, and waste management. Communication networks make data flow between sensors, control systems, and data centers easier. Control systems analyze data and make real-time choices to control and optimize physical processes, such as adjusting traffic signals, regulating energy use, and optimizing garbage collection routes.
- *Data centers and cloud computing: data-driven decision-making (DDDK).* Large-scale data centers process and analyze massive amounts of sensor data, offering actionable insights. Big data analytics and artificial intelligence (AI) technologies aid in extracting significant patterns and trends from data, allowing for more informed decisions in urban planning and resource optimization.
- *Machine learning and optimization algorithms for resource optimization (OR).* It allows the city to make better use of resources like electricity, water, and transportation, resulting in increased efficiency and sustainability. Intelligent grids are used to better efficiently manage and distribute energy resources.
- *Mobile applications and citizen portals for citizen engagement (CM).* It gives citizens real-time access to information and services and the capacity to participate actively in governance and decision-making processes. Social media integration enables real-time communication and feedback loops between citizens and city officials.
- *Cybersecurity measures for security and resilience (SR).* It was implemented to protect data, networks, and control systems against cyberattacks and to ensure the resilience of smart city infrastructure.
- *Dynamic interaction (DI).* Bidirectional information flow between the physical and cyber components results in a dynamic and responsive urban environment. Feedback circuits, real-time continuous monitoring, analysis, and response to changing situations are present in the DI.

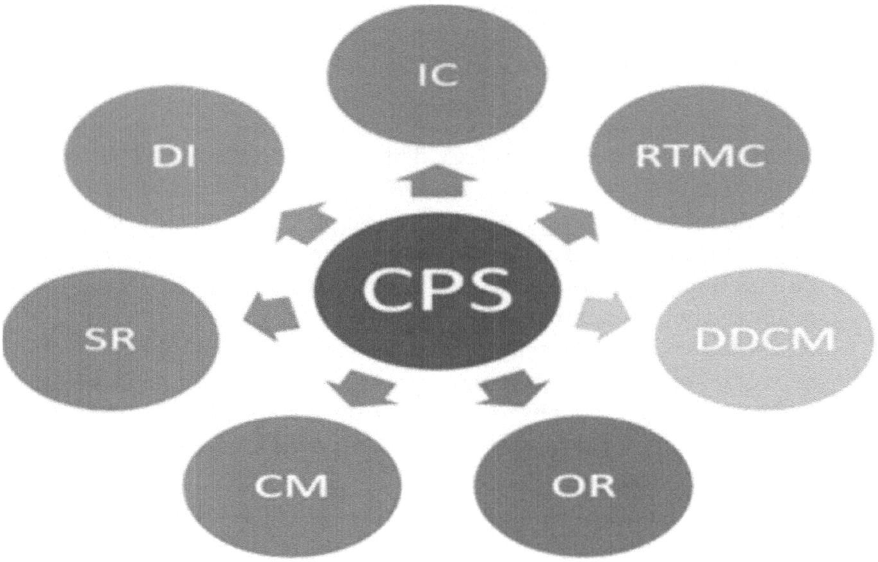

Figure 5.1 Components of CPS.

5.1.2 Applications of smart cities

Everyone uses different methods and technology, and the applications vary and spread to all facets of city life. New concepts and solutions are always being developed through ongoing research and development. However, there has not been a significant connection made between the wealth of results from these endeavors and the moral issues they raise.

Smart city applications are classified according to their usage, communication, and interactions between devices, devices to humans, and external and internal smart devices. Some of the smart city's applications are shown in Figure 5.2.

Some of the smart city applications are as follows:

- *Smart industries.* The industry's interest in integrating ICT in the production environment has increased as a result of recent exponential growth in technology developments, including big data, cloud services, machine intelligence, and 5G. When industrial machinery and ICT are used in tandem, there are opportunities to increase output, reduce waste, increase efficiency, and improve working conditions in the manufacturing sector. The phrase "smart manufacturing" describes incorporating sophisticated data analytics, computing power, sensing inputs, and always-on networking into traditional manufacturing processes. It should be able to take advantage of new opportunities

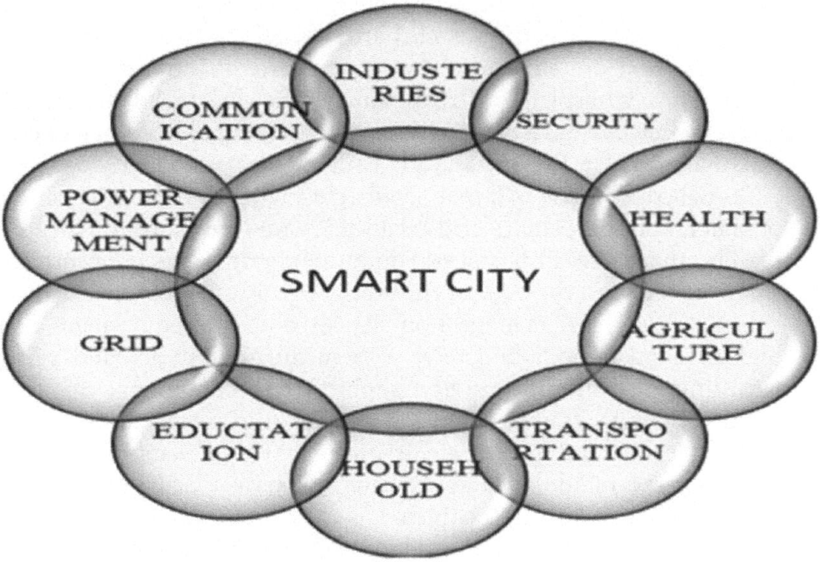

Figure 5.2 Different smart city applications.

to accelerate development, reduce waste, and enhance transparency and traceability with the help of these technologies. Adding new technologies into the process can make production more productive and efficient. Costs are reduced, delivered goods are of greater quality, and waste is reduced. End goods are removed from the conveyor belt more quickly, and downtime on the same conveyor line declines. The throughput of productions can be greatly increased by combining digital and 5G technology.

- *Smart power management*. In the present era, the use of battery-powered equipment is increasing, and effective power management equipment (PME) is needed for maximum performance. As PMEs evolve, they demand more processing power, making power consumption reduction a critical concern. Although many solutions focus on improving embedded system design, few consider the input signal's time-varying nature. This chapter offers a signal-driven sampling scheme: driven sampling scheme adapts its acquisition rate based on variations in the input signal, resulting in significant computational power savings. This strategy prioritizes intelligent energy allocation.
- *Smart grid*. A smart grid is an improved electrical infrastructure that includes a variety of operational and energy-saving techniques. Smart meters, smart distribution boards, load control switches, and smart appliances are examples of advanced grid infrastructures. Furthermore,

they can incorporate renewable energy sources and energy storage solutions, such as electric vehicle batteries. They also emphasize energy efficiency and are supported by reliable broadband connectivity for monitoring, with plenty of redundancy. The smart grid requires electronic management of electricity production and delivery. The policies for smart grids in Europe and the United States are clearly organized. Although the phrase is largely connected with its technological components, adopting smart grid technology involves substantially overhauling the electrical business. Some smart grid issues concentrate on smart meters and the equipment that goes with them.

Smart grids can manage non-critical domestic equipment during high-demand periods and restore them during low-demand periods. Real-time grid monitoring is a sophisticated component of modern electrical networks that enables real-time observation of power flow, voltage levels, and potential disruptions. This constant monitoring improves the reliability and efficiency of power distribution by discovering and correcting problems as they arise. Utilities can acquire granular insights about the grid's performance and possible areas of concern by employing cutting-edge sensors and software.

Remote grid maintenance supplements real-time monitoring by allowing utilities to handle faults without assigning staff to a specific spot. Many minor grid disturbances or faults can be corrected from a central location using sophisticated control systems. This not only saves time and money but also lowers service interruptions for customers. In an era of increasing electricity consumption, these advances are critical to guaranteeing a continuous and stable power supply. By combining real-time monitoring with remote maintenance capabilities, electrical networks become more resilient, adaptable, and ready for future challenges.

- *Smart education.* Virtual learning, digitalization, and augmented reality (AR) have transformed the educational landscape, causing profound changes in how people receive and process information. These breakthroughs have shattered traditional learning barriers, allowing for flexible, personalized educational experiences suited to the specific needs of individual learners. Furthermore, blended learning, which blends traditional classroom methods with digital features, has grown in popularity, aided by enormous data and advanced analytics. This amount of data allows educators to create customized classes and track students' development more precisely. There is a clear shift in concentration as we enter this new era of learning. Rather than simply digitizing the classroom, a growing emphasis is being placed on empirical learning.

  This implies that students are encouraged to learn through hands-on experiences and real-world applications, resulting in a deeper comprehension and more meaningful engagement with the subject matter.

Immersive teaching and learning go beyond standard techniques by engaging students profoundly through technology such as virtual reality (VR) and AR. These tools immerse students in various contexts or scenarios, giving them firsthand experience and developing a greater knowledge of the subject matter. On the other hand, remote interactive learning allows students and educators to connect from anywhere in the world, using digital platforms for live discussions, collaborative projects, and real-time feedback. This strategy not only gives flexibility but also broadens students' minds by exposing them to multiple ideas and knowledge from various geographical locations.

- *Smart logistics.* Fully automated warehousing has transformed product storage and retrieval by utilizing advanced robotics and artificial intelligence to improve productivity and eliminate human error. These cutting-edge warehouses interact seamlessly with self-driving transportation systems, allowing items to be loaded onto self-driving trucks with minimal manual intervention. This autonomous transportation promises to deliver products in a safer, more consistent, and timely manner, addressing the issues associated with human-driven logistics. Drone delivery has also emerged as a game-changing solution for last-mile deliveries, particularly in congested urban areas or isolated locales. These drones are quick, cutting delivery delays and ensuring things get to the customer's door as soon as possible.

  These innovative methods are complemented with real-time commodities tracking, ensuring transparency across the supply chain. Customers and organizations can now track their shipment's exact location and status, creating trust and allowing for a proactive response to any potential interruptions. Taken together, these innovations pave the path for a more efficient and customer-centric logistics future.
- *Smart household.* Incorporating the Internet of Things (IoT) into furniture has ushered in a new era of intelligent living spaces. Furniture IoT allows interconnected goods, such as sofas, beds, and cabinets, providing functionality beyond their usual functions. Consider a chair that adjusts its padding based on the user's preferences, or a table that can charge your electronics wirelessly. Furthermore, managing these furniture pieces remotely via smartphones or voice commands improves convenience and personalization in homes and offices. In addition, immersive entertainment is changing how we watch movies, play video games, and consume other forms of information. With the introduction of VR and AR, viewers are no longer passive observers but active players in the story. This immersion blurs the distinction between reality and the virtual world, creating a more immersive and compelling entertainment experience. Furniture IoT and immersive entertainment, when coupled, can create a living area that is useful and gives unprecedented entertainment value.

- *Smart transport.* Modern automobiles have evolved into sophisticated sensor platforms in today's quickly evolving technology ecosystem. They constantly absorb and process information from their environment using powerful on-board processors. These computers are then used to help with a variety of functions, like navigation, pollution control, and traffic management. One of the issues is that real-time processing of this massive amount of data necessitates highly powerful computers, which often raise the price of premium automobiles equipped with modern driver-aid systems.

  However, a potential solution exists: using the Internet to dump this information onto cloud systems, allowing the huge processing workload to be handled. This is where IoT comes into play. IoT not only assists vehicles in gathering extra data but also supplements traffic information already gathered by traffic management centers.

  The vehicular cloud computing paradigm exemplifies this, offering an intriguing opportunity to test the potential of future 5G networks. Vehicular-to-everything (V2X) communication, in which vehicles may talk with other vehicles (V2V), roadside infrastructures, pedestrians, and almost any element in a smart city, is a critical component of this ecosystem. With this backdrop, some current developments in vehicular communication include remote/self-driving capabilities, infotainment on high-speed trains, AR-assisted navigation, and intelligent traffic planning, all to improve user safety, mobility, and comfort.
- *Smart security.* By delivering efficient and cost-effective solutions, 5G technology is revolutionizing public security and law enforcement; 5G patrolling robots improve public safety while lowering labor expenses dramatically. AR mobile policing uses 5G to provide officers with smart devices for better-coordinated responses. Using 5G-powered unmanned aerial vehicles (UAVs) ensures thorough patrolling, even in difficult terrain. Furthermore, integrating 5G with smart environmental protection tools strengthens supervision. Real-time ultra-high-definition (UHD) monitoring, in conjunction with sophisticated robot and drone patrols, offers a future of enhanced security and surveillance efficacy.
- *Smart agriculture.* The IoT holds great promise for revolutionizing agriculture. Farmers can improve crop growth using less water and fertilizer by combining sensors with wireless Internet in crop fields. Remote livestock and farm equipment monitoring simplifies operations even further, resulting in more cost-effective production. Furthermore, 5G technology enables improved water body monitoring, allowing for multi-dimensional management and increasing agricultural efficiency.
- *Smart health.* The 5G technology is driving transformative transformation in the health-care industry, opening the door for a more egalitarian and widely accessible system. The introduction of 5G-powered

telemedicine eliminates time and distance obstacles, bridging the gap between doctors and patients and providing on-time consultations. Furthermore, 5G improves outdoor first aid responses, ensuring life-saving actions promptly. Furthermore, 5G is important in epidemic prevention by enabling exact monitoring, facilitating rapid communication, and dramatically decreasing potential hazards through prompt interventions and data-driven initiatives.

Implementing health care in a smart city is a transformative endeavor that leverages technology, data, and connectivity to enhance health-care delivery, improve patient outcomes, and optimize resource utilization. Smart cities, characterized by their integration of digital technologies and data-driven systems, offer a unique environment for integrating health-care services. By utilizing innovative solutions such as telehealth, wearable devices, data analytics, and artificial intelligence, health care in a smart city aims to provide personalized, accessible, and efficient care to residents. However, the successful implementation of health care in a smart city is not without its challenges and issues.

Health care in a smart city operates as a CPS, merging digital technologies with physical health-care infrastructure. Wearable gadgets, sensors, and medical equipment are used in the system to collect real-time health data from individuals. Interconnected networks send this data to health-care facilities and data centers for examination. Automated systems and AI algorithms contribute to patient health monitoring, illness outbreak prediction, and resource allocation in health care. Telemedicine and remote patient monitoring improve access to health-care services even further, leading to a more responsive and efficient health-care system within the context of a smart city.

This chapter explores the challenges and issues encountered while implementing health-care CPS in a smart city. It delves into the infrastructure and connectivity challenges, technological complexities, legal and ethical considerations, human factors, and financial constraints that arise during this process. Furthermore, the chapter will present strategies, case studies, and best practices that can help overcome these challenges and optimize the implementation of health care in a smart city.

5.2 CYBER-PHYSICAL SYSTEM BASED HEALTH-CARE SYSTEM

A cyber-physical system–based health-care system revolutionizes the delivery of medical services by integrating advanced technology with physical health-care infrastructure. Through interconnected networks, sensors, and actuators, these systems enable real-time monitoring of patient health, facilitate remote diagnosis, and optimize treatment processes. By collecting and analyzing vast amounts of data from wearable devices, medical equipment, and electronic health records, health-care providers can offer personalized and proactive care tailored to individual patient needs. Moreover, these systems enhance collaboration among health-care professionals, streamline administrative tasks, and improve resource allocation, ultimately leading to more efficient and effective health-care delivery. With cyber-physical systems at the forefront, health care becomes not just reactive but predictive, preventive, and patient-centered, ushering in a new era of transformative health-care solutions. Figure 5.3 provides an insightful overview of a cyber-physical health-care system (CPHS), depicting the intricate network of interconnected components facilitating modern health-care delivery. At its core, the system revolves around patients and their interaction with various medical devices equipped with sensors for data collection. This data is seamlessly transmitted to a central system, where it undergoes processing and analysis. Health-care providers leverage the processed information to make informed decisions and provide personalized care to patients. The figure also highlights crucial aspects, such as security measures, regulatory compliance, and integration with electronic health records, underscoring the importance of safeguarding

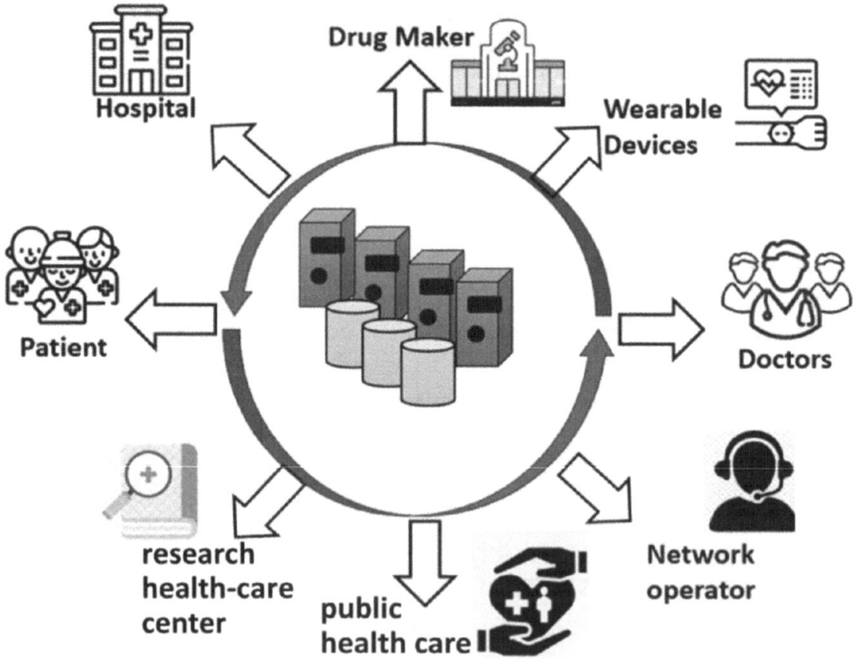

Figure 5.3 Overview of cyber-physical health-care system.

patient data and ensuring interoperability within existing health-care infrastructure. Through a feedback loop, patients receive valuable insights and guidance, fostering a collaborative approach toward improving health outcomes. This comprehensive depiction encapsulates the complexity and potential of CPHS in revolutionizing health-care delivery.

A number of studies are being conducted on the application of smart health, some of which are included in what follows. Hamza et al. [1] aim to identify the main issues facing each participant in the smart health-care system. In the state-of-the-art literature, eight major stakeholders in smart health care were revealed to face a total of 27 difficulties. Verma et al.'s [2] case study on health-care cyber-physical systems outlines the system's characteristics, the role of various technologies in their development, and the key obstacles to their successful implementation. Many safety-related systems are evolving into CPSs, integrating information technologies in their control architecture and modifying the interactions among automation and human operators, as Guzman et al. [3] discussed. Since a significant amount of data needs to be handled intelligently, a machine learning technique is essential for the effective implementation of IoT-powered wireless sensor networks (WSNs) for this purpose. How AI-powered IoT and WSNs are used in the health-care industry is covered in detail by TM Ghazal et al. [4]. Residents of smart cities now enjoy better lives and healthier bodies because of integration. The privacy of patient health information and the security of nearby mobile health users are two security issues that the integration has exposed to the health-care business. The usage of blockchain, however, is a promising technology that will allow the health-care sector to address security issues in smart cities. Blockchain technology has allowed patients' information to be stored in the health-care system safely and securely, according to J. Qiu et al. [5].

Ahmad et al. [6] discuss smart cities, health-care system analysis, and technical elements, particularly network technology. To analyze and make decisions for future planning, Reddy et al. [7] propose an architecture for smart cities, where the authors propose a distributed smart transport service (DSTS) model that effectively manages road traffic of traditional vehicles and intelligent vehicles to improve the QoS of the smart city. The authors also envision improving the QoS of smart transportation while utilizing a context-aware computing approach that reduces fog node data transfer.

To analyze security and privacy in the context of smart cities for health-care applications, S. Alromaihi et al. [8] make two contributions. As a result, an overview of various IoT applications and their digital weaknesses is provided on the one hand. On the other hand, a thorough analysis of potential solutions to address the issue of cyberattacks is given. In Kamruzzaman et al. [9], the PRISMA flowchart is proposed as a visual representation of the selection process. It has been determined that applying AI, ML, DL, edge AI,

IoMT, 6G, and cloud computing can tackle these growing health-care difficulties. However, just a few places have adopted these most recent innovations, and the results have improved.

Devices used for health monitoring have limited power and connectivity capabilities. The devices are outfitted with potent microprocessors to process the received data and conduct intelligent, decisive actions. Selective data gathering is a method to avoid accelerated energy dissipation and confined communication. S. S. Bhunia et al. [10] proposed a fuzzy-assisted data collection and alarm system for health-care services. Roy et al. [11] envisioned addressing this delay requirement of such unified IoT applications by considering applying lessons acquired from context-aware computing, notably context sharing among interdependent vertical IoT apps. This will minimize system delay by implementing context sharing among fog nodes.

The key uses of smart cities are highlighted in Al-Turjman et al. [12], which also discusses the important privacy and security concerns in the design of the apps for smart cities. It also discusses current approaches to the privacy and security of information-centric smart city applications and outlines upcoming research problems for performance enhancement. The needs, architecture, and components of Society 5.0 are detailed in P. Mishra et al. [13]. The world has made significant progress with the cutting-edge Society 5.0 and its connection to Industry 4.0/5.0. Additionally, the role of Society 5.0 in the UN's Sustainable Development Goals is thoroughly explained. Demands for medical services offered by smart devices in smart cities complicate the environment for analyzing medical data.

Boyi Xu et al. [14] proposed a health-care data analysis system for regional medical unions built to assist physicians from various hospitals in evaluating patient's health status in a comprehensive data view to address the aforementioned issues. Physiological index values are used to extract behavioral patterns. Social network data creates tags to identify the popular subjects people in each area are interested in. Experiments show the system's viability in assisting health-care data analysis. Behera et al. [15] presented a health-care application in the fog. A level monitoring task scheduling (LMTS) algorithm is suggested to respond promptly to delay-sensitive jobs with the least delay and network utilization. The results obtained from simulating the proposed algorithm using the CloudSim simulator showed that the proposed model is effective.

The context-aware addition to mobile health in smart cities is the new idea of smart health introduced by A. Solanas et al. [16] and summarizes the primary domains of knowledge utilized in constructing this novel notion. In [17], Reddy et al. examine the viability of energy minimization at the fog layer using context-aware fog nodes with intelligent sleep and wake-up periods. Using a genetic algorithm (GA), it suggests a virtual machine management technique for efficiently allocating service requests with a small number of active fog nodes. Thereafter, a reinforcement learning (RL) approach is included to optimize the duty cycle of fog nodes.

Rafiq et al. [18] review the IoT's current characteristics, architecture, communication infrastructure, and applications. This overview discusses communication protocol characteristics and applications, along with the number of IoT applications and problems in creating smart cities. Improvements to the smart grid, health applications, transportation applications, and smart city services are also covered. Reducing energy consumption, improving communities' economy and quality of life while simultaneously protecting the environment, and helping people use and adapt to modern information and communication technology (RTMCT) more effectively are elaborated by Prawiyogi et al. [19]. A health-care system is developed gradually and systematically. Numerous authors have studied these stages of development in-depth. Understanding the true nature of the health-care development stages from 1.0 to 4.0 can be aided by the findings described by Ahmad et al. [20]. The literature review focuses on various approaches, technology, and applications for health care in smart cities. There are numerous gaps in implementing health care in smart cities, which are explored in Section 5.3.

5.3 CASE STUDIES

This section presents some case studies analyzed from different perspectives, including implementation, evaluation, prospects, etc.

5.3.1 A few key development for implementing cyber-physical health-care system

Analyzing case studies and best practices can give important insights into effective health-care implementations in smart cities. The examples that follow showcase real-world situations and show how to overcome obstacles and get results that are beneficial:

- *Barcelona, Spain.* Barcelona has received recognition as a prominent smart city in health-care innovation. The "CityOS" platform, which unifies health-care data from various sources, including hospitals, clinics, and wearables, into a centralized system, was put into place by the city. With the help of this platform, proactive health-care treatments can be supported by real-time monitoring, data analytics, and predictive modeling. Barcelona's success can be attributed to its collaborative approach, which unites academic institutions, health-care providers, and technology firms in order to promote innovation and accelerate the adoption of smart health-care solutions.

- *Singapore.* Smart health care is a key element of Singapore's "Smart Nation" plan. The city-state has implemented several initiatives and technologies to enhance health-care delivery. For instance, the "National Electronic Health Record" system allows health-care organizations to share patient data, making care coordination easier and seamless. Singapore uses mobile health apps and telemedicine to deliver accessible and practical health-care services. Strong governance, strategic planning, and investment in research and development can be credited for Singapore's smart health-care implementation success.
- *Medellín, Colombia.* Through the incorporation of technology in a CPS, Medellín has changed its health-care system. The city established a telemedicine program that uses video chats to link isolated populations with medical experts. Additionally, Medellín makes use of wearables and IoT devices for remote patient monitoring, allowing for the early identification of health issues and prompt intervention. The city's focus on closing the digital divide and supplying health care to neglected communities shows a commitment to ensuring fair access to health care.
- *Copenhagen, Denmark.* In Copenhagen's smart city initiatives, user-centered design, and citizen interaction stand as paramount considerations. Anchored within the framework of cyber-physical systems, the city has introduced a digital health platform empowering residents to seamlessly access their medical records, manage appointments, and receive personalized health guidance. Complementing this, Copenhagen places significant emphasis on enhancing digital literacy among both the populace and health-care professionals to maximize the efficacy of smart health-care technologies. This concerted effort toward user acceptability and engagement underpins the success of their smart health-care deployment, aligning with the interconnected nature of cyber-physical systems in urban innovation.

Several top practices are highlighted in these case studies. The important aspects are described next.

- *Partnerships and collaboration.* Successful implementations necessitate collaboration among health-care organizations, technology vendors, government agencies, and academic institutions in order to drive innovation, share expertise, and efficiently utilize resources. A collaboration between a local hospital, a technology company, and a university may result in the development of a telemedicine platform that enhances patient care.
- *User-centered design.* Prioritizing users' wishes, preferences, and involvement results in smart health-care solutions that are rational,

simple to use, and tailored to the context and population. To encourage adoption and regular usage, a health-care app with a user-friendly interface and features that address the special needs of senior users should be designed.

- *Data integration and interoperability*. To ensure coordinated care delivery and comprehensive patient information, tools and platforms that enable easy data exchange and interoperability among health-care providers and systems must be implemented, using electronic health records (EHRs) that can be viewed and updated by all health-care professionals involved in a patient's care.
- *Digital literacy and training*. Investing in digital literacy initiatives and training programs aids in the acceptance and effective use of smart health-care technologies by citizens and health-care professionals, educating patients on how to utilize health monitoring applications for self-care and conducting workshops for health-care personnel to familiarize them with new technologies.
- *Equitable access*. Closing gaps in health-care access through programs that bridge the digital divide, prioritize underserved communities, and promote inclusivity in the use of intelligent health care. Using mobile health clinics to reach out to remote communities with few health-care options, and ensuring that technology serves a varied range of populations.
- *Regulatory frameworks*. To enable the responsible and compliant adoption of smart health-care solutions, clear regulatory frameworks and guidelines that address legal, ethical, and privacy issues must be established, developing policies and procedures that control the collection, storage, and exchange of patient data in order to protect patient privacy and maintain compliance with health-care regulations.

Other cities and health-care organizations can learn from successful implementations and modify their approaches to overcome obstacles and achieve successful outcomes when implementing health care in a smart city by reviewing these case studies and following these best practices.

5.3.2 Implementation challenges for cyber-physical health-care system

In cyber-physical systems, stakeholders are poised to forge a robust and enduring health-care ecosystem by proactively addressing the complexities inherent in this domain. By cultivating an acute awareness of these challenges and engaging in collaborative problem-solving endeavors, stakeholders can pave the way for a transformative evolution in health-care provisioning. Such concerted efforts hold the promise of revolutionizing health-care

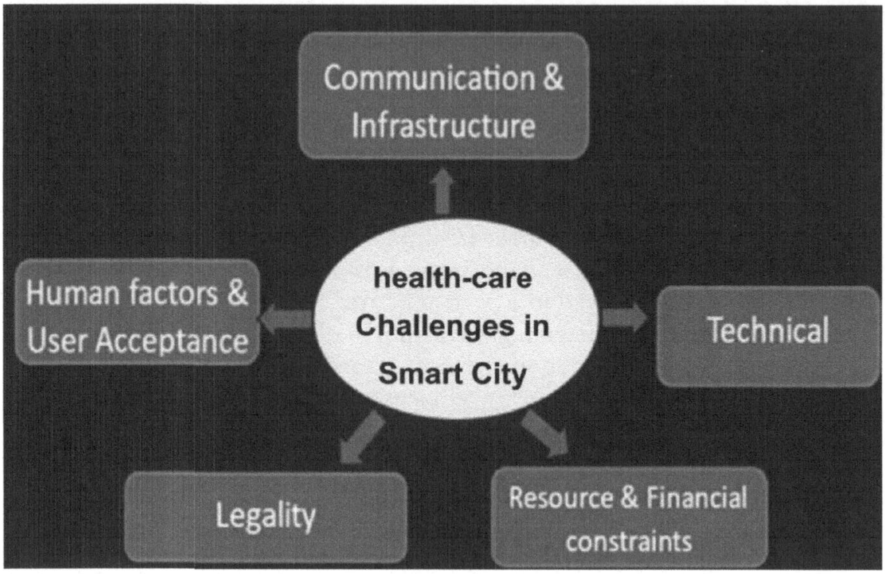

Figure 5.4 Health-care challenges cyber-physical system.

delivery paradigms, enhancing population health outcomes, and fostering an overarching improvement in citizens' well-being [21]. The integration of health-care services with cyber-physical system initiatives stands poised to catalyze this paradigm shift, ushering in a new era of health-care innovation and accessibility [22]. Despite the manifold benefits, the journey toward implementing health care within a cyber-physical framework is not without its obstacles, as depicted in Figure 5.4.

5.3.2.1 *Infrastructure and connectivity challenges*

In cyber-physical systems, the successful integration of health care hinges upon robust infrastructure and seamless connectivity. Various challenges and obstacles must be addressed to ensure the smooth delivery of health-care services. The following list delves into key connectivity and infrastructure issues encountered during the integration of health care within cyber-physical systems.

- *Infrastructure challenges.* For cyber-physical systems, reliability is a crucial parameter, especially for the health-care sector. Figure 5.5 illustrates the dependence of health-care systems on hardware components, such as health sensors and actuators; software dependencies on systems computing patients' health status; and network dependencies on communication networks for patient data transfer.

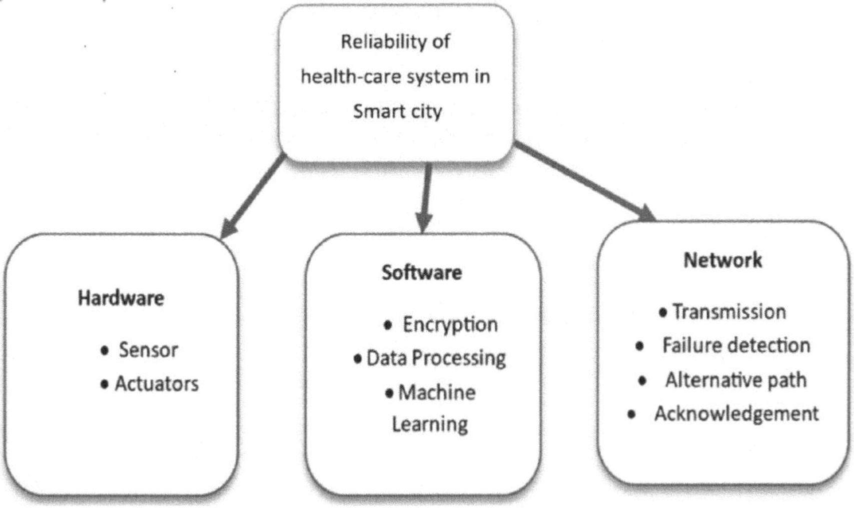

Figure 5.5 Reliability factors in the health-care system in CPS.

Embedding health-care systems within smart cities necessitates resource optimization and the cultivation of self-adaptive capabilities to ensure the delivery of efficient, reliable, and enhanced services. Autonomous systems should possess the ability to discern failures within different closed-loop system components and enact suitable responses to fulfill their functions. Health-care systems within smart cities ought to incorporate self-adaptive elements, drawing insights from past experiences to inform present actions. Addressing challenges in service quality can be done through the deployment of intelligent technologies, such as health-care robots capable of self-organization within dynamic environments [23].

- *Lack of network infrastructure.* The lack of a sufficient network infrastructure is one of the main obstacles to implementing health care in a CPS. To enable seamless communication between diverse health-care equipment, systems, and stakeholders, health-care systems require dependable and high-bandwidth networks. However, the current network infrastructure in many cities may be insufficient or antiquated, resulting in connectivity problems, sluggish data transmission, and unreliable service. The data-intensive requirements of smart health-care systems necessitate updating and increasing network infrastructure.
- *Health-care system integration.* Interoperability poses a significant barrier to the implementation of health care within a cyber-physical system. The utilization of diverse standards, protocols, and data formats across different health-care systems, devices, and applications complicates data sharing and seamless integration. Achieving interoperability

is crucial for ensuring efficient care coordination, facilitating data exchange among systems, and unlocking the full capabilities of intelligent health-care solutions. Addressing this challenge requires the establishment of common standards and protocols, the development of interoperability frameworks, and the promotion of channels for data interchange. This concerted effort is essential to enable the seamless flow of information and enhance the effectiveness of health-care delivery within cyber-physical systems [24].

5.3.2.2 *Technological challenges*

Utilizing cutting-edge technologies to improve health-care delivery, patient outcomes, and resource efficiency is essential to the implementation of health care in a CPS. The list that follows go through some of the major technological obstacles that smart cities' implementation of health care must overcome.

- *Wearables and IoT device integration.* In the domain of cyber-physical systems, a notable technological challenge lies in the integration and management of numerous IoT devices and wearables. These devices play a pivotal role in collecting and transmitting real-time health data, enabling functionalities such as remote monitoring, personalized treatment, and early detection of illnesses. However, the integration of multiple IoT devices presents challenges, such as ensuring data interoperability, managing device connectivity, and addressing scalability issues. To tackle this challenge effectively, it is imperative to establish robust frameworks for IoT device integration, standardization of data formats, and efficient device management within the cyber-physical system ecosystem.
- *Big data analytics and management.* Smart health-care infrastructures generate vast quantities of health-care data encompassing patient records, sensor data, and population health statistics [25]. Managing and analyzing this big data present a significant technological hurdle. To harness valuable insights from this immense volume of data, health-care organizations must deploy scalable data storage solutions, implement efficient data retrieval mechanisms, and leverage state-of-the-art analytics technologies. Machine learning and artificial intelligence emerge as indispensable tools in data analytics, facilitating the identification of patterns, prediction of health outcomes, and enhancement of health-care interventions.

5.3.3 Hacking (breaching) risks and cybersecurity

The integration of advanced technologies and interconnected systems in smart health care introduces cybersecurity risks and vulnerabilities. Given the high value and attractiveness of health-care data to cybercriminals, it becomes susceptible to cyberattacks and data breaches. Robust cybersecurity measures, including stringent access controls, regular security assessments, encryption protocols, and continuous monitoring, are essential to safeguard health-care systems, IoT devices, and patient data against such threats. Mitigating these risks requires the implementation of strict access controls, regular training sessions, and fostering a culture of cybersecurity awareness.

- *Delay/latency*. In real-time applications such as health care, low latency is crucial. Delays in transmitting patient data can disrupt the telemonitoring cycle of cyber-physical systems, potentially leading to delayed medication and treatment. Fault latency, measuring the time delay between fault occurrence and recognition, must be minimal in health care to ensure prompt management and enhance system reliability [26].

5.3.4 Challenges in ethics and law

To ensure responsible and ethical use of technology, preserve patient rights, and uphold public trust, several legal and ethical concerns are raised by deploying health care in a CPS. These issues must be resolved. The following parts go through some of the major moral and legal issues that arise when implementing health care in a CPS.

- *Governance and compliance with regulations*. Following numerous legal and regulatory frameworks is necessary to implement health care in a CPS. Compliance with licensing requirements, health-care standards, data protection, and privacy laws is essential. To ensure responsible adoption and preserve patient rights, health-care organizations, technology suppliers, and governments must navigate complex regulatory frameworks, ensure compliance, and build governance systems.
- *Ownership of data and consent*. A significant amount of patient data is produced by the integration of smart health-care systems. Establishing data ownership and secure informed consent for data collection, storage, and sharing can be difficult. Health-care organizations must implement transparent permission procedures that respect patient autonomy and privacy and clear policies on data ownership. Mechanisms for people to revoke consent and take back control of their data should also be implemented.
- *AI's potential for health care: ethical consequences*. There are moral questions raised by the application of artificial intelligence (AI) in health care. AI algorithms can potentially create biases, affect clinical judgment, and present ethical issues. The use of ethical frameworks, ensuring algorithm transparency and explainability, correcting biases, and creating channels for human oversight, is essential. Collaboration between health-care practitioners, lawmakers, and technology creators must create rules that put patient welfare, justice, and equality first.

5.3.5 Financial and resource constraints

Health-care implementation in a CPS necessitates large financial outlays and effective resource management. Sufficient funds and resources must be available for the successful integration of smart health-care solutions. But there are a lot of obstacles and limitations in this field. The primary financial and resource limitations encountered when implementing health care in a CPS are covered in the following:

- *Exorbitant implementation costs.* Implementing smart health-care solutions entails significant up-front costs, including infrastructure creation, technology procurement, system integration, and employee training. Expenses include acquiring and maintaining advanced medical equipment, IoT infrastructure, data storage, and analytics capabilities. To optimize investments and reduce implementation expenses, health-care organizations and local governments should devise cost-effective strategies and secure reliable financing sources.
- *Challenges with return on investment (ROI).* It might be difficult to prove that smart health-care solutions are cost- and ROI-effective. Smart health care has advantages that might not be immediately observable or quantifiable, such as better patient outcomes, cost reductions, and effective resource utilization. Evidence-based results, thorough cost–benefit analyses, and long-term assessment frameworks are required to demonstrate the value of smart health care and support financial commitments to stakeholders and funders.
- *Resource allocation and sustainability.* Critical factors to consider are resource allocation optimization and guaranteeing the long-term viability of smart health-care initiatives. Infrastructure, technology, and human resources investments must be balanced with other competing objectives by health-care organizations and local government agencies. To maximize the effectiveness and sustainability of efforts, effective budget allocation techniques, stakeholder collaboration, and regular monitoring and evaluation of smart health-care programmers are crucial.
- *Innovative public–private partnerships in health care.* Collaborations between public and private entities offer a promising avenue to surmount resource and financial constraints in the realm of smart health care. By pooling knowledge, funding, and assets, such partnerships accelerate the adoption of innovative health-care solutions. Cultivating robust relationships, aligning incentives, and delineating responsibilities are pivotal strategies for fostering innovation, enhancing resource efficiency, and navigating financial limitations.

Strategic planning, collaboration, and innovative approaches are imperative in addressing resource and financial constraints. Health-care institutions and governmental bodies must explore diverse funding avenues, including government grants, private investments, and charitable contributions. Overcoming financial barriers necessitates the establishment of public–private partnerships, pursuit of external funding sources, and exploration of unconventional financing mechanisms. Moreover, prioritizing resource optimization, implementing sustainable cost-saving strategies, and conducting comprehensive cost–benefit analyses ensure prudent resource utilization in the integration of health care within cyber-physical systems.

5.4 STRATEGIC APPROACHES FOR IMPLEMENTING CPS

In the realm of cyber-physical systems, the implementation of health care demands strategic collaboration and concerted efforts from diverse stakeholders. Overcoming the hurdles and ensuring seamless health-care integration within such systems necessitate a multifaceted approach. This section elucidates strategies to surmount obstacles. Leveraging cutting-edge communication and computational technologies presents promising avenues to tackle the myriad challenges across sectors, particularly in health care. Advanced communication infrastructures like 5G networks and Internet of Things (IoT) devices facilitate uninterrupted connectivity and real-time data exchange among health-care entities. This fosters remote monitoring, telemedicine, and efficient communication channels between health-care providers and patients. Moreover, computational advancements, such as artificial intelligence (AI) and machine learning algorithms, sift through extensive health-care datasets to glean invaluable insights, forecast disease outbreaks, tailor treatment regimens, and optimize resource allocation. Cloud computing architectures furnish scalable frameworks for securely storing and processing health-care data, allowing seamless access and sharing among providers across disparate locations. Additionally, blockchain technology ensures the integrity and security of health-care data through decentralized and immutable ledger systems. This amalgamation of technologies empowers health-care systems to surmount challenges pertaining to accessibility, efficiency, accuracy, and security, thereby enhancing patient outcomes and revolutionizing health-care service delivery.

- *Strengthening network infrastructure.* In cyber-physical health-care systems, ensuring seamless connectivity and efficient data transmission necessitates the acquisition of robust network infrastructure. Collaborative efforts between local government bodies and telecommunications enterprises are imperative to enhance network performance, expand coverage, and ensure reliable connectivity citywide. Meeting the demands for real-time data transmission among health-care devices, systems, and stakeholders entails substantial investments in high-speed Internet, wireless networks, and communication infrastructure. Undoubtedly, the latest advancements in communication and computational technologies hold immense potential in mitigating the challenges inherent in cyber-physical health-care systems. Cutting-edge communication technologies such as 5G networks offer the capability for high-speed and low-latency data transmission, facilitating real-time monitoring and remote patient care. Furthermore, the integration of edge computing augments the processing capabilities of health-care devices and sensors, enabling expedited decision-making and diminishing reliance on centralized infrastructure.

- *Communication and computational technology.* Within cyber-physical health-care systems, artificial intelligence and machine learning algorithms play a pivotal role in processing vast datasets, unveiling patterns, forecasting health trajectories, and tailoring treatment strategies. These technologies extend their utility by automating mundane tasks, affording health-care practitioners more bandwidth to focus on intricate patient care endeavors. Furthermore, blockchain technology emerges as a cornerstone for fortifying the security and veracity of health-care data. By furnishing a decentralized and tamper-resistant ledger, blockchain ensures the sanctity of patient information, shielding it from unauthorized access and upholding data privacy and confidentiality standards. The amalgamation of these cutting-edge technologies within cyber-physical health-care ecosystems holds the promise of catalyzing a paradigm shift in health-care delivery mechanisms, elevating patient outcomes, and augmenting the overall efficacy and efficiency of health-care services.
- *Ensuring data security and privacy.* To preserve public confidence and safeguard sensitive health information, it is essential to implement robust data security procedures and privacy safeguards. To reduce cybersecurity threats, health-care organizations should use encryption methods, put access controls in place, and often update security procedures. It is crucial to adhere to data protection laws, such as the General Data Protection Regulation (GDPR). Users' trust can be increased by being transparent and unambiguous about data collection, storage, and usage procedures.
- *Promoting interoperability and standardization.* To facilitate data interchange and deliver coordinated care, health-care systems, devices, and platforms must be interoperable. Data formats, protocols, and interfaces must be standardized to maintain compatibility and interoperability. The creation and adoption of common interoperability standards and frameworks should be the focus of cooperative efforts among health-care providers, technology companies, and

governments. This makes it possible to integrate various health-care systems, coordinate care, and share data effectively.

- *Enhancing accessibility and equity in health care.* For the equitable deployment of smart health care in a CPS, it is crucial to address gaps in access to health-care services and technology. By giving access to technology, digital literacy programs, and assistance to marginalized communities, efforts should be made to close the digital divide. Incorporating the community's different needs and preferences into the design and implementation of smart health-care solutions should be a top priority for health-care organizations and city authorities.
- *Developing robust technological solutions.* To ensure the creation of dependable and user-friendly smart health-care solutions, collaboration among health-care entities, technology innovators, and researchers is indispensable. Adopting user-centered design principles becomes imperative to ensure intuitive user interfaces, smooth user experiences, and robust engagement. Continuous innovation, research, and development endeavors are vital to enhance the accuracy, reliability, and usability of technologies such as IoT devices, artificial intelligence algorithms, and data analytics platforms.
- *Regulatory reforms and policy frameworks.* The adoption of smart health care should be governed by transparent regulatory frameworks that cover all relevant legal, ethical, and privacy issues. Guidance on data protection, consent, liability, and accountability should be provided through these frameworks. Creating policies that enable responsible innovation, safeguard patient rights, and assure regulatory compliance can be made easier by cooperation between legislators, health-care organizations, and legal professionals.
- *Promoting digital literacy and training.* Facilitating effective utilization of smart health-care devices requires funding digital literacy programs and training initiatives. Collaboration among health-care providers, governmental organizations, and community associations is vital to provide education and training on digital health literacy, data privacy, and technology usage. These programs should target both individuals and health-care professionals to enhance their knowledge and proficiency in utilizing smart health-care solutions.
- *Addressing financial and resource constraints.* Health-care organizations and municipal authorities can explore avenues such as public–private partnerships, seek external funding, and employ innovative financing methods to navigate resource and budget constraints. Collaboration with corporate entities, philanthropic bodies, and technology firms presents an opportunity to leverage expertise, capital, and resources in advancing smart health-care initiatives. Maximizing the utility of available resources demands the implementation of

sustainable cost-saving measures, rigorous cost–benefit assessments, and optimal resource allocation.

By embracing these principles, health-care institutions, local administrations, technology providers, and policymakers can surmount challenges and ensure the effective integration of health care within cyber-physical systems. Collaboration, stakeholder engagement, and a user-centric approach are pivotal in addressing infrastructure and connectivity issues, technological complexities, legal and ethical considerations, human factors, and financial constraints. Through collective effort, technology holds the promise of enhancing patient outcomes, elevating health-care delivery standards, and steering toward a healthier and more interconnected future.

5.5 CONCLUSION

The incorporation of health care into cyber-physical systems within smart cities holds promise for transformative advancements in the delivery, outcomes, and ecosystem structure. Essential to this success is overcoming infrastructure limitations, ensuring robust data security, and fostering interoperability. Equally crucial is the establishment of ethical frameworks to address legal compliance, data ownership, and AI ethics. Human factors and user acceptance are central considerations, demanding efforts to enhance digital literacy and cultivate trust. Sustainable implementation necessitates creative approaches, public–private collaborations, and judicious resource optimization. Stakeholders are urged to address challenges through strategies such as infrastructure reinforcement and the cultivation of partnerships. Successful implementations yield valuable insights, underscoring the significance of collaboration among health-care entities, technology providers, policymakers, and communities. Prioritizing user needs and resource optimization can bolster initiatives, fostering improved access to health care. In navigating this dynamic landscape, staying informed and fostering innovation are paramount as technology continues to evolve. Drawing lessons from past experiences and embracing innovation can pave the way for patient-centered health care in the context of smart cities.

REFERENCES

[1] Hamza, M., & Akbar, M. A. (2022). Smart Healthcare System Implementation Challenges: A Stakeholder Perspective. arXiv preprint arXiv:2208.12641.
[2] Verma, R. (2022). Smart City Healthcare Cyber Physical System: Characteristics, Technologies and Challenges. Wireless Personal Communications, 122, 1413–1433. https://doi.org/10.1007/s11277-021-08955-6

[3] Carreras Guzman, N. H., Wied, M., Kozine, I., & Lundteigen, M. A. (2020). Conceptualizing the Key Features of Cyberphysical Systems in a Multi-Layered Representation for Safety and Security Analysis. Systems Engineering, 23, 189–210. https://doi.org/10.1002/sys.21509
[4] Ghazal, T. M., Hasan, M. K., Alshurideh, M. T., Alzoubi, H. M., Ahmad, M., Akbar, S. S., Al Kurdi, B., & Akour, I. A. (2021). IoT for Smart Cities: Machine Learning Approaches in Smart Healthcare—A Review. Future Internet, 13(8), 218. https://doi.org/10.3390/fi13080218
[5] Qiu, J., Liang, X., Shetty, S., & Bowden, D. (2018). Towards Secure and Smart Healthcare in Smart Cities Using Blockchain. 2018 IEEE International Smart Cities Conference (ISC2), Kansas City, MO, pp. 1–4. https://doi.org/10.1109/ISC2.2018.8656914
[6] Ahmad, K. A. B., Khujamatov, H., Akhmedov, N., Bajuri, M. Y., Ahmad, M. N., & Ahmadian, A. (2022). Emerging Trends and Evolutions for Smart City Healthcare Systems. Sustainable Cities and Society, 80, 103695.
[7] Reddy, K. H. K., Goswami, R. S., & Roy, D. S. (2023). An Artificial Intelligence Approach to Enabled Smart Service Towards Futuristic Smart Cities. In *Handbook of Research on Applications of AI, Digital Twin, and Internet of Things for Sustainable Development* (pp. 12–29). IGI Global.
[8] Alromaihi, S., Elmedany, W., & Balakrishna, C. (2018). Cyber Security Challenges of Deploying IoT in Smart Cities for Healthcare Applications. 2018 6th International Conference on Future Internet of Things and Cloud Workshops (FiCloudW), Barcelona, pp. 140–145. https://doi.org/10.1109/W-FiCloud.2018.00028
[9] Kamruzzaman, M. M. (2021, December). New Opportunities, Challenges, and Applications of Edge-AI for Connected Healthcare in Smart Cities. 2021 IEEE Globecom Workshops (GC Wkshps), IEEE, pp. 1–6.
[10] Bhunia, S. S., Dhar, S. K., & Mukherjee, N. (2014). iHealth: A Fuzzy Approach for Provisioning Intelligent Health-Care System in Smart City. 2014 IEEE 10th International Conference on Wireless and Mobile Computing, Networking and Communications (WiMob), Larnaca, pp. 187–193. https://doi.org/10.1109/WiMOB.2014.6962169
[11] Roy, D. S., Behera, R. K., Reddy, K. H. K., & Buyya, R. (2018). A Context-Aware Fog Enabled Scheme for Real-Time Cross-Vertical IoT Applications. IEEE Internet of Things Journal, 6(2), 2400–2412.
[12] Al-Turjman, F., Zahmatkesh, H., & Shahroze, R. (2022). An Overview of Security and Privacy in Smart Cities' IoT Communications. Transactions on Emerging Telecommunications Technologies, 33(3), e3677.
[13] Mishra, P., Thakur, P., & Singh, G. (2022). Sustainable Smart City to Society 5.0: State-of-the-Art and Research Challenges. SAIEE Africa Research Journal, 113(4), 152–164. https://doi.org/10.23919/SAIEE.2022.9945865
[14] Xu, B., Li, L., Hu, D., Wu, B., Ye, C., & Cai, H. (2018). Healthcare Data Analysis System for Regional Medical Union in Smart City. Journal of Management Analytics, 5(4), 334–349. https://doi.org/10.1080/23270012.2018.1490211
[15] Behera, R. K., Patro, A., Reddy, K. H. K., & Roy, D. S. (2022). An Efficient Fog Layer Task Scheduling Algorithm for Multi-Tiered IoT Healthcare Systems. International Journal of Reliable and Quality E-Healthcare (IJRQEH), 11(4), 1–11.

[16] Solanas, A. et al. (2014). Smart Health: A Context-Aware Health Paradigm Within Smart Cities. IEEE Communications Magazine, 52(8), 74–81. https://doi.org/10.1109/MCOM.2014.6871673
[17] Reddy, K. H. K., Luhach, A. K., Pradhan, B., Dash, J. K., & Roy, D. S. (2020). A Genetic Algorithm for Energy Efficient Fog Layer Resource Management in Context-Aware Smart Cities. Sustainable Cities and Society, 63, 102428.
[18] Rafiq, I., Mahmood, A., Razzaq, S., Jafri, S. H. M., & Aziz, I. (2023). IoT Applications and Challenges in Smart Cities and Services. The Journal of Engineering, 1–25. https://doi.org/10.1049/tje2.12262
[19] Prawiyogi, A. G., Purnama, S., & Meria, L. (2022). Smart Cities Using Machine Learning and Intelligent Applications. International Transactions on Artificial Intelligence, 1(1), 102–116.
[20] Ahmad, K. A. B., Khujamatov, H., Akhmedov, N., Bajuri, M. Y., Ahmad, M. N., & Ahmadian, A. (2022). Emerging Trends and Evolutions for Smart City Healthcare Systems. Sustainable Cities and Society, 80, 103695.
[21] Alnuaimi, A. S., & Alneyadi, H. (2017). Challenges in Implementing Smart Healthcare Systems in Smart Cities. 2017 3rd International Conference on Future Internet of Things and Cloud (FiCloud) (IEEE), pp. 74–81.
[22] Lee, M., Liew, J. C., Lim, H. B., & Chuah, S. P. (2016). Internet of Things for Smart Cities: Issues and Challenges. Journal of Network and Computer Applications, 66, 120–134.
[23] Hashem, I. A. T., Yaqoob, I., Anuar, N. B., Mokhtar, S., Gani, A., & Khan, S. U. (2016). The Rise of "Big Data" on Cloud Computing: Review and Open Research Issues. Information Systems, 47, 98–115.
[24] Bu, D., & Fu, H. (2019). Challenges and Countermeasures for the Development of Smart Healthcare in Smart Cities. 2019 4th International Conference on Smart City and Intelligent Building (ICSCIB) (IEEE), pp. 73–76.
[25] PwC. (2018). Smart Health in the Smart City: Overcoming the Challenges of Implementing Smart Health Services. Retrieved from www.pwc.in/assets/pdfs/publications/2018/smart-health-in-the-smart-city.pdf
[26] Singh, R. K., & Adhikary, A. (2018). Healthcare in Smart Cities: Opportunities and Challenges. In *Smart Cities and Homes* (pp. 77–94). Springer.

Chapter 6

Enabling smart manufacturing through cloud computing and cyber-physical systems

Isaac O. Olalere

LIST OF ABBREVIATIONS

AE	acoustic emission
AI	artificial intelligence
ANOVA	analysis of variance
CNC	computer numerical control
CPS	cyber-physical system
DWT	discrete wavelet transform
EMD	empirical mode decomposition
FFT	fast Fourier transform
FT	Fourier transform
GA	genetic algorithm
HHT	Hilbert–Huang transform
IIOT	industrial Internet of Things
IIRA	industrial Internet reference architecture
IMF	intrinsic mode function
IOMT	Internet of Manufacturing Things
IoT	Internet of Things
ISO	International Standard Organization
KNN	K-nearest neighbor
LAN	local area network
ML	machine learning
RAMI 4.0	Reference Architectural Model Industry 4.0
RW	roulette wheel
R_a	arithmetic mean
R_{ku}	kurtosis
R_{max}	maximum valley depth
R_q	maximum height of roughness
R_{sk}	skewness
R_t	maximum roughness depth
R_y	maximum valley depth
R_z	average maximum height of profile
R_z	root mean square deviation
SEM	scanning electron microscope
SM	smart manufacturing

DOI: 10.1201/9781003559993-6

SOA	service-oriented architecture
SVM	support vector machine
TCM	tool condition monitoring
TGAN	tabular generative adversarial networks
TWCM	tool and condition monitoring
VB	flank wear
Vc	cutting speed
WLAN	wireless local area network
WPT	wavelet packet transform
WSN	wireless sensors network

6.1 INTRODUCTION AND MOTIVATION

Smart manufacturing (SM) has emerged since the introduction of Industry 4.0, and it consists of resource sharing and networking, predictive engineering, and material and data analysis [1]. This has the capability of increasing the efficiency and effectiveness of the production and manufacturing processes. SM, enabled with Industry 4.0 technologies, such as the industrial Internet of Things (IIoT), smart sensors, and cyber-physical systems (CPS), is gradually reducing human operations and replacing them with computerized systems capable of varying system responses to different situations and requirements. This has a significant positive impact on the product quality, resource utilization, production cost, and performance of most machines and systems. IIoT and CPS have enabled intelligent manufacturing through the network of connected sensors, actuators, and controllers to machines and production processes. Complex systems and processes with remote areas and locations in manufacturing facilities are being transformed into SM through the possibilities of a wireless sensor network (WSN). This makes the connection of sensors to machines and industrial facilities seamless and overcomes the obstructive nature of the location being monitored. The intelligent manufacturing system, also known as SM, optimizes production using advanced information and manufacturing technologies [2]. Intelligent manufacturing can be classified into three processes, which are digital manufacturing; digitally networked manufacturing, which applies to a network of interconnected manufacturing outlay; and new-generation intelligent manufacturing, which integrates advanced manufacturing with artificial intelligence (AI) [3].

Over time, CPSs have employed a network of sensors that consider the embedded service-oriented architecture (SOA) [4], which has widened its application in the manufacturing sector. For instance, smart machining manufacturing is evolving over conventional methods in precision manufacturing, with a focus on improving productivity by optimizing the facility outlay. There are two main concerns with machining operation: the first concern is the tool condition, and the second concern is the quality of the product output. The condition of the tool affects the quality of the product, and the quality specifications of the manufactured product determine if the product will be accepted or will need to be reworked or scrapped. This directly increases the cost of production, as both the tool and workpiece must be replaced. According to Rudek [5], quite a few machining stations have embraced a just-in-time maintenance policy that schedules cutting tool replacement at a specified time before the useful life. While the approach may seem to reduce machine downtime, the drawback, however, is that the running cost increases because the cutting tool is not fully utilized [6]. This has endeared many research methods to investigate smart manufacturing that factors in product quality requirements, cutting tool conditions, and manufacturing conditions into production using Industry 4.0 revolution technology, such as smart sensors and IoT devices, machine learning (ML) methods, and cloud computing. Integrating manufacturing enterprises in the physical world with virtual enterprises in cyberspace requires cyber-physical systems. This has further expanded the world IIoT to the Internet of Manufacturing Things (IoMT), as stated by Yang [7]. In the machining manufacturing industry, there are quite some challenges that limit the capabilities and application of CPS for SM. Some of them is the nature of the manufacturing outlay, which is most times harsh or obstructive for the installation and functioning of IoT devices. Mabkhot [8] highlighted some requirements to meet for a smart factory, which are modularity, interoperability, decentralization, virtualization, service orientation, and real-time capability. The design of the system components in a machining station, which is explained as the ability to separate and reconfigure system components seamlessly and quickly, and the interoperability, which entails the exchange of information between manufacturing enterprises powered through CPS-enabled connections are required for a smart factory.

6.2 CONTRIBUTION AND OVERVIEW OF THE STUDY

This section discusses the contribution and detailed insights of the study regarding intelligent machining in a manufacturing company. The study focuses on a turning operation which involves a rotation workpiece and a stationary cutting tool at the machining workstation. Several studies have considered milling operations that consist of a stationary workpiece and a rotating cutting tool.

6.2.1 Contribution of the study

The study develops an SM system in the machining station by applying Industry 4.0 and CPS technology. Two critical components are significant at a machining station for optimal productivity. The cutting tool and the workpiece determine the cost of production, as the former determines if the product manufactured from the latter would be scrapped, reworked, or accepted. Due to the obstructive nature of the machining operation, it is daunting to evaluate the condition of the cutting tool during operation, and measuring the condition and the quality output of the workpiece by intermittently stopping the machining operation increases the machine downtime. The inability to capture the condition of the cutting tool implies higher scrapped workpieces or an under-utilized cutting tool. This chapter therefore develops the SM system that captures pertinent conditions during machining for intelligent detection, diagnosis, and knowledge-based decisions using Industry 4.0 and cyber-physical systems technology.

6.2.2 Machining operation and the monitored parameters

The cutting tool and the workpiece are the important components at a machining station that require being optimized for smart or intelligent manufacturing. Establishing a sensor network that monitors the condition of the cutting tool and the workpiece using CPS is required to measure some influential parameters of the cutting tool and the workpiece. Some studies have used some measured parameters from the cutting tool, workpiece, and machine tool to support machining station decision-making. Workpiece condition has been determined by its surface roughness parameter R_a [9], which is the arithmetic mean roughness, while the cutting tool is assessed by the wear, crack, or chip parameters. Even though the arithmetic mean surface roughness R_a parameters of the workpiece can indicate a quality characteristic that is essential in terms of customer and product specification, more detailed parameters can be extracted for intelligent condition diagnosis of both the workpiece and the cutting tool during machining operation. There are several other surface roughness parameters, such as the total height of roughness profile R_t, maximum roughness depth R_{max}, maximum valley depth R_y, root mean square deviation R_q, average maximum height of profile R_z, kurtosis R_{ku}, skewness R_{sk}, and others, which can be measured to evaluate the quality condition of a material's surface. Determining the parameters that correlate to varying conditions of the tool and workpiece is fundamental to the choice of the devices and the conditions that the CPS is built on. These surface parameters are measured both for the workpiece and the cutting tool edge and surface conditions and can determine the condition of both the workpiece and the cutting tool.

6.2.3 Evaluating significant parameters for tool and workpiece conditions monitoring

The sensitive surface roughness device used captured 20 parameters both from the surface profile of the workpiece and the cutting tool, determining the essential parameters that can differentiate different conditions based on the condition of the machine tool and the remaining life of the tool or quality output of the workpiece. Das [10] showed that with decreased tool wear obtained from the minimized coefficient of friction between the tool and workpiece, the workpiece structure changes from coarse to fine. This therefore indicates a correlation between the condition of the cutting tool edge/surface and the surface quality output of the workpiece material. If these can be quantified through parameter measurements, then the corresponding effect can be studied and, therefore, an appropriate sensing device is used for capturing and monitoring the device. Figure 6.1 illustrates the schematic of parameter measurement during machining to extract knowledge on the condition of the workpiece and the cutting tool. Establishing this is very important to applying CPS in the manufacturing outlay for proper event and condition capturing, diagnosis, and apt decision-making.

To adopt CPS for SM using IoMT for machining operations, some requirements must be met, as illustrated in Figure 6.1. Since the machining operation is obstructive, there are limitations to the application of sensor networks that can be installed to capture the tool condition as well as the condition of the workpiece during operation. The measured parameters from both the cutting tool and the workpiece should, at different thresholds, be able to generate signals that can be sensed with a sensor. These signals could be one or more of either vibration, acoustics, temperature, force field, etc., which can be sensed using the sensor.

The SEM images in Figure 6.2 show the edge and surface wear condition of the indexable tungsten cutting tool, with CCMT09T3034 carbide insert. The edge and surface wear parameters increase as the tool life decreases. Figure 6.2 shows the SEM images of four different cutting tools, classified into new, good, rough, and worn-out tools using ISO 3685:1993 standards, as also illustrated in Table 6.1.

These tools create different surface wear signatures on the workpiece and generate different signals during the machining operation. There are known conditions that contribute to the wear condition of a cutting tool during machining operation. A study carried out by Khan [11] showed that operating parameters directly affect cutting tool wear, considering cutting velocity, feed rate, cutting depth, and the cutting tool texture pattern. The study

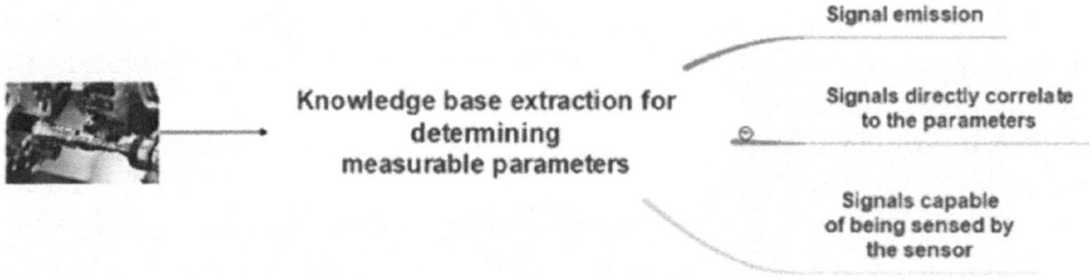

Figure 6.1 Schematics of signal capturing requirements for CPS at a machining station.

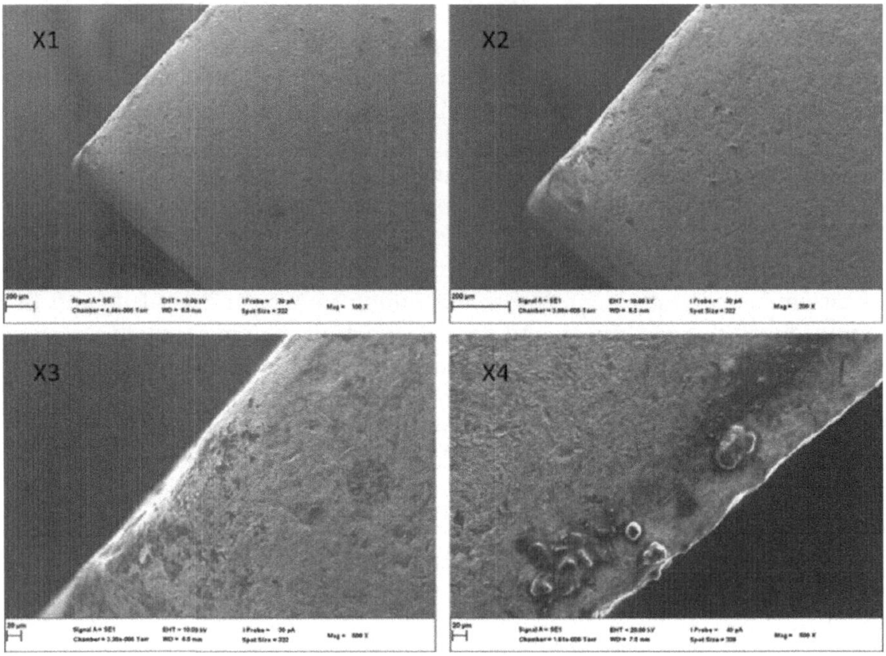

Figure 6.2 SEM images of cutting tool types based on degree of wear.

Table 6.1 Tool classification based on flank wear

Tool label	*Tool caption*	*Flank wear*
X1	New	0
X2	Good	0–0.15
X3	Rough	0.15–0.25
X4	Worn	0.25–0.35

asserted that tool wear was found to have increased as the cutting depth and feed rate increased. A similar study conducted by Kumar [12] showed that there is a correlation between both cutting speed and depth and flank wear, noting that the flank wear was lower than the standard limit of 0.2 mm, with the maximum surface roughness of 0.9 μm.

Another study was conducted on cutting tool wear morphology by optimizing tool life reliability considering two operating parameters, cutting speed/velocity and feed rate. The experimental study showed a resulting indication of varying tool wear, such as crater wear, resulting in cutting-edge damage at V_c = 175 m/min and f = 0.02 m/r; tool fracture at V_c = 55 m/min and

$f = 0.3$ m/r; tool chipping and tipping at $V_c = 85$ m/min and $f = 0.2$ mm/r; and tool breakage at $V_c = 175$ m/min and $f = 0.05$ m/r and different cutting speed and feed rates [13]. The study indicates that cutting parameters influence the cutting tool wear. According to Aralikatti's [14] study, tool conditions can be classified using vibration and cutting force signals into "healthy," "worn flank," "broken insert," and "extended tool." To vary the tool condition to the signal emission during operation, it is therefore important to keep the machine operating parameters constant. Therefore, the condition of the cutting tool and the workpiece, which is measured by surface/edge parameters, can be correlated with the signals emitted during the operation while keeping the operating parameters constant. The SM system is, hence, possible through capturing the signals emanating during the operation using IoMT devices.

Several signals are emitted during machining operation, such as vibration, acoustics, and heat, which may be monitored to diagnose the conditions of the cutting tool and the workpiece material. The study by Kiew [15] showed that at different depths of cut and feed rate, the correlation between tool wear and the machine vibration signal varies. Even though another study by Ochoa [16] found that installing the acoustic emission (AE) sensor on the toolholder as against the workpiece could provide a more reliable TCM system, in contrast, another study by McIntyre [17] argued that AE sensors are incapable of distinguishing between the new and worn-out tool when it is installed on the toolholder or the workpiece. An investigation by Kang [18] to determine the effect of vibration on surface roughness in a finished turning operation concluded that, depending on the magnitude of vibration, combined with the vibration-free workpiece profile, the average roughness may increase continuously or fluctuate randomly. Şahinoğlu, Karabulut, and Güllü [19] investigated the correlation between both the vibration of the spindle and surface roughness in the turning of aluminum alloy AI 7075 and examined the effect of cutting parameters on surface roughness and spindle vibration. The result showed that feed rate directly affects both the spindle vibration, the workpiece's surface profile roughness, and the tool condition. During turning operation, the radial vibration of the spindle has a strong effect on the surface roughness of the workpiece and tool condition [20]. Vibration signal is hence very important in detecting and diagnosing the condition of the cutting tool and the workpiece surface finish. Acoustic signal is also emitted during the machining operation; however, due to the excessive external sound/noise from the machine being in operation, it is very challenging to discriminate acoustic signals emitted from the interface of the cutting tool with the workpiece surface. Thermal/heat signal is another signal that is generated during the machining operation; however, capturing this signal during operation and remotely may be a lot more challenging.

6.3 ENABLING SMART MANUFACTURING THROUGH IOT DEVICES AND SENSOR DEPLOYMENT

Upgrading a manufacturing process or system to smart manufacturing through the application and implementation of CPS requires a painstaking approach. One of the core steps in implementing CPS for SM is the integration and installation of sensors for condition monitoring during operation. It is therefore important and of enormous interest to determine the suitability of an appropriate sensor deployment for signal capturing for on-condition monitoring of the machining operation.

6.3.1 Smart manufacturing architecture with cyber-physical system design at the machining station

The implementation of CPS at a machining station requires a stepwise process that follows sequentially from start to finish. More recently, the Industry 4.0 (I4.0) paradigm has adopted meeting emerging DIN SPEC 91345 norm for SM [21]. The Reference Architectural Model 4.0 (RAMI 4.0) for factories is gradually migrating from conventional automation systems to cyber-physical production systems [22]. This has been illustrated to have a three-dimensional layered hierarchical model comprising of the layers, life cycle value stream, and hierarchy levels. Another prominent architecture according to the standardization/IEC/IEEE 42010 standard is the industrial Internet reference architecture (IIRA) developed by the industry IoT consortium considering the business, usage, functional, and implementation viewpoint [23]. There are, however, shared conceptual similarities in both the RAMI 4.0 and IIRA models in terms of interoperability [24]. This includes the business domain, application and operation domain, information domain, network connectivity, and physical systems. The architecture for smart machining manufacturing adapts these basic similarities, with no limitations to any of these layers of hierarchy. The CPS's architecture for the smart machining system can be explained based on the application and operation and the implementation strategy, which automatically cuts through the physical system, network connectivity, information from processed data, application, and operation. The first section of the framework deals with knowledge base extraction from the machining operation that identifies the signals, configuration, and measurements from the manufacturing system. Implementing an SM system at a machining station requires considering the physical asset/system for signal sensing, processing, real-time data acquisition and analytics capabilities, and communication networks.

The developed architectural framework for CPS implementation for smart machining operation encompasses the CPS's 5Cs framework for manufacturing based on standards, namely, configuration, conversion of data to information, smart connection, cyber computation, and cognition [25]. Extracting signals that capture data that depicts the condition of the machining operation is carefully implemented through the selection of appropriate sensor networks and controllers for efficient network communication. Efficient communication ensures that the feedback system to the operator of the operating station is seamless. Data processing and cognition of the system deal with the processing, interpretation, and transmission of information as feedback to the system. All these actions and operations within the CPS require great coherence and synergy with real time, therefore making configuration a pertinent step in the implementation. The condition at the machining station during operation is therefore monitored to identify trends depicting significant conditions in the measured parameters. The overview of the framework for implementing CPS at the machining station in this study is illustrated in Figure 6.3. A suitable sensor is selected for signal capturing, and the knowledge base to be extracted from the signal is determined by classification or regression trend to be captured during online operation or condition. Communication is established through the connection of the gateway to the sensor network and the configuration of the network to a server. Data processing, analysis, and feedback are initiated from the cloud server, and actionable information is routed to the actuators or machine operator at the machining station. The condition of the component of the machining station is thereby, in real time, determined as the sequence of these reactions occurs at intervals, known as heartbeat.

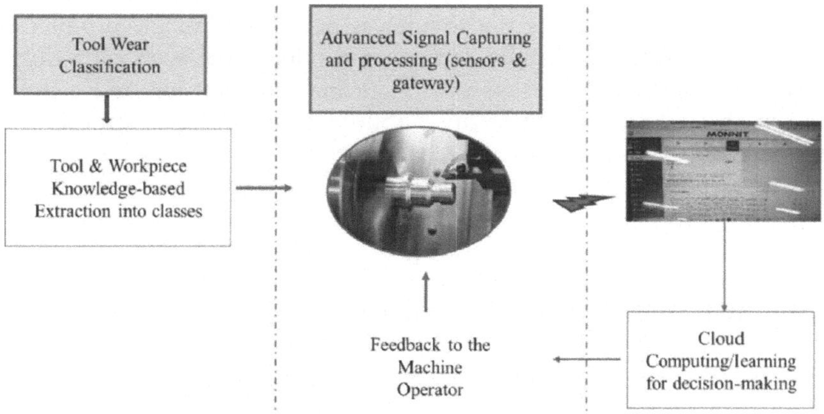

Figure 6.3 The implemented architectural framework for the CPS for SM.

6.3.2 Deployment of sensors and IoT controllers

An early study by Dureja [26] evaluated the relationship between machining parameters, workpiece roughness, and tool flank wear by applying analysis of variance (ANOVA) analysis and an interaction graph to show that there is a correlation and influence between these factors in a machining operation. At the machining station, tool and workpiece conditions can be monitored through direct or indirect methods. A contact instrument can be used for evaluating the wear and the surface roughness condition of the tool and workpiece area of estimation. Direct tool wear and workpiece roughness monitoring relies on directly measuring and analyzing the tool wear and workpiece profile by using high-precision instruments, such as contact detectors or industrial cameras [27]. When compared to the indirect method, it produces a more accurate measurement of the observations. However, direct methods come with some limitations, such as lighting conditions, cutting fluid, and chip interference during machining; hence, the method is only applied for tool wear and workpiece roughness measurement offline [28]. The indirect method measures tool wear and workpiece roughness by monitoring alterations or changes in some parameters, or signals, such as vibration, cutting temperature, force, acoustics, and others, by constructing the relationship model between characteristic signals and tool wear and indirectly inferring the tool wear and workpiece roughness state [18]. Even though the implementation of the indirect method is computationally more challenging, advancements in sensing technology, cloud computing, and the increasing number of different sensors and modern CNC machines with in-built sensors have altogether enhanced this approach. This approach is a data-driven method that evaluates the tool and workpiece conditions by the intelligent analysis of signals captured during the machining operation. Tool wear directly increases the roughness parameters of the product due to reduced machining accuracy. It further results in the size being outside the tolerance limit, an increase in the cutting temperature, and a noticeable increase in vibration. These cause abnormal operating conditions of the machine, which affect the workpiece with a high-accuracy requirement [29]; hence, the captured data in the indirect method is deployed for the prognosis of tool and workpiece condition.

Several factors could result in the vibration of the machine tool during operation. One of these factors that could result in the vibration of the machine tool is a broken or worn cutting tool during the cutting operation, and the same can happen with a faulty jaw, which also induces the vibration of the spindle that affects both the condition of the tool and the workpiece [30]. Since tool wear directly increases the cutting temperature and the vibration of the machine tool, leading to an abnormal operation of the processing equipment, this negatively impacts the accuracy requirement of the workpiece [29]. A careful study of each possible cause and the vibration threshold from each corresponding condition can be used in the condition diagnosis during the machining operation. The condition of the cutting tool and the workpiece can therefore be diagnosed by capturing, processing, and analyzing the vibration signal during operation. Similarly, acoustic signals were used by Kuram and Ozcelik [31] to optimize tool wear, surface roughness, and cutting force measured in the machining of Ti6Al4V titanium alloy and Inconel 718 workpiece materials using Taguchi's signal-to-noise ratio. The study captured the acoustic signals generated during the machining process and applied the regression model on fitting and predicting tool wear, surface roughness, and the cutting force in the machining of Ti6Al4V. Other signals have been captured during the machining process using different sensors, such as temperature and motor current [32, 33]; however, vibration signals have been widely used because of their potential, ability, and robustness in sensing and analyzing different conditions of the system during operation. The vibration signals emitted during operation were captured through IoT-enabled sensors (advanced vibration sensor from iMonnit) and gateways. The vibration data was captured on the cloud server, enabling remote access and robust data analysis. The experiment (turning operation) was performed while keeping some operating parameters constant, such as speed, feed rate, cutting force, etc. Each cutting tool class was used for the turning operation, and the label indicating the tool type was recorded against the captured vibration signal using a long-range wireless advanced vibration meter.

6.3.3 Cloud data mining and processing and analysis

Resource sharing, data acquisition, and analysis with CPS are mainly achieved through cloud server configuration for data hosting, analysis, and information transmission. Signals are captured at the machining station using varying sensors, depending on the nature of the signal, the robustness of the application, and the location of the region of installation. Due to the limitations of the cutting zone which limit the accuracy of most signal types, vibration signal is mostly adopted for diagnosing the condition of both the cutting tool and the workpiece during operation. In this study, vibration signals during operation were captured using IoT-enabled sensors (advanced vibration sensors) and a gateway for data transmission to the cloud server.

The sensor adapter gateway helps with signal transmission from the smart vibration sensor to the remote cloud server that is accessible online, bringing about the implementation of the CPS for the smart machining process. The basic functions needed to control and manage smart sensors were highlighted according to ISO/IEC/IEEE 21450:2010, while a reliable and interchangeable control system was expressed by IEC 61131 and IEC 61499. Suitable standards for industrial wireless communication, measurement,

Table 6.2 Wireless CPS communication and networking protocols

Network type	*Com protocol*	*Features/coverage*	*Freq./Data rate*	*CPS app layer*
Wireless	ZigBee	• Energy saving, wireless LANs (WLANs) • Up to 100 m (standard-dependent)	2.4 GHz/250 Kbps	WSANs M2M
	Bluetooth	• Cable connection replacement • Up to 100 m (version-dependent)	2.4 GHz/1 Mbps	WSANs M2M
	Wi-Fi	• Data networks, LAN, WLAN • Up to 250 outdoors (standard-dependent)	2.4–5 GHz/1–150 Mbps (standard-dependent)	WSANs M2M IoT (all)
	WiMax	• Metropolitan area network (MAN) • Up to 56 km	2–66 GHz/2–75 Mbps	WSANs M2M IoT (all)
	Cellular 3G, 4G/ LTE, 5G	• Wide-area networks (WANs), digital data packets • 1 several km (cell radius–dependent)	• 3G: 800–1,900 MHz • 4G/LTE: 700–2,500 MHz • 5G/LTE: 600–6 GHz • 5G/mmWave: 24–86 GHz	WSANs M2M IoT (all)

monitoring, and control are both the IEC 62591:2016 (Wireless HART™) and IEC 62601:2015 (WIA-PA) [25]. The sensor adapter gateway is built based on these control system standards for its CPS principal functionality for communication with the cloud server. Table 6.2 illustrates the wireless CPS communication and the networking protocols for the framework [34]. The sensor deployed for the research study conducted was wireless; hence, the communication and networking protocol falls under the Wi-Fi communication protocol, as illustrated in Table 6.2.

6.3.4 Data analysis and processing

In this study, tools classification was based on flank wear, and the new tool is assumed to have 0 flank wear. The good tool is taken to have the same range with the normal wear (VB = 0–015 mm), the rough tool has medium wear (VB = 0.16–0.29 mm), and the flank wear for the worn tool is taken as the same with the critical wear range (VB > 3.0 mm), according to ISO 3685:1993.

There are two classes of signals captured from the installed monitoring devices, namely, the steady-state signals and the dynamic/transient signals. The steady-state signal is captured when the condition of the machine is stable during operation, while the signals captured when the machining operation is unstable are known as the dynamic/transient signal. These two signals captured during the machining operation are analyzed using different methods and techniques for extracting intelligent information that identifies the distinctive state of the process. Different signal processing methods can be adopted for the TWCM system. There is the time domain analysis and frequency domain techniques. While the former evaluates physical signals and mathematical functions with respect to time [35], the latter analyzes signals or mathematical functions with respect to frequency instead of time [36]. Signals can also be converted from either the time domain to the frequency domain, or vice versa, with an operator called transforms.

Some studies on the TCM system have adopted the Fourier transform (FT), which converts time function into an integral of sine waves of various frequencies; it is, however, not efficient in analyzing non-periodic and non-stationary signals. Therefore, other types of transforms have been developed [37]. Wavelet packet transform (WPT) models have shown a better result as a computational approach for time–frequency signal conversion, and this has resulted in wider acceptance for tool condition monitoring research [38, 39]. A comparative study on bearing degradation prediction by Bhavsar and Vakharia [40] using discrete wavelet transform (DWT), tabular generative adversarial networks (TGAN), and ML models also showed that DWT is an efficient signal processing tool in decomposing signals that are non-stationary signals. Another transform model for decomposing signals and computing time–frequency conversion is the Hilbert–Huang transform [41]. The vibration signal captured from the machining station is a non-stationary and nonlinear signal, which requires some signal processing techniques to use the signal for analysis. An ideal transform for linear and stationary signals is the fast Fourier transform (FFT) because of the uniform trigonometric function, while the alternative transform for extracting time–frequency resolutions of signals is the wavelet transform (WT) because of its unique sparse representation. The empirical mode decomposition (EMD) method is a better approach for analyzing nonlinear and non-stationary signals, such as the vibration signal captured at the machining station. The signal is decomposed into a finite number of IMFs (real part) and the residual (imaginary part), as illustrated in equation 1.

$$f(t) = \sum_{i} imf_i + res \tag{6.1}$$

The EMD function evaluates the local extrema of the signal and fits the maxima (E_{upper}(t)) and the minima (E_{lower} (t)) to an individual envelope. The mean of the upper and lower envelope is determined using equation 2.

$$E_{mean}(t) = \frac{E_{upper}(t) + E_{lower}(t)}{2} \tag{6.2}$$

The residual part of the signal is evaluated using equation 3.

$$res(t) = f(t) + E_{mean}(t) \tag{6.3}$$

Also, the stopping criteria are evaluated using equation 4, since the process is finite.

$$\sum t = \frac{\left(res(t) - f(t)\right)^2}{f(t)^2} < \in \tag{6.4}$$

The process of decomposition ends when the residual approaches a monotonic function. The features that discriminate and classify the condition during the machining process can be obtained from the intrinsic mode function (IMF) component of the decomposed signal. These instantaneous properties, such as the energy and frequency, are evaluated by applying the HHT on the IMFs. For each IMF x_i, the HHT function computes the components as indicated in equation 5.

$$x_i = f(i) + iH\{f(t)\} = A(t)e^{i\varphi(t)} \tag{6.5}$$

A unique imaginary component H{f(t)} and real component $f(i)$ are in equation 5. The instantaneous energy is derived from equation 6, and the instantaneous frequency from equation 7.

$$\rho = |A(t)|^2 \tag{6.6}$$

$$\omega(t) = \frac{d\varphi(t)}{dt} \tag{6.7}$$

where $A(t)$ is the amplitude and $\varphi(t)$ is the phase of the signal. The feature vectors are mainly the instantaneous frequency and the energy of the models, because the instantaneous energy is very dependent on the amplitude. A total of 12 features was computed from the IMFs of the decomposed signals. These features are deployed for developing the classification model. Figure 6.4 shows the transformed decomposed IMFs after applying HHT, while the residual components are discarded.

6.3.5 Model development and optimization

Enabling smart manufacturing using CPS requires processing and analyzing monitored signals and cloud computing for decision-making. The features vector extracted from the signals is used to develop a classification model

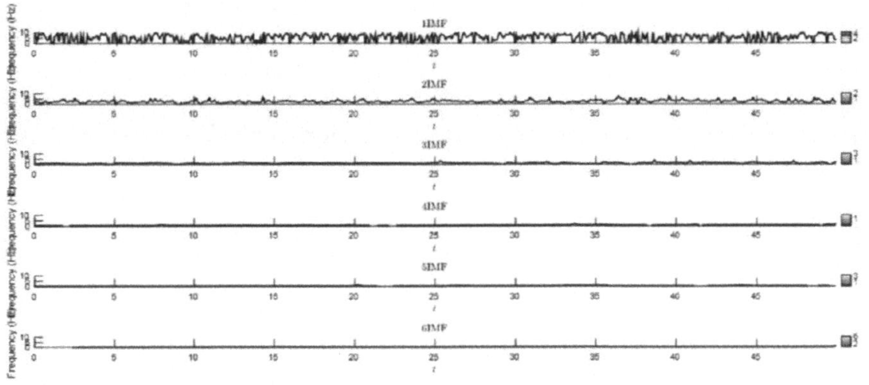

Figure 6.4 Applying HT on the IMFs evaluated from EMD function.

that identifies the tool class, as illustrated in Table 6.1, and, hence, its corresponding workpiece quality output. This therefore indicates that the number of scraps produced during machining due to product damage (surface quality exceeding the required standards) can be reduced as well as the machine downtime because the tool condition as well as the workpiece can be accurately monitored during production. Feature classification is essentially the last data processing task performed on the signal to predict the condition of the cutting tool and the workpiece for optimization during turning operation. This implies that the product surface quality requirement can be closely monitored during machining operation. That means the cutting tool's life can be used optimally based on the product quality requirement, which is one of the benefits of adopting smart manufacturing systems for machining operations.

To reduce the computation cost based on the number of features used in developing the model, a feature selection algorithm can be used to optimize the most important features that are critical for the classification algorithm. A genetic algorithm (GA) model using the roulette wheel (RW) method is used for feature selection, as illustrated in equation 8.

$$F_p = \frac{F_i}{\sum_{i=1}^{n} F_i} \tag{6.8}$$

where F_p is the fitness probability of the ith chromosome, and F_i is the fitness value of the ith chromosome.

A total of 4 features out of 12 features is selected after feature selection using the GA model, which makes the classification process computationally less expensive, as it reduces the time and space required for the computation process. These four features are then used for developing the classification model for the smart tool and workpiece condition monitoring. Deploying the machine learning AI algorithm for smart manufacturing is a data-driven approach which relies on the level of data processing and optimization done to increase the level of accuracy of the model. One of the powerful techniques used for data classification and regression analysis is support vector machine (SVM) because of their good theoretical foundation and generalization capacity [42]. However, the study by Altaf [43] adopted KNN classifiers for classifying both EMD and FFT features extracted from bearing vibration signals for detecting bearing fault without any statistical information, with the method yielding an accuracy percentage of 96.64%. The result of this classification algorithm showed good performance, even though the study applied FFT on the decomposed signal as opposed to this study that applied HHT on the decomposed signal. This study therefore developed the model using the KNN algorithm. The biggest challenges faced when applying ML algorithms and techniques for classification problems are bias and misclassification errors. Therefore, to avoid these issues, k-fold cross-validation techniques are applied. This study therefore considered both fivefold and tenfold cross-validation, and the model with the best performance based on the error loss in classification was selected after direct comparison. The overall error loss for the models can be evaluated using equation 9.

$$E = \frac{1}{k}\sum_{i=1}^{k} E_i \tag{6.9}$$

where the number of folds considered and the error loss are k and E, respectively. The overall error loss when fivefold cross-validation is applied to the KNN model is 0.0318, while the value is 0.343 for the tenfold cross-validation model. This result indicates that the fivefold cross-validation model performs better than the tenfold cross-validation model. Furthermore, the error loss was also evaluated on the models without feature selection to check for the performance when the feature selection algorithm is used against when it is not. The overall error loss when the feature selection algorithm is not applied with the fivefold cross-validation model is 0.2202, while the error loss for the tenfold cross-validation model is 0.2172. The result shows that the fivefold cross-validation KNN model has the best performance when the feature selection algorithm is implemented.

The performance of the fitted KNN model is further improved through a hyperparameter optimization that minimizes the fivefold cross-validation loss through automatic optimization using the "expected-improvement-plus" acquisition function and random seed for reproducibility. The objective function of the optimization is the classification error loss of the model, while the constraints are the distance metrics and the number of neighbors of the model. Figure 6.5 illustrates the objective function of the model after

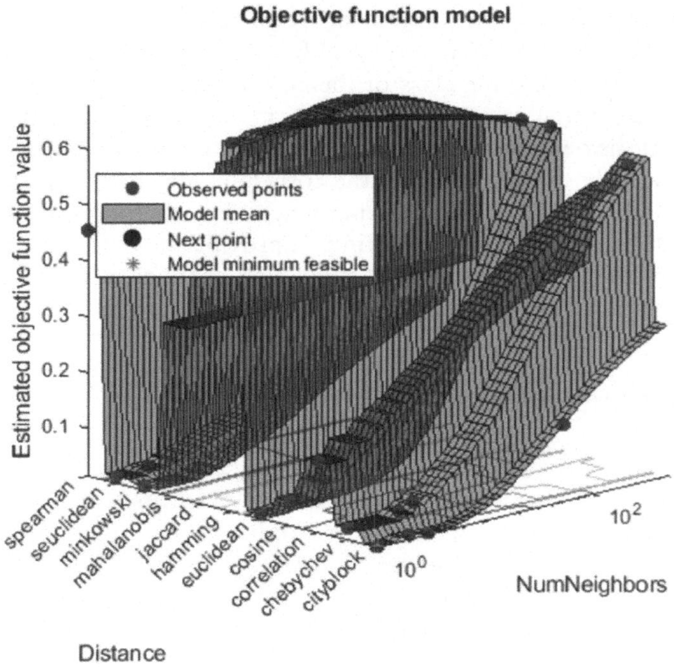

Figure 6.5 Hyperparameter optimization objective function model.

performing. The objective function of the model specified "cityblock" distance metrics to perform optimally using the "kdtree" neighbor searcher method. The estimated objective function value of the optimized model is 0.01416, while the observed objective function value is 0.0140, which indicated the model had been improved. The estimated function evaluation time and the actual function evaluation time were very close, being 0.14396 and 0.13112, respectively, while the "NumNeighbor" was 4. *Cityblock* is a special case of Minkowski distance where p is equal to 1 and evaluated by applying equation 10.

$$d_{st} = \sqrt[p]{\sum_{j=1}^{n} \left| X_{sj} - y_{ij} \right|^p} \tag{6.10}$$

The optimized model is therefore used for decision-making, which is decided within the same interval as determined by the heartbeat settings of the IIoT gateway for data transmission and processing. Depending on the robustness and sophistication of the cloud server for the CPS application, the heartbeat may be set to 1 s, 10 s, 30 s, 60 s, etc. This implies that the heartbeat of the CPS application for smart manufacturing can be determined by the sensitivity or severity of the condition being measured. The decision is made by identifying the class of the cutting tool and the corresponding workpiece condition and making the information available to the operator for smart decision-making during the machining operation. Therefore, the conditions of the cutting tool and the corresponding workpiece profile can be monitored and determined during machining operation, thus reducing machine downtime due to monitoring, improving the quality output of the process, and reducing scraps due, while also enabling remote monitoring access through CPS implementation.

6.4 CONCLUSION AND FUTURE DIRECTION

Machining operation is an important manufacturing process as it is undertaken by both companies manufacturing new products and companies that produce parts for maintenance services. Studies have shown that monitoring tool conditions to prevent failure and damage also prevents lower-bound approach, which is the damage done to the workpiece because of sudden damage to the cutting tool. This phenomenon also increases the downtime at a machining station, as production time is lost due to the replacement of the cutting tool and the workpiece, and in cases where the event causes machine fault, time is lost to machine repair or machine reconfiguration. This generally increases the cost of production due to lost production time, increased cost of cutting tool replacement, increased cost of scrapped workpieces, and cost of machine tool repair or reconfiguration. The study hence develops a smart machining manufacturing system that is enabled by CPS for monitoring the condition of the tool and the workpiece during operation. SM enabled through CPS combines the capabilities of smart sensors, a gateway for signal/data transmission to the cloud server, cloud computing, data processing and analysis, and a feedback system. Signal processing and analysis are done using ML models, such as KNN and SVM, for condition classifications. The best model is optimized using parameter optimization to give better performance. The results indicate that KNN with a fivefold cross-validation model performs well in classifying the condition of the cutting tool and the workpiece during operation. The efficiency and robustness of the CPS implementation for smart machining manufacturing can be improved through the implementation of structured and thorough data processing and analysis, efficient cloud server and computing, and data measurement (signal capturing) through smart sensors.

Further research work can be done in optimizing turning operation by integrating the condition of the machine tool and the workpiece into the SM system. This would enhance the robustness of the SM system for smart production.

REFERENCES

[1] Kusiak, A., *Fundamentals of smart manufacturing: A multi-thread perspective*. Annual Reviews in Control, 2019. **47**: pp. 214–220.

[2] Zheng, P., et al., *Smart manufacturing systems for Industry 4.0: Conceptual framework, scenarios, and future perspectives*. Frontiers of Mechanical Engineering, 2018. **13**: pp. 137–150.

[3] Leng, J., et al., *Blockchain-secured smart manufacturing in Industry 4.0: A survey*. IEEE Transactions on Systems, Man, and Cybernetics: Systems, 2020. **51**(1): pp. 237–252.

[4] Siddesh, G. M., et al., *Cyber-physical systems: A computational perspective*. 2015: CRC Press.

[5] Rudek, R., *A generic optimization framework for scheduling problems under machine deterioration and maintenance activities*. Computers & Industrial Engineering, 2022. **174**: p. 108800.

[6] Olalere, I. O. and O. A. Olanrewaju, *Tool and workpiece condition classification using Empirical Mode Decomposition (EMD) with Hilbert–Huang Transform (HHT) of vibration signals and machine learning models*. Applied Sciences, 2023. **13**(4): p. 2248.

[7] Yang, H., et al., *The Internet of Things for smart manufacturing: A review*. IISE Transactions, 2019. **51**(11): pp. 1190–1216.

[8] Mabkhot, M. M., et al., *Requirements of the smart factory system: A survey and perspective*. Machines, 2018. **6**(2): p. 23.

[9] Pathiranagama, G. J. and H. Namazi, *Fractal-based analysis of the effect of machining parameters on surface finish of workpiece in turning operation*. Fractals, 2019. **27**(4): p. 1950043.

[10] Das, A., et al., *Experimental investigation into machinability of hardened AISI D6 steel using newly developed AlTiSiN coated carbide tools under sustainable finish dry hard turning*. Proceedings of the Institution of Mechanical Engineers, Part E: Journal of Process Mechanical Engineering, 2022. **236**(5): pp. 1889–1905.

[11] Khan, W. Z., et al., *Industrial Internet of Things: Recent advances, enabling technologies and open challenges*. Computers & Electrical Engineering, 2020. **81**: p. 106522.

[12] Roy, S., et al., *Cutting tool failure and surface finish analysis in pulsating MQL-assisted hard turning*. Journal of Failure Analysis and Prevention, 2020. **20**: pp. 1274–1291.

[13] Liu, E., et al., *Experimental study of cutting-parameter and tool life reliability optimization in inconel 625 machining based on wear map approach*. Journal of Manufacturing Processes, 2020. **53**: pp. 34–42.

[14] Aralikatti, S. S., et al., *Comparative study on tool fault diagnosis methods using vibration signals and cutting force signals by machine learning technique*. Structural Durability & Health Monitoring, 2020. **14**(2): p. 127.

[15] Kiew, C. L., et al., *Complexity-based analysis of the relation between tool wear and machine vibration in turning operation*. Fractals, 2020. **28**(1): p. 2050018.

[16] Ochoa, L. E. E., et al., *New approach based on autoencoders to monitor the tool wear condition in HSM*. IFAC-PapersOnLine, 2019. **52**(11): pp. 206–211.

[17] McIntyre, S., et al., *Best paper award*. IEEE Transactions on Haptics, 2016. **9**(3): p. 293.
[18] Kang, L., et al., *Tool wear monitoring using generalized regression neural network*. Advances in Mechanical Engineering, 2019. **11**(5).
[19] Şahinoğlu, A., Ş. Karabulut, and A. Güllü, *Study on spindle vibration and surface finish in turning of Al 7075*. Solid State Phenomena, 2017. **261**: pp. 321–327.
[20] Abdullahi, Y. U. and S. A. Oke, *Optimizing the machining process of IS 2062 E250 steel plates with the boring operation using a hybrid Taguchi-pareto box Behnken-teaching learning-based algorithm*. International Journal of Industrial Engineering and Engineering Management, 2022. **4**(2): pp. 49–64.
[21] Cruz Salazar, L. A., et al., *Cyber-physical production systems architecture based on multi-agent's design pattern—comparison of selected approaches mapping four agent patterns*. The International Journal of Advanced Manufacturing Technology, 2019. **105**(9): pp. 4005–4034.
[22] Cheng, Y., et al., *Cyber-physical integration for moving digital factories forward towards smart manufacturing: A survey*. The International Journal of Advanced Manufacturing Technology, 2018. **97**: pp. 1209–1221.
[23] Lin, S.-W., et al., *Industrial internet reference architecture*. Industrial Internet Consortium (IIC), Technical Report, 2015. https://www.iiconsortium.org/pdf/SHI-WAN%20LIN_IIRA-v1%208-release-20170125.pdf
[24] Leitão, P., et al., *Alignment of the IEEE industrial agents recommended practice standard with the reference architectures RAMI4. 0, IIRA, and SGAM*. IEEE Open Journal of the Industrial Electronics Society, 2023. **4**: pp. 98–111.
[25] Ahmadi, A., et al., *A review of CPS 5 components architecture for manufacturing based on standards*. in *2017 11th International Conference on Software, Knowledge, Information Management and Applications (SKIMA) (IEEE)*. 2017.
[26] Dureja, J., et al., *Design optimization of cutting conditions and analysis of their effect on tool wear and surface roughness during hard turning of AISI-H11 steel with a coated—Mixed ceramic tool*. Proceedings of the Institution of Mechanical Engineers, Part B: Journal of Engineering Manufacture, 2009. **223**(11): pp. 1441–1453.
[27] Zhu, K. and X. Yu, *The monitoring of micro milling tool wear conditions by wear area estimation*. Mechanical Systems and Signal Processing, 2017. **93**: pp. 80–91.
[28] Jantunen, E., *A summary of methods applied to tool condition monitoring in drilling*. International Journal of Machine Tools and Manufacture, 2002. **42**(9): pp. 997–1010.
[29] Cheng, Y., et al., *Tool wear intelligent monitoring techniques in cutting: A review*. Journal of Mechanical Science and Technology, 2023. **37**(1): pp. 289–303.
[30] Amici, C., et al., *Multi-sensor validation approach of an end-effector-based robot for the rehabilitation of the upper and lower limb*. Electronics, 2020. **9**(11): p. 1751.
[31] Kuram, E. and B. Ozcelik, *Optimization of machining parameters during micro-milling of Ti6Al4V titanium alloy and Inconel 718 materials using Taguchi method*. Proceedings of the Institution of Mechanical Engineers, Part B: Journal of Engineering Manufacture, 2017. **231**(2): pp. 228–242.

[32] Zhou, Y. and W. Xue, *Review of tool condition monitoring methods in milling processes*. The International Journal of Advanced Manufacturing Technology, 2018. **96**: pp. 2509–2523.
[33] Mohanraj, T., et al., *Tool condition monitoring techniques in milling process—A review*. Journal of Materials Research and Technology, 2020. **9**(1): pp. 1032–1042.
[34] Banjanin, M. K., M. Stojčić, and D. Drajić, *Software networks in the logical architecture of the cyber-physical traffic system*. in *Science and Higher Education in Function of Sustainable Development–SED 2021*. 2021.
[35] Vasilevskyi, O. M., et al., *Evaluation of dynamic measurement uncertainty in the time domain in the application to high speed rotating machinery*. International Journal of Metrology and Quality Engineering, 2017. **8**: p. 25.
[36] Chen, Y., R. Bian, and W. Ding, *A fault diagnosis method of CNC machine tool spindle based on deep transfer learning*. Computer Information Analytics and Intelligent Systems, 2019: pp. 136–140.
[37] Hurley, C., *Wavelet: Analysis and methods*. 2018: Scientific e-Resources.
[38] Chen, Y., et al., *An intelligent chatter detection method based on EEMD and feature selection with multi-channel vibration signals*. Measurement, 2018. **127**: pp. 356–365.
[39] Pahuja, R. and M. Ramulu, *Surface quality monitoring in abrasive water jet machining of Ti6Al4V–CFRP stacks through wavelet packet analysis of acoustic emission signals*. The International Journal of Advanced Manufacturing Technology, 2019. **104**: pp. 4091–4104.
[40] Bhavsar, K. and V. Vakharia, *Prediction of remaining useful life (RUL) of bearing using exponential degradation model*, in *Recent advancements in mechanical engineering: Select proceedings of ICRAME 2021* (pp. 439–447). 2022: Springer.
[41] Vazirizade, S. M., A. Bakhshi, and O. Bahar, *Online nonlinear structural damage detection using Hilbert Huang transform and artificial neural networks*. Scientia Iranica, 2019. **26**(3): pp. 1266–1279.
[42] Cervantes, J., et al., *A comprehensive survey on support vector machine classification: Applications, challenges and trends*. Neurocomputing, 2020. **408**: pp. 189–215.
[43] Altaf, M., et al., *A new statistical features based approach for bearing fault diagnosis using vibration signals*. Sensors, 2022. **22**(5): p. 2012.

Chapter 7

An open architecture for cyber-physical systems 2.0 From a holonic perspective for application in sustainable manufacturing

M. J. Ávila-Gutiérrez and F. Aguayo-González

LIST OF ABBREVIATIONS

AR	augmented reality
BIM	building information modeling
CAD	computer-aided design
CAE	computer-aided engineering
CAM	computer-aided manufacturing
CAS	complex adaptive systems
CBPS	cyber-biophysical systems
CPMS	cyber-physical manufacturing system
CPS 1.0	cyber-physical systems 1.0
CPS 2.0	cyber-physical systems 2.0
CPS 3.0	cyber-physical systems 3.0
CPS 4.0	cyber-physical systems 4.0
CPS X.0	cyber-physical systems X.0
CPS	cyber-physical systems
FVN	functional virtualized networks
H-CPS	human cyber-physical systems
HMI	human–machine interaction
I4.0	Industry 4.0
I5.0	Industry 5.0
IoT	Internet of Things
KETs	key enabling technologies
LoRa	long-range communications
PLCs	programmable logic controllers
PLM	product life cycle management
RTOS	real-time operating systems
SDGs	Sustainable Development Goals
SDN	software-defined networks
TIA	totally integrated automation
UA	Unified Architecture
VR	virtual reality
VUCA	volatile, uncertain, changing, and ambiguous

DOI: 10.1201/9781003559993-7

7.1 INTRODUCTION

Cyber-physical systems (CPS) stand as one of the key enabling technologies that have catalyzed the advent of Industry 4.0 (I4.0), presenting substantial challenges on the path toward realizing Industry 5.0 (I5.0) [1, 2]. In general terms, *CPS* can be defined as the convergence between the real world or physical model (encompassing machines, products, and systems) and the virtual world or digital model (digital twin) of these elements [3]. This convergence is materialized through a cybernetic system that links the real system and its digital twin in real-time connectivity, with intelligence and proactive control that integrates communication, computation, and control capabilities. Its purpose is to achieve functional and sustainable behavior in environments that vary in terms of their degree of intelligence. The exact definition and configuration of CPS may vary depending on the specific application, scope, and evolutionary generation to which CPS X.0 belongs, giving rise to diverse technological solutions and architectures.

In a more detailed definition, considering implementation technologies, a *CPS* can be described as a physical mechanism or system designed (or natural entity), whether electronic, electromechanical, or other, that is monitored and controlled through remote computational algorithms hosted in the cloud and even locally, as is the case with autonomous vehicles. These algorithms are based on simulations conducted in the digital twin with historical and real-time data obtained from the physical system, deriving surrogate models that are implemented through control loops at the Edge, fog, or cloud, leveraging telematic networks.

Key enabling technologies (KETs) in these CPSs encompass fields such as robotics, drones, big data management, cloud computing, augmented reality (AR), virtual reality (VR), artificial intelligence (AI), Internet of Things (IoT), software-defined networks (SDN), functional virtualized networks (FVN), intelligent telematic networks, collaborative robotics, 5G, 6G, and blockchain, among others [1, 4].

To emphasize the evolution of industrial automation underlying the concept of Industry 4.0 and the emergence of CPS, Table 7.1 is presented, highlighting one of the distinctive features of these CPSs: the use of virtual models (digital twin) and their symbiotic integration with the real world through communication, computation, and control to optimize functionality through remote and/or local control loops.

CPS has undergone significant advancements in design engineering, embracing life cycle concepts and integrating physical models with digital twins. A cybernetic system is crucial in connecting real products with their digital counterparts, ensuring desired CPS behavior. CPS architectures must be open and adaptable to accommodate evolving technologies

Table 7.1 Industrial revolutions and the emergence of CPS 2.0 and beyond

Industrial revolution	*Technology*	*Control*	*Model*	*Model use*	*Connectivity, intelligence, and virtuality*
First industrial revolution	Mechanization (mechanical, hydraulic, thermal); non-existence of CPS	Inherent to the mechanism	Inexistent or Euclidean tool graphs	Machine construction	Isolated; no intelligence; only real part
Second industrial revolution	Automatism (electrical, electronic); non-existence of CPS	Differentiated driving force mechanism or circuit and control circuit, logic	Numerical and graphical models with slide rules and Euclidean tools; continuous and discrete event models	Machine construction, maintenance, and renovation	Isolated machine; intelligence embedded in the control circuit; only real part
Third industrial revolution	Automation (electronics and informatics); increased range of automatisms; CPS 0.0	Differentiated analog and digital control of sensorization and effective actuator actions; analog and digital control	Numerical models and computer graphics, simulation; CAD, CAE, CAM; continuous logic and statistical models	Construction, maintenance, and renovation; beginning of model-driven engineering	Local connection, local networks; local intelligence; real and virtual models not connected
Fourth industrial revolution	Cyber-physical system + human operator (sociotechnical system); CPS 1.0	By embedded and surrogated models in the edge from fog and cloud; hierarchical control, distributed control, predictive control	Virtual models in BIM and PLM environments for construction purposes, and digital twin functions in other life cycle phases; predictive models	Construction of the product, system, machine, or device; integration as a digital twin in the construction of the CPS once the real system has been built	Local and global connectivity; local intelligence (individual) and global intelligence in edge (collective); real and virtual models connected through a cyber-physical system of connectivity, communication, and intelligence
Fifth industrial revolution	Cyber-physical system ecosystem; CPS 2.0 or the next generation of cyber-physical systems (NG-CPS)	Distributed and self-organized control by CPS, holarchy control by cybernetic system	Cyber-physical models integrated in the metaverse, CPS holarchy	Generation of proactive, evolutionary, and adaptive experiments	Interactive and connected multiverse

and increasing diversity, transitioning from CPS 1.0 to CPS 2.0. CPS applications have expanded across various sectors, leading to new business models like sterilization driven by data collection and processing within digital models, impacting fields from health care to smart cities and education.

CPS 1.0, CPS 2.0, and their evolutions, such as CPS 3.0 and CPS 4.0, have been studied [5].

The first generation of CPS is characterized by static design and operation without changes throughout its life cycle. These systems employ conventional control mechanisms that can regulate known parameters but require human intervention in case of faults or environmental changes.

The second generation of CPS introduces the ability to switch between predefined control modes based on established criteria. These systems use data-driven control, leveraging information from the system and environment to optimize each operational mode. These are CPSs with proactive intelligence, supported by technologies such as intelligent telematic networks 5G and 6G, SDN, FVN, IoT, long-range communications (LoRa), wireless sensor networks, and metaverse networks. Furthermore, they benefit from emerging computing technologies, like big data, cloud computing, blockchain, deep learning, machine learning, and containers, among others.

The third generation of CPS exhibits adaptability within predefined limits during operation. These systems have self-learning capabilities and can modify control algorithms based on operational feedback.

The fourth generation of CPS is highly adaptable and autonomous, capable of generating and adapting control strategies in response to largely unknown changes, potentially beyond predefined safety limits. These systems employ advanced machine learning and artificial intelligence to autonomously adapt to new situations and challenges.

Figure 7.1 shows the different generations of SCP and their main characteristics.

The aforementioned situation highlights the need to establish a framework for designing and developing hosting platforms for CPSs that can address the life cycle engineering of diversity in CPS 1.0, CPS 2.0, and future generations. This framework should enable intelligent, evolutionary, sustainable, and environmentally friendly systems that are scalable and coevolving toward future digital environments like the metaverse.

Based on everything mentioned earlier, formulated are several research questions (RQ) related to the life cycle engineering of CPS, which operates as a complex, intelligent, and coevolving system in volatile, uncertain, changing, and ambiguous (VUCA) environments:

- RQ1: How can we characterize the diversity and complexity of CPS 2.0 and its evolutions to establish a framework for standardization by national and international organizations?

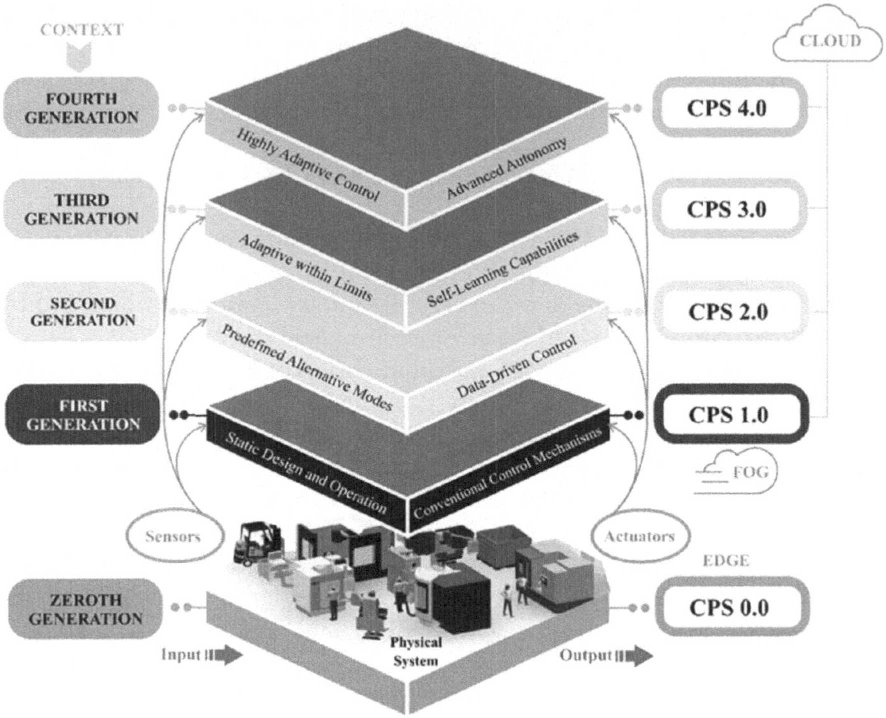

Figure 7.1 Different CPS generations.

- RQ2: How can we define and characterize a framework for the life cycle engineering of CPS X.0, considering their coevolving operation, ecological compatibility in VUCA environments, and scalability in forming CPS ecosystems on hosting platforms?
- RQ3: Can an open architecture for CPS 2.0 systems and hosting platforms for CPS X.0 ecosystems be established?
- RQ4: Can a platform for managing the life cycle engineering of CPS 2.0 be designed and controlled by models?

The answers to these questions will be developed throughout this chapter and organized based on three main objectives. The first objective is to define and characterize CPS 2.0 systems, establish their classification and typologies, and review proposed architectures for their design and development from the perspective of life cycle engineering [6, 7]. This will focus on physical and virtual hybridization, real-time linkage, and synchronization through communication elements, computing engineering, control loops, and cloud, fog, and edge telematics. Additionally, there is an interest in formulating an architecture inspired by the biosociety for CPS 2.0 systems to support the required variety of complex sociotechnical (or natural) and adaptive systems in a coevolutionary manner [8, 9].

Once the need for a reference architecture for CPS 2.0 systems that enables their coevolution with complex sociotechnical systems has been identified, the second objective is to explore the characteristics of holonic systems as systems inspired by the biosociety and carriers of the required variety. These holonic systems emerge as a solution to support CPS 2.0 systems, providing the necessary diversity and enabling efficient collaboration between physical and digital elements [10–12]. The third objective is to realize CPS 2.0 systems for holonic manufacturing systems, as CAS is inspired by the biosociety [13, 14]. This realization should be compatible with sustainable manufacturing systems geared toward the Sustainable Development Goals (SDGs) of the 2030 agenda [15, 16]. In this process, various elements are emphasized, such as life cycle, physical dimension, virtual dimension, and self-regulation through surrogate models, aiming to achieve adaptive systems with the required variety.

The central contribution of this chapter lies in presenting an architecture for CPS 2.0 and its evolutionary CPS X.0 that integrates research and proposals for CPS and digital twin architectures in a unified manner, aiming for their projection in the development of CPS standards. Additionally, it introduces an architecture for the life cycle engineering of holonic manufacturing specifically designed for CPS 2.0 systems. This architecture is based on the concept of holons, which are autonomous entities capable of interacting and collaborating toward shared objectives. Each holon can be a machine, a robot, a sensor, or even a human operator, equipped with decision-making capabilities at the local level. Each holon possesses intelligence and competencies to perform specific tasks, collaborating with other holons to achieve common goals. Communication and coordination among the holons are achieved through advanced technologies, such as the IoT and cloud computing [8, 17].

In summary, this chapter is structured around exploring CPS 2.0 and the subsequent evolutions, CPS X.0, emphasizing their self-adapting nature. A pivotal aspect is formulating a framework deeply rooted in cybernetic principles designed to manage the required variety in CPS 2.0 effectively. Delving into the design realm, the holonic reference architecture takes center stage, providing insights into the holistic nature of holons, their impact on products and processes, and their dynamic, adaptive characteristics. The discussion extends to the holonic engineering environment, elucidating its role in the structured design and development of holonic CPSs. The chapter culminates with exploring holonic cyber-physical manufacturing systems 2.0, presenting a comprehensive understanding of the evolving landscape of CPS 2.0 and CPS X.0.

7.2 CPS 2.0 AND ITS EVOLUTIONS (CPS X.0) AS SELF-ADAPTING SYSTEMS

In this section, the characterization of CPS2.0 is described in detail through a review of the architectures formulated in research [18–20], which support both CPS 2.0 architectures and their evolutionary CPS X.0 architectures [21, 22].

It is important to highlight that the choice of architectures and platforms most used in CPS 1.0 and CPS 2.0 will depend on the specific requirements and needs of each application and industrial sector. The ongoing evolution of these technologies remains an active area of research and development.

Among the architectures and platforms associated with CPS 1.0, we can find thus: (1) *Distributed control architecture* [23, 24]. In CPS 1.0, a distributed control architecture is employed, where sensors and actuators are connected to a central controller. This central controller processes information and makes decisions to control and regulate the system. An example is the use of programmable logic controllers (PLCs) in the manufacturing industry. (2) *Real-time communication middleware* [25]. To ensure real-time communication among components of a cyber-physical system 1.0, specialized communication middleware is used. A common standard used in industrial automation is OPC UA (Unified Architecture), which facilitates interoperability and communication between devices and systems from various manufacturers. (3) *Industrial control platforms* [26]. Leading companies in industrial automation, such as Siemens with its TIA Portal (Totally Integrated Automation) platform, Rockwell Automation with its Studio 5000 platform, and Schneider Electric with its EcoStruxure platform, offer comprehensive automation and control solutions for CPS 1.0. (4) *Embedded systems and real-time operating systems (RTOS)* [27, 28]. To execute critical tasks in CPS 1.0, real-time operating systems (RTOS) and embedded hardware are used. RTOSs, such as VxWorks and FreeRTOS, ensure a quick and deterministic response, which is necessary for industrial and control applications.

The architectures and platforms associated with CPS 2.0 include:

- *Service-oriented architecture (SOA)* [29, 30]. SOA is used in CPS 2.0 to create more modular and flexible systems. Components are interoperable services that can be connected and reconfigured as needed. This facilitates adaptation to dynamic environments and the incorporation of new devices and functionalities.
- *IoT platforms* [31, 32]. IoT platforms like AWS IoT, Microsoft Azure IoT, and Google Cloud IoT enable the connection of physical devices to the cloud. They provide data management, analysis, security, and scalability tools in IoT environments, which is crucial for data collection and processing in CPS 2.0.
- *Blockchain* [33]. Blockchain technology is employed in CPS 2.0 to ensure the security and integrity of data and transactions. Blockchain,

through its immutability and decentralization, is used to verify the origin and authenticity of data in distributed systems.

- *Edge computing* [34, 35]. Edge computing involves processing data near the source instead of sending all data to the cloud. This reduces latency and improves efficiency in CPS 2.0, especially in applications that require rapid responses, such as industrial automation and autonomous vehicles.
- *Simulation platforms and digital twins* [36]. These platforms allow the modeling and simulation of CPS before physical implementation. Digital twins are virtual replicas of physical systems that facilitate the design, testing, and optimization of CPS 2.0. Examples include MathWorks' Simulink platform and cloud-based simulation platforms.
- *5G and 6G networks* [37]. Next-generation communication networks like 5G and 6G are essential for the connectivity of CPS 2.0. These networks offer ultra-fast data transmission speeds and low latency, enabling real-time data transmission and precise device coordination.

Starting from the architectures and platforms of CPS and continuing with the premise that CPS 1.0 and CPS 2.0 have roots in control engineering and cybernetics, it is logical to follow the characteristic approach of *cybernetics*, defined as the "science that studies communication and automatic regulation systems in living beings and applies them to mechanical, electronic, and computer systems that appear in them." This definition expands the concept of bioinspired design beyond the biological, encompassing psychological and social aspects.

In the context of the proposal formulated for CPS 2.0 and its evolutions, such as cyber-biophysical systems (CBPS) [38] with applications in agronomic engineering [39–41] and human CPS [42–45], we proceed to integrate the variety of CPS referring to living systems, including both living beings and humans, into the characterization of CPS 2.0 and their projection into CPS X.0. This is done from a bio-psychological perspective that is incorporated into the architectures proposed so far and is carried out under the cybernetic approach.

Human cyber-physical systems (H-CPS) are systems in which physical elements and human components, or the human–machine interaction (HMI) [44, 46], are closely and collaboratively integrated. These systems are characterized by the convergence of physical and digital technologies with the active involvement of humans in various capacities, such as operators, users, designers, or supervisors. The definition of H-CPS focuses on how humans interact and collaborate with CPS, and how these systems are designed and operated to optimize this interaction [47, 48]. In the context of manufacturing, they give rise to sociotechnical manufacturing systems, in which humans are H-CPS.

These systems have a wide range of applications in various fields, ranging from medicine and health care [20, 49] to the automotive industry [50], transportation [51], manufacturing [46], collaborative robotics [45], smart cities [52], and the entertainment industry [53], among others. Effective collaboration between humans and CPS is essential to address complex challenges and enable safe automation in various domains. It is crucial to note that the specific definition and characteristics of an H-CPS may vary depending on the context and the application in which they are implemented.

Figure 7.2 illustrates the principles of bionic design for cybernetic implementation, using humans as a model for a cyber-physical system, with three main components:

- *Physical component.* This involves the human biological body equipped with sensory perception capabilities, including exteroceptors for the external environment and interceptors for the internal environment. It includes effectors and actuators (muscles, immune system, endocrine system) for interaction with surroundings. The biological system exhibits various levels of intelligence, including instinctive-reactive, cognitive, and deliberative aspects.
- *Virtual body (digital twin).* The virtual body exists as a mental representation within the brain, reflecting sensations and perceptions of the physical body. Generated through the somatosensory cortex, it processes information from exteroceptors and interceptors, creating a digital replica enabling the use of mental models, brain prediction, and deliberation.
- *Cybernetic system.* Also known as the Bayesian brain, the cybernetic system encompasses communication, computation (cognition), and control. Operating in the cyber-real dimension, it interacts with the real world within a virtual domain. This system utilizes sensor information and memory of past experiences to process and respond to the

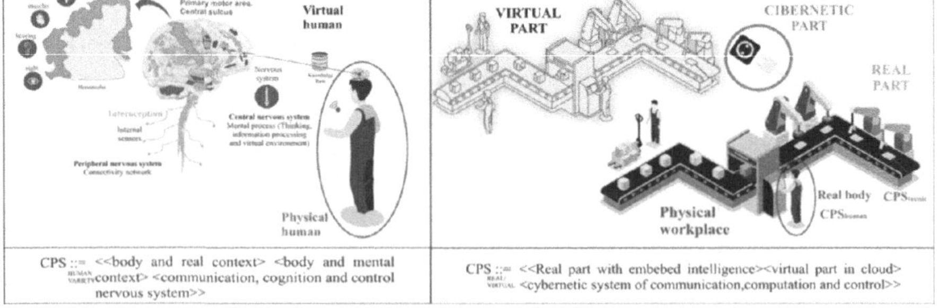

Figure 7.2 Conceptualization of CPS X.0 under cybernetic design.

environment, creating digital representations for the body and surroundings. It serves as a source of inspiration for the entire cyber-physical system, drawing from bioinspired principles.

In this way, integrating the bioinspired potential of the human CPS allows us to provide an initial characterization of CPS 2.0 and its evolutions as a triad consisting of the real part, the virtual part, and the cybernetic part.

The architecture for CPS 2.0 and its evolution to CPS X.0 is crafted for diverse CPS agents, spanning physical, biological, psychological, and social aspects. Rooted in cybernetic principles, it extends from earlier structures. Analyzing Figure 7.3, the following correlations emerge:

- *Real body* corresponds to the collaboration and cooperation domains in real-time operation.
- *Virtual body* symbolizes virtualization in fog and edge environments of the collaboration and cooperation domain.

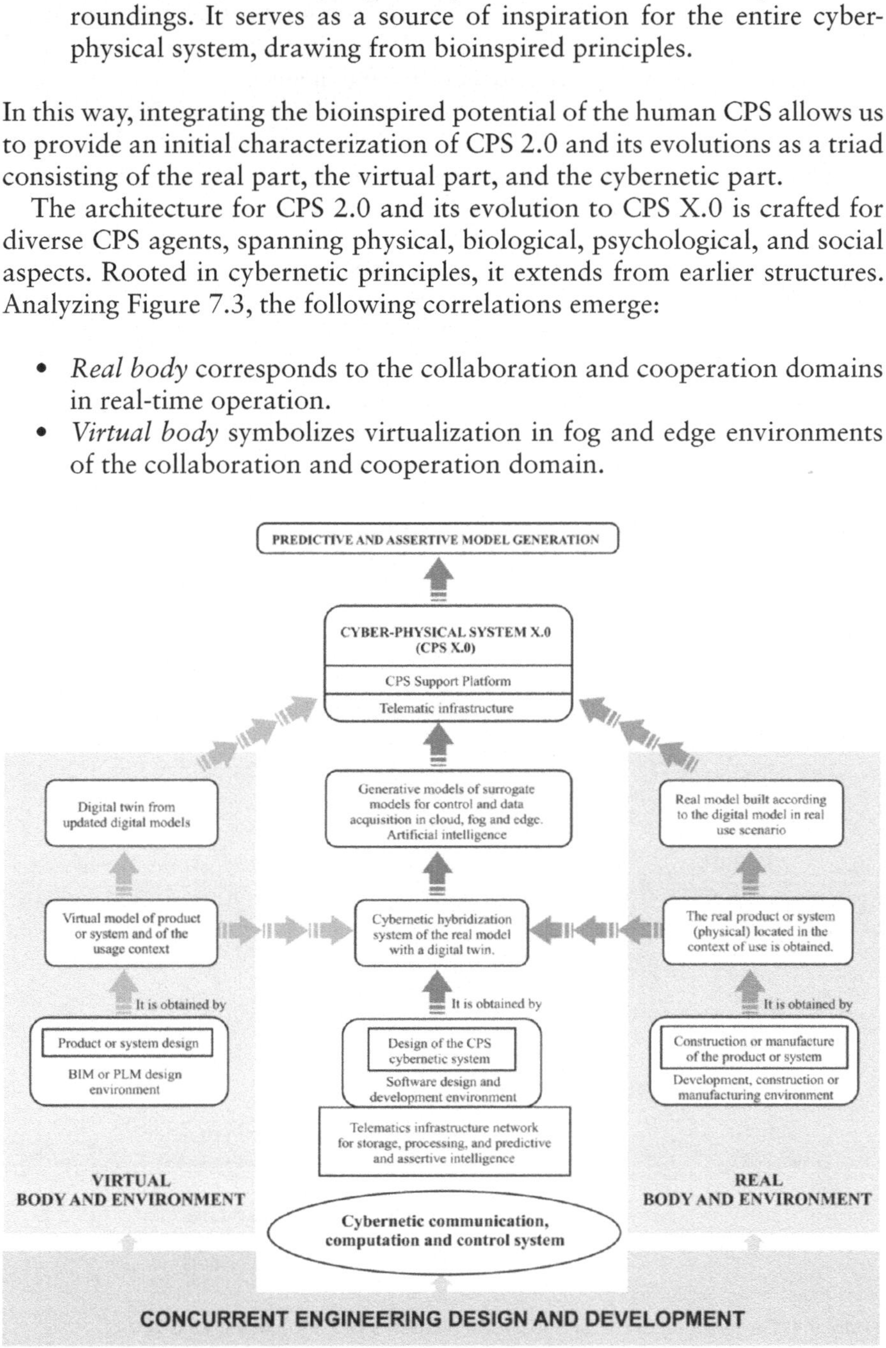

Figure 7.3 Schematic architecture for CPS 2.0 and its evolutions (CPS X.0).

- *Cybernetics system* encompasses the life cycle, surrogate model, homeostatic control loops, telematic communication network, data processing, AI technologies, simulation strategies, surrogate model training, and predictive and assertive intelligence.

Bioinspired design principles from cybernetics contribute to developing complexity dimensions for each element in CPS 2.0 architecture throughout its life cycle, which is illustrated in Figure 7.4. These dimensions integrate into the design process, bridging virtual models from BIM (building information modelling) and PLM (product life cycle management) environments with real product, process, or construction models. In the context of CPS 1.0, CPS 2.0, and their evolution (CPS X.0), the structure is conceptualized as a triad: real, virtual, and cybernetic systems. This aligns with established CPS frameworks and architectural proposals.

CPS X.0 encompasses all CPS evolutions, emphasizing diversity, establishing a framework for CPS diversity in architecture and support platforms is crucial, as well as identifying intergenerational diversity within CPS for a comprehensive understanding of their evolution.

To explore the diversity of CPS 2.0 and CPS X.0 and their applicability in a variety of contexts, various categories can be considered, encompassing applications such as intelligent exoskeletons aimed at enhancing strength,

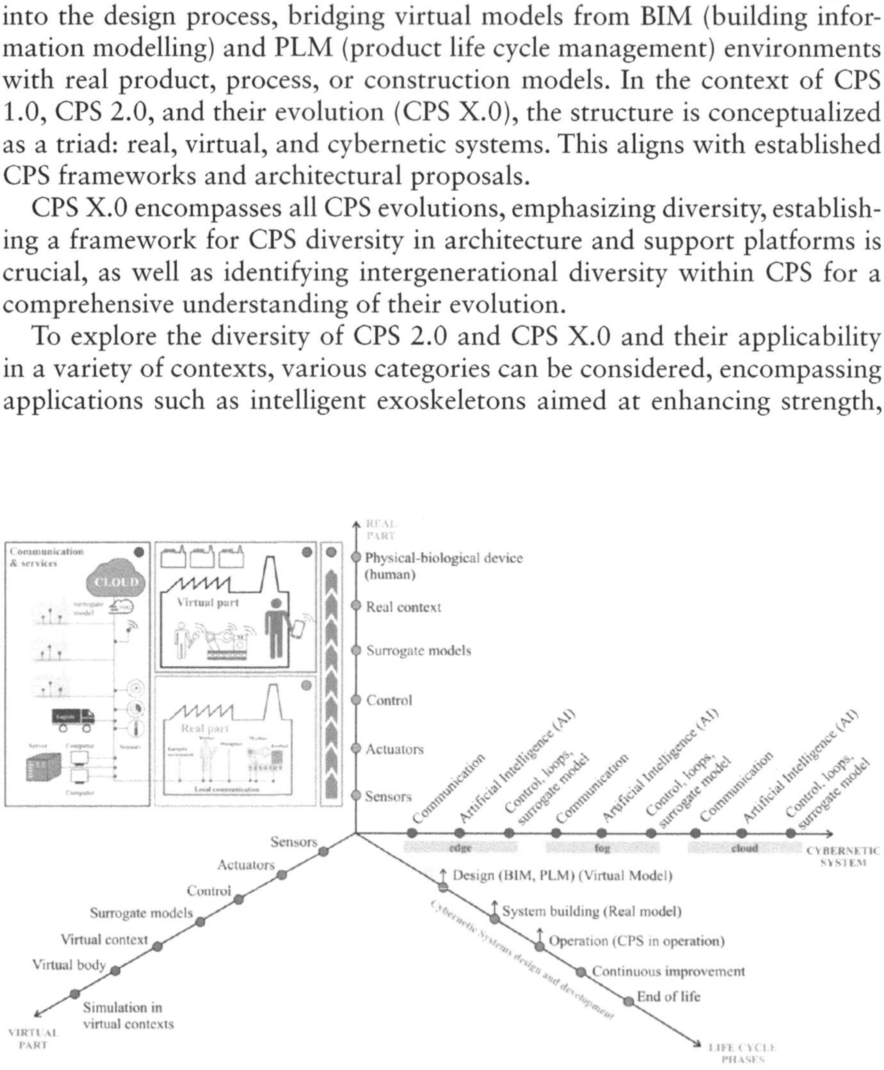

Figure 7.4 Dimensions of the architectural elements of CPS 2.0 and CPS X.0.

reducing fatigue, or restoring mobility [54, 55], among others. The required diversity for these CPSs arises from cybernetic, physical, and biological domains.

This diversity of CPS 2.0 and CPS X.0 covers various applications and fields, from industry to agriculture, human interaction, and social collaboration. The flexibility and adaptability of these systems enable them to address complex challenges and leverage the diversity of scenarios in which they can be applied [1].

Examples of this diversity can be found in sociotechnical systems that operate in VUCA environments [56, 57], such as:

- *Sociotechnical systems for sustainable manufacturing*, which focus on product manufacturing while considering environmental, social, and economic factors. Variety arises from the need to adapt to different production conditions and sustainability regulations.
- *Sociotechnical systems for elderly care*, where variety arises from the changing needs and capabilities of older individuals who require care and assistance. CPS must adapt to individual situations and the evolving health of users.
- *Sociotechnical systems for tourism*, as the tourism industry encompasses a wide variety of destinations and experiences. CPS in this domain must adapt to travelers' preferences, local conditions, and the constantly changing market trends.
- *Sociotechnical systems for agriculture*, as modern agriculture faces challenges such as efficient resource management, automation, and adaptation to climate change. Variety comes from the diversity of crops, weather conditions, and agricultural practices.
- *Sociotechnical systems for the military*, where CPS must adapt to different operational scenarios, missions, and threats. The variety includes the diversity of equipment, tactics, and combat environments.

The telematics network plays a crucial role in CPS by enhancing interaction between real and virtual components. It enables:

- *Real-time communication*. Facilitating continuous real-time data acquisition from the physical model to maintain an accurate view of the environment.
- *Control and communication loops*. Enabling constant feedback between the virtual model and the real world, supporting real-time adaptation and decision-making.
- *Transmission of intelligence*. Serving as the medium for transmitting intelligence required for CPS operation, including scenario generation, predictive modeling, and adaptive surrogate models.

The cybernetic component of CPS is categorized into three levels based on the operating context: edge, fog, and cloud, each with specific functions.

- *Cybernetic CPS (edge).* This level is closest to the cyber-physical system and focuses on local connectivity, communication, and computation. It utilizes local networks for real-time communication with the cyber-physical system and implements control through surrogate models. Local artificial intelligence (AI) tools help adapt surrogate model parameters to the local context.
- *Cybernetic CPS (fog).* This dimension involves the fog-based cybernetic CPS, which consists of a common surrogate model instantiated with generic parameters across different local environments. It manages the interconnection of these environments, transforms surrogate models from the cloud to suit specific parameters or domains, and serves as a gateway for real-time data and temporary storage.
- *Cybernetic CPS (cloud).* This level handles telematics network connectivity, data ingestion, database storage, data processing using big data techniques, and simulation generation for optimization. It utilizes global telematics networks, processes data for various agents and departments, and manages data storage in structured or unstructured databases.

Effective management of virtual model variety is crucial across the CPS life cycle, starting with the initial design in BIM or PLM environments. Virtual models are the foundation for constructing the real system, transitioning into a dynamic digital twin continually updated in real time. The virtual part hybridizes with the real system through the cybernetic system, facilitating connectivity, artificial intelligence–driven data processing, simulation generation, optimization, and decision-making. This approach aligns with model-based engineering principles, supporting efficient planning, design, and construction of CPS with varying levels of detail.

Variety management in the virtual system of a CPS is vital for adapting to diverse contexts within complex sociotechnical systems. There are several categories of virtual system variety in CPS:

- *Unary variety CPS.* These systems lack a physical counterpart and solely consist of virtual components. They virtualize functions, like telematic devices, in a virtual cloud network.
- *Artificial variety CPS.* These systems begin with digital model design in BIM and PLM environments, later integrating the real part to form a complete CPS.
- *Living variety CPS.* These CPSs involve virtualizing living entities and creating digital twins representing living beings and their behavior in digital environments.

- *Psychological variety CPS*. These systems virtualize aspects of human psychological life and the body, often using avatars in specific universes or metaverses, with wearables aiding in sensing psychological aspects.
- *Variety of social activity CPS*. This dimension involves virtualizing social systems and group interactions for various purposes, enabling interactions across multiple-dimensional universes and metaverses.

Variety in the virtual system of a CPS refers to the diversity of approaches, contexts, and types of virtual models used to represent different aspects of the cyber-physical system. Each category of variety has its characteristics and purposes, and the choice of variety depends on the specific needs and applications of a CPS X.0. The detailed specification of CPS X.0 encompasses aspects of both the real system and the virtual system, in addition to the cybernetic component that connects and manages them throughout their life cycle. Given the variety of the triple that defines a CPS X.0, we proceed to carry out the complete SNF specification.

CPS X.0 ::=≪Real System><Virtual System><Cybernetic System≫ (7.1)

<Real System$_{CPS}$>::=≪Real context><Physical Devices>
<Real sensor><Real actuator><Control><Surrogate model> (7.1.1)
<Interface><Virtual System≫

<Virtual System$_{CPS}$>::=≪Virtual context><Virtual Devices>
<Virtual sensor><Virtual actuator><Control> (7.1.2)
<Surrogate model><Interface≫

<Cybernetic System$_{CPS}$>::=≪Telematic Network>
<Communication and loop policy><Data storage>
<Artificial Intelligence techniques><Simulation and (7.1.3)
Optimization scenarios><Surrogate model>
<Lifecycle management$_{CPS}$≫

7.3 FORMULATION OF A FRAMEWORK FOR REQUIRED VARIETY IN CPS 2.0

A CPS 2.0 framework aligned with the Sustainable Development Goals of the 2030 Agenda is characterized by key principles:

- *Holonic paradigm*. Organizing systems into autonomous entities called "holons" allows independent functioning or collaboration for common goals, providing adaptability to address CPS 2.0's complexity.

- *Cybernetic principles.* Derived from cybernetics, the framework incorporates principles enabling CPS to adapt, self-organize, self-learn, be self-aware, predict, and exhibit resilient behavior. Crucial for managing complex systems in volatile environments.
- *Emergent properties.* The framework supports the emergence and evolution of properties beyond individual CPS components, playing a fundamental role in adaptation and real-time decision-making.
- *Evolutionary processes.* Integration of evolutionary processes like allostasis ensures self-regulation for balance and stability in changing environments, enhancing CPS robustness and sustainability.
- *Fractal complexity.* Recognizing complexity arising from seemingly simple systems exhibiting intricate behaviors at various scales is essential for understanding CPS behavior.

These characteristics lay the foundation for designing, developing, and managing CPS 2.0 and CPS X.0 ecosystems, fostering intelligent, sustainable, and adaptable complex systems. The framework provides a solid foundation for engineering the life cycle of CPS 2.0, facilitating the management of the required variety to adapt to diverse applications and contexts. This aligns with the sustainability goals of Agenda 2030, promoting intelligent and resilient solutions to contemporary challenges.

7.3.1 Cybernetics principles

Cybernetics plays a pivotal role in engineering the life cycle of CPS 2.0 and its evolutionary versions, CPS X.0, providing a conceptual framework for designing intelligent and adaptable complex systems. Two pertinent levels of cybernetics are identified in this context:

- *First-order cybernetics, or first cybernetics.* This initial stage focuses on the homeostatic mechanisms of cybernetic systems, emphasizing how systems maintain balance and stability through negative feedback. In CPS 2.0, this translates to the ability to adapt to the environment through self-regulation, facilitated by surrogate models for internal and coevolutionary regulation.
- *Second-order cybernetics, or second cybernetics.* This stage, linked to morphogenesis [58, 59], goes beyond homeostasis, introducing the idea that systems can evolve and adapt through positive feedback mechanisms. In CPS 2.0, second-order cybernetics is particularly concerned with the evolution of systems throughout their life cycle, allowing adaptation and evolution without self-destruction, essential for sustainability.

Both levels of cybernetics are applied in the design, development, and management of CPS 2.0 and CPS X.0. First-order cybernetics focuses on creating systems that self-regulate and adapt through negative feedback, while second-order cybernetics relates to the evolution and adaptation of systems through positive feedback and morphogenesis.

Key cybernetic principles include feedback, adaptation, self-organization, self-learning, self-awareness, self-prediction, resilient behavior, and control and regulation. These principles are fundamental for designing intelligent and adaptable systems like CPS 2.0, allowing them to gather information, adjust, change in response to new conditions, structure themselves efficiently, learn from experiences, understand their own state, predict future behavior, maintain functionality despite disruptions, and control and regulate operations.

7.4 DESIGN OF OPEN ARCHITECTURES FOR CPS 2.0 AND CPS X.0 FROM HOLONIC PERSPECTIVE

To tackle the complexity of CPS 2.0 and its evolutions, CPS X.0, a comprehensive approach is necessary. This involves:

- *Variety across domains and scales.* CPS 2.0 and CPS X.0 must be adaptable across various domains (physical, biological, psychological, social) and operate effectively at different complexity levels and scales.
- *Ontology and complexity science.* Establishing an ontological framework rooted in natural systems diversity is crucial. Complexity science, incorporating concepts like complex adaptive systems, autopoiesis, interpoiesis, simplicity, and fractality, aids in modeling CPS behavior.
- *Life cycle engineering methodology.* Designing CPS 2.0 and CPS X.0 requires a methodology with intelligent controllers. These controllers, employing bioinspired design, ensure adaptability, resilience, optimization, and harmonious operations, essential for coevolution and compatibility with natural ecosystems.

Frameworks for CPS 2.0 and CPS X.0 must effectively handle diverse domains and environments. Key considerations include: (1) *Required variety* [60]. Frameworks must meet the varied requirements of CPS 2.0 across physical, biological, psychological, social, and cultural domains. Addressing variety at multiple levels and scales is crucial. (2) *Holonic paradigm.* The holonic paradigm is a promising approach for managing complexity in CPS 2.0. Holonic systems, acting independently or as part of larger systems, provide a strong foundation for handling diverse challenges. (3) *Reverse engineering on holonic proto-model.* Designing CPS X.0 can benefit from reverse engineering existing systems or holonic proto-models. This involves analyzing these models to create compatible and sustainable CPS tailored to specific environments.

In summary, handling the required variety in CPS 2.0 and CPS X.0 is a key challenge. The holonic paradigm and reverse engineering are effective strategies for ensuring compatibility and sustainability in diverse environments.

7.4.1 Holonic reference architecture for CPS

The holonic paradigm, based on Koestler's concept of "holon" [10–12], is fundamental for understanding and organizing complexity in CPS. A *holon* is an entity that can function both as an autonomous whole and as part of a larger entity, as shown in Figure 7.5. This approach is applied to CPS 2.0 and their evolutionary, CPS X.0, to design flexible and adaptable systems in complex environments. The holonic paradigm allows CPS to operate autonomously and collaboratively, which is essential in changing environments.

$$\langle \text{Holon} \rangle ::= \frac{\langle \text{Holon}_{/W} \rangle \langle \text{Holon}_{/P} \rangle}{\langle \text{Holon}_{/W} \rangle \langle \text{Holon}_{/P} \rangle} \langle \text{Product} \rangle \langle \text{Process} \rangle \langle \text{Evolution} \rangle \langle \text{Adaptation} \rangle \tag{7.2}$$

Each holon is characterized by a set of properties, which are discussed in what follows [61]:

- A holon is WHOLE/PART. A holon can be a part of another holon or integrate different holons or holarchies in a harmonious interaction, forming collaborative ($n + 1$) or cooperative ($n - 1$) domains.

  $$\langle \text{Holon} \rangle ::= \langle\!\langle \text{Holon}_{/W} \rangle \langle \text{Holon}_{/P} \rangle\!\rangle \tag{7.3}$$

  El Holon$_{/W}$. It can be modeled as a sevenfold integrated by the following elements:

  $$\langle \text{Holon}_{/W} \rangle ::= \langle\!\langle \text{Identifier} \rangle \langle \text{Inputs} \rangle \langle \text{Outputs} \rangle \langle \text{Set of internal states} \rangle \langle \text{Evolution function} \rangle \langle \text{Operational function} \rangle \langle \text{domain} \rangle\!\rangle \tag{7.3.1}$$

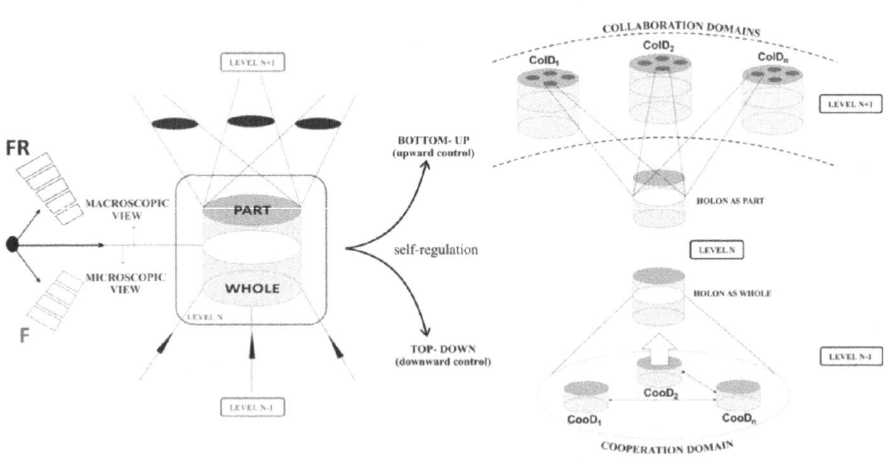

Figure 7.5 Properties of the holon as part of a higher-level $n + 1$ holarchy (macroscopic vision) and as any member of lower-level $n - 1$ holarchies (microscopic vision).

El Holon$_{/P}$. It refers to the specification of the holon from the perspective of the part that is integrated into different holons/holarchies. This can be modelled as threefold:

$$\begin{aligned}&<\text{Holon}_{/P}>::=\ll\text{Holon}_{/P}><\text{Operational function}>\\&<\text{Collaboration domain}_{n+1}><\text{Rols}\gg\end{aligned}\qquad(7.3.2)$$

- A holon is an AUTONOMOUS entity. It can create and control the execution of its own rules or strategies autonomously and self-control, utilizing the variety with which it is equipped. A holon is autonomous if and only if its behavior (ϕi) depends on its state and its perceptions, that is:

$$\Psi = \mathrm{s}_i \times \mathrm{p}_i \rightarrow s_i \times \mathrm{a}_i \qquad (7.4)$$

s_i: possible holon states h_i
p_i: perceptions or inputs of the holon h_i
a_i: actions that the holon is equipped with h_i

- A holon is a COLLABORATIVE entity. It is integrated into one or more higher-level holarchies, called collaborative domains, forming holarchies of level $n + 1$. This is a property of the holon in terms of its expression as a part.

$$\Pi\left(\text{Holon}_{/P},\mathrm{p}_i\right) \rightarrow \text{Collaboration domain} \qquad (7.5)$$

- Π. A perception/entry function that determines the perceptions/entries of the holon.

$$\begin{aligned}&\text{Collaboration=}\\&<\text{Set of relations with the Holon}_{/W}\text{ of level n+1}>\end{aligned}\qquad(7.5.1)$$

$$\begin{aligned}&<\text{Collaboration domain}>::=\ll\text{ID}>\\&<\text{R.Collaboration domain}><\text{Rol}\gg\end{aligned}\qquad(7.5.2)$$

- *ID*. Identity and goals of the $n + 1$ level holon from the collaboration domain that integrates the holon as part of level n.
- *R.Collaboration domain*. Required resources for the collaborative domain, such as communication, coordination, control, operation, etc.
- *Rol*. The "roles" assigned to the holon in the collaborative domain and which it plays as the entire $n + 1$ level holon i. It is the specification of functional requirements (FRi), domains and areas of responsibility of the role (Dr_i), realizations associated with the role, set of activities and operations that integrate it (R_i), and competencies associated with the role and its associated knowledge classes (C_i).

$$<\text{Rol}>::=\ll\text{FR}_i><\text{Dr}_i><\text{R}_i><\text{C}_i\gg \qquad (7.5.2.1)$$

From the perspective of the collaboration domain, the number of domains deployed at level $n + 1$ (for example, collaboration D.1, collaboration D.2, . . .) provides a degree of variety that can be amplified or attenuated depending on the control desired over it.

- A holon is a COOPERATIVE entity. It integrates other holons and processes where a set of entities work together to create acceptable plans to perform a function. Emergent properties are derived from cooperation. It is a property of the holon as a whole. This property must synthesize processes and products with the required variety and minimal static and dynamic complexity.

$$\Pi\left(\text{Holon}_{/\text{W}},\text{p}_i\right) \rightarrow \text{Cooperation domain} \tag{7.6}$$

$$\text{Cooperation} = < \text{Set Holon}_{/\text{P}} \text{ of level n} - 1 > \tag{7.6.1}$$

$$\begin{aligned} &<\text{Cooperation domain}> ::= \ll \text{ID}> \\ &<\text{R.Cooperation domain}><\text{Rol}\gg \end{aligned} \tag{7.6.2}$$

- *ID*. Identity and goals of the $n - 1$ level holon of the cooperation domain that integrates the holon as an n-level whole.
- *R.Cooperation domain*. Resources required by the cooperation domain, such as communication, coordination, control, operation, etc.
- *Rol*. The "roles" assigned to the holon in the cooperative domain and performed as part of the $n - 1$ level holon i. It is the specification of functions or processes (F_i), domains and areas of responsibility of the role (Dr_i), realizations associated with the role, set of activities and operations that integrate it (R_i), and competencies associated with the role and its associated knowledge classes (C_i).

$$<\text{Rol}> ::= \ll F_i><Dr_i><R_i><C_i\gg \tag{7.6.2.1}$$

From the cooperation domain perspective, each of the perspectives deployed at that level ($n - 1$) generates a degree of variety that must be dealt with according to the perspective we are in (for example, information perspective, activity perspective, material flow perspective, etc.).

- A holon is SELF-ASSERTIVE. It has the capacity to impose on other holons forms of interaction, plans, strategies, ideas, criteria, or thoughts to develop plans. When performance, perception, and state are not coupled to other holons, $\Psi_i = s_i \times p_i \rightarrow s_i \times a_i$, performance depends solely on its own state s_i and its perception p_i. In this way, from its perception, the holon determines its new state $s_i = \Phi_i\left(s_i \times p_i\right)$

and its next action $a_i = \Phi_i(s_i \times p_i)^2$ in a decoupled manner from other holons. Modes of cooperation generate different types of functions that result in the required cooperation and the balance of self-assertion according to the type of task.

- The holon is SELF-REGULATING. It can change how it cooperates to perform a function, which provides resilience to the required variety. It can be internal or external to the holon. The holon, as an autonomous entity in harmonic interaction with the environment, has a set of required variety and self-regulatory mechanisms, based on feedback from the collaborative domains of the $n + 1$ holarchies and the cooperative domains belonging to the $n - 1$ holarchies. From these, the holon establishes self-adjustment strategies on the cooperation domain by means of top-down control strategies, and on the collaboration domain based on bottom-up control strategies.
- The holon has an integrated life cycle. The holon's life cycle is structured in three contained dimensions, shown in Figure 7.6: the stages of the life cycle, the perspectives of complexity, and the degrees of specificity of the holon as a particular, partial, or general entity, depending on its level of concreteness or specialization.

7.4.1.1 Holon as whole and part

The concept of a *holon* refers to an entity that can be seen as both an independent whole and a part of a higher-level entity, resembling a complex adaptive system that operates in VUCA environments. Its dual structure and nature are illustrated in Figure 7.7.

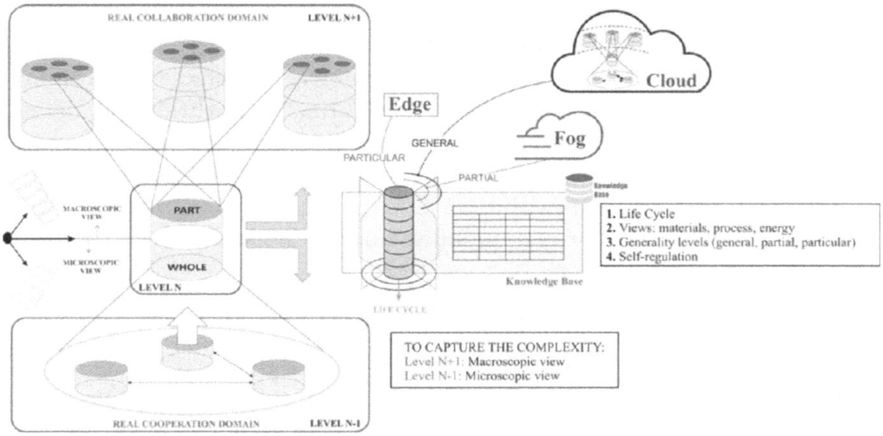

Figure 7.6 Dimensions of the holonic variety on level *n* as an interface of level *n* and *n* + 1 or *n* − 1.

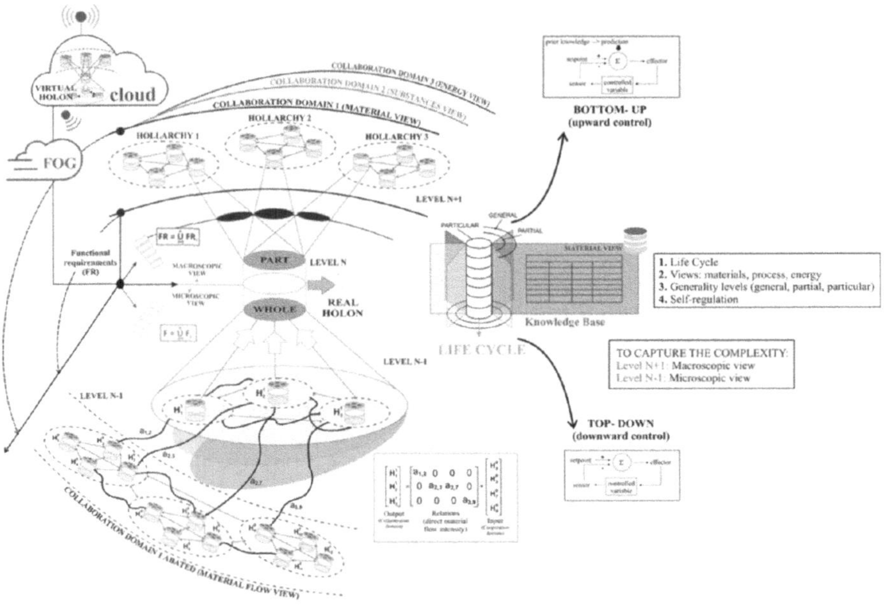

Figure 7.7 The holon is "part" and "whole."

Figure 7.7 is composed of:

- Its hierarchical structure, which can give rise to other structures, such as hierarchies, heterarchies, or reaches, representing the interactions between them (denoted as *aij*).
- Each holon can function as a "part" by integrating with other multi-level holons that make up its domain of collaboration. Each domain of collaboration can have different perspectives of complexity, addressing aspects such as materials, energy, information, hybrids, and others.
- Additionally, each holon can function as a "whole" by integrating into a *holarchy*, which is a set of holons in cooperative interaction, oriented toward stability and evolution. Emerging properties in this context can be formalized through input and output holarchies.

7.4.1.2 Holon as product and process

The holon is a dynamic entity influenced by internal forces, resulting in both bottom-up and top-down dynamics. These dynamics impact the domains of collaboration and cooperation to which it belongs, as shown in Figure 7.7.

The holon can be seen as a "process" emerging from the interaction of two domains: one from the networks or holarchies of the cooperation domains (bottom-up dynamics), and the other from the collaboration domain (top-down dynamics). These domains change, creating a dynamic entity that serves as both an actor and a network component.

Furthermore, the holon is an evolving entity, an "actor network," following a life cycle encompassing design, manufacturing, operation, re-engineering, and end-of-life stages. Throughout these stages, the holon integrates various complexity perspectives and generality levels into its operations. It also incorporates knowledge engineering, learning strategies, artificial intelligence, and other approaches to support its development and evolution.

7.4.1.3 Holon as an evolutionary and adaptive entity

The holon possesses evolutionary and adaptive capabilities that enable self-production and evolution across generations. This process involves integrating epistemic knowledge related to CAS, leading to holonic entities and processes characterized by fractal features, spanning from simplicity to complexity.

The holonic entity establishes control loops within both the collaboration (horizontal) and cooperation (vertical) domains. These control loops rely on information derived from virtual models of the holon and its cooperation and collaboration domains, facilitated by AI-enhanced simulations.

To achieve these holonic entity properties, the incorporation of a framework rooted in the concept of CAS is proposed. This framework ensures the essential attributes required for CPS 2.0 and their subsequent developments. Additionally, the utilization of theoretical frameworks like the Bayesian brain is suggested, among others.

Figure 7.7 illustrates two homeostatic feedback loops, one ascending (bottom-up) and one descending (top-down), aimed at attaining equilibrium, self-production, and adaptation. These regulatory loops leverage surrogate models generated through cloud-based artificial intelligence techniques, enabling the simulation of scenarios with digital holon models, adaptable to various CPS types or their varieties.

7.4.1.4 Holonic engineering environment for the design and development of holonic CPSs

One fundamental aspect related to CPS 2.0 or CPS X.0 holons is the creation of suitable environments to develop the life cycle engineering of these holonic systems. These environments should address multiple scales and aspects, allowing the design of holonic CPS at any level of systemic aggregation and forming CPS ecosystems.

The life cycle engineering environment is composed of a holarchy that includes:

- An entity dedicated to the life cycle engineering of the CPS holonic system.

- A set of proto-models representing use cases of holonic CPS that can be instantiated in various contexts, such as manufacturing, health care, energy, and construction. These proto-models are adaptable to different levels of intelligence, from CPS 0.0 to CPS 2.0—CPS X.0.
- A set of holonic proto-models that can be instantiated in different domains, universes, or multiverses, operating at various levels of intelligence, from CPS 0.0 to CPS 2.0—CPS X.0.

In the following, we will provide an example of how this conceptualization is applied in the manufacturing domain, considering the level of intelligence and self-regulation of CPS 2.0 and considerations regarding container-based technology and microservices architectures.

7.5 HOLONIC CYBER-PHYSICAL MANUFACTURING SYSTEMS 2.0

A cyber-physical manufacturing system (CPMS) facilitates the implementation of life cycle engineering in such systems. CPS 2.0 and its subsequent versions, known as CPS X.0, are composed of a real physical object, whether designed or natural, and its corresponding digital replica, or digital twin. These elements are interconnected through a cybernetic system that governs their dynamics and promotes their coevolution with the environment in an eco-compatible manner. Thus, a CPS 2.0 can be conceptualized as a triad, following this notation:

$$\begin{array}{c}\text{CPS 2.0}::=\ll\text{Real Manufacturing Object}>\\ <\text{Digital Manufacturing Twin}><\text{Cybernetic System}\gg\end{array} \qquad (7.7)$$

Manufacturing objects can encompass a wide range of elements, from industrial machines, factory facilities, and industrial products to businesses related to the industry.

Digital twins, on the other hand, represent models of these objects and manufacturing systems and can vary in complexity depending on the modeled entity. This can include equations describing physical behaviors, historical datasets, schematics, and diagrams, or even realistic 3D models, for specific simulations.

The cybernetic system that connects the real object and its digital twin has communication, computation, and control capabilities. Its performance varies according to operational needs and network infrastructure. The system operates at the edge, where surrogate model parameters are adjusted and adapted to new situations, or in the cloud, where historical and real-time data are used to conduct more complex simulations and derive evolutionary

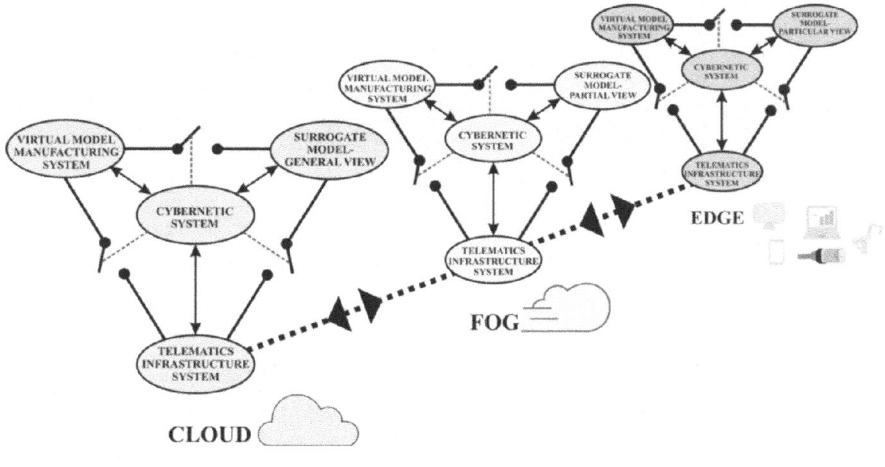

Figure 7.8 Cyber-physical systems supported by the telematic network at the edge, fog, and cloud levels.

surrogate models with parameters that can be instantiated in regional and operational contexts in both edge and fog environments, as shown in Figure 7.8.

The proposed cyber-physical system (CPS) architecture introduces the concept of a digital twin and emphasizes scalability and hierarchical or federated integration. This approach effectively manages complexity and variety in systems, addressing predefined variety and natural variability. The vision includes the potential creation of an industrial metaverse through the federation of CPS and CPS ecology platforms.

Aligned with the holonic approach inspired by natural systems, the CPS 2.0 design incorporates holonic principles. Central to the proposal is the life cycle engineering of CPS 2.0, focusing on sustainable sociotechnical systems aligned with the 2030 Agenda's Sustainable Development Goals (SDGs). This involves addressing specific life cycles within manufacturing CPS 2.0, such as coevolutionary and adaptive life cycles, product holon life cycle, manufacturing process holon life cycles, and industrial building life cycle.

Figure 7.9 illustrates the holon of an industrial plant, divided into collaboration, central, and cooperation domains. These domains represent different levels of the holarchies of companies, final products, processes, machines, and design engineering, forming a cohesive manufacturing plant. The central holon acts as an interface, containing the life cycle with detailed knowledge views, including real and virtual components, feedback loops, and instances of digital twins and CPS in edge, fog, and cloud layers.

The proposed structure demonstrates high scalability in granularity and geographic scope, accommodating various levels of detail and adapting to diverse needs and contexts.

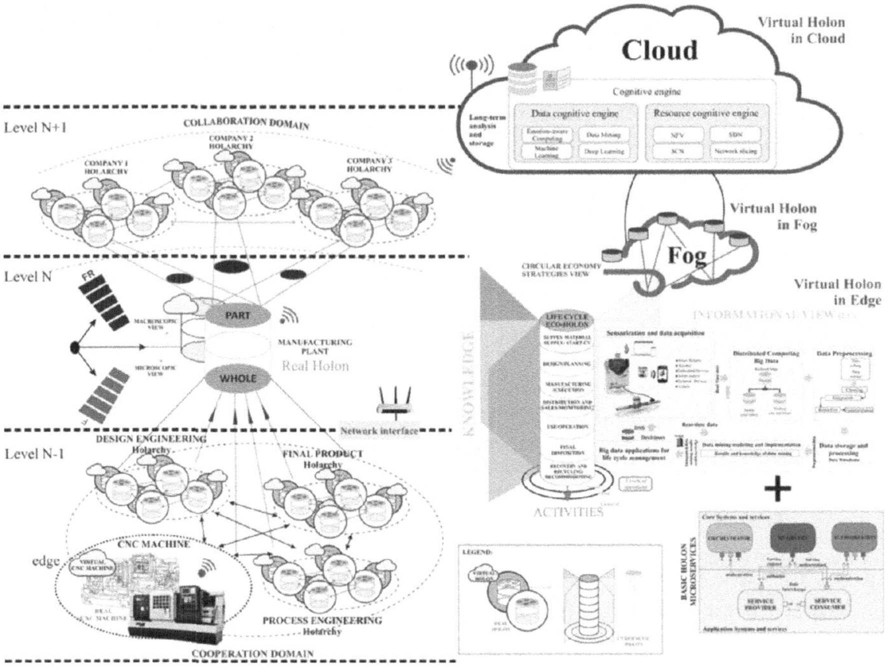

Figure 7.9 Holon life cycle industrial plant.

The design and construction of a holonic cyber-physical system 2.0, whether for an industrial plant as a system of holonic CPS or for an individual holonic CPS, can occur in different situations. In a "new project" scenario, the process begins from scratch, with holon life cycles starting at the most basic level of granularity. Concurrent engineering is employed, allowing the simultaneous design and construction of both the machine and the cyber-physical system, including instances of edge, fog, and cloud.

On the other hand, a "redesign project" involves transforming an existing real system into a holonic CPS 2.0. This scenario applies to industrial plants already under construction or in operation. Various models are generated, ranging from mathematical representations to realistic 3D models obtained through 3D scanning and photogrammetry. Simultaneously, the design of the cyber-physical system considers communication requirements, computation aspects, data processing applications, and storage systems, incorporating technologies such as artificial intelligence and structured/unstructured databases. These two situations offer flexible contexts for developing holonic CPS 2.0, adapting to specific needs and conditions.

7.6 CONCLUSIONS

This chapter thoroughly explores CPS 2.0 and its subsequent evolutions, CPS X.0, emphasizing their intricate nature and the imperative to address their inherent diversity. The holonic paradigm for complexity emerges as a key consideration, allowing entities to function autonomously or as part of larger wholes, providing a fundamental framework for the flexibility and adaptability of CPS 2.0 in complex and dynamic environments. The proposed holonic architecture exhibits essential features, including managing CPS complexity, adaptability to changing manufacturing environments, scalability for integrating new components, and efficiency in agile decision-making and process optimization. Life cycle engineering, incorporating intelligent controllers and bioinspired approaches, plays a crucial role in ensuring adaptability, resilience, and operational efficiency.

CPS 2.0 demonstrates broad applicability across sectors like manufacturing, health care, and energy, addressing various evolutionary types from CPS 0.0 to CPS X.0. Future research directions are identified, focusing on practical implementation in industrial settings, cybersecurity considerations due to interconnected critical systems, and exploration of emerging technologies, like quantum computing and nanotechnology. The concept of an industrial metaverse, achieved through the federation of CPS and CPS ecology platforms, emerges as a promising avenue for further research, exploring collaboration and evolution in a broader context. In summary, this chapter establishes a solid foundation for comprehending CPS 2.0 and CPS X.0, outlining future research priorities in practical implementation, security, and sector-specific applications.

REFERENCES

[1] Valette, E., El-Haouzi, H., and Demesure, G., 2023. Industry 5.0 and its technologies: A systematic literature review upon the human place into IoT-and CPS-based industrial systems. *Computers & Industrial Engineering*, 184, pp. 109426.

[2] Lee, J., Bagheri, B., and Kao, H. A., 2015. A cyber-physical systems architecture for Industry 4.0-based manufacturing systems. *Manufacturing Letters*, 3, pp. 18–23. https://doi.org/10.1016/j.mfglet.2014.12.001

[3] Sadiku, M., Wang, Y., and Cui, S., 2017. Cyber-physical systems: A literature review. *European Scientific Journal*, 13(36), pp. 1857–7881. https://doi.org/10.19044/esj.2017.v13n36p52

[4] Fei, X., Shah, N., Verba, N., Chao, K. M., Sanchez-Anguix, V., Lewandowski, J., James, A., and Usman, Z., 2019. CPS data streams analytics based on machine learning for cloud and fog computing: A survey. *Future Generation Computer Systems*, 90, pp. 435–450. https://doi.org/10.1016/j.future.2018.06.042

[5] Tavcar, J., and Horvath, I., 2019. A review of the principles of designing smart cyber-physical systems for run-time adaptation: Learned lessons and open issues. *IEEE Transactions on Systems, Man, and Cybernetics: Systems*, 49(1), pp. 145–158. https://doi.org/10.1109/TSMC.2018.2814539

[6] Chen, H., 2017. Applications of cyber-physical system: A literature review. *Journal of Industrial Integration and Management*, 2(3), p. 1750012. https://doi.org/10.1142/S2424862217500129

[7] Tan, Y., Goddard, S., and Pérez, L. C., 2008. A prototype architecture for cyber-physical systems. *ACM SIGBED Review*, 5(1), pp. 1–2. https://doi.org/10.1145/1366283.1366309

[8] Siddesh, G., Deka, G., Gopalaiyengar, K., and Patnaik, L., 2015. Cyber-physical systems: A computational perspective. In *Cyber-Physical Systems: A Computational Perspective*. CRC Press.

[9] Shi, J., Wan, J., Yan, H., and Suo, H., 2011. A survey of cyber-physical systems. *International Conference on Wireless Communications and Signal Processing (WCSP)*, pp. 1–6.

[10] Koestler, A., 1967. *The Ghost in the Machine*. Hutchinson.

[11] Koestler, A., 1979. Janus: A summing up. *Bulletin of the Atomic Scientists*, 35(3), pp. 4–4.

[12] Koestler, A., 1964. *The Act of Creation*. Macmillan.

[13] Velte, C., Wilfahrt, A., Müller, R., and Steinhilper, R., 2017. Complexity in a life cycle perspective. *The 24th CIRP Conference on Life Cycle Engineering*, 61, pp. 104–109. https://doi.org/10.1016/j.procir.2016.11.253

[14] Efthymiou, K., Pagoropoulos, A., Papakostas, N., Mourtzis, D., and Chryssolouris, G., 2012. Manufacturing systems complexity review: Challenges and outlook. *Procedia CIRP*, 3(1), pp. 644–649. https://doi.org/10.1016/j.procir.2012.07.110

[15] Goldstein, R., 2021. Desafíos del Desarrollo Sostenible en la Nueva Normalidad. Coherencia de Políticas para la Agenda 2030 y los ODS en la Década de Acción. In *La Administración Pública en tiempos disruptivos. Diego Pando (compilador)* (Issue January, pp. 171–177).

[16] Boto-Álvarez, A., and García-Fernández, R., 2020. Implementation of the 2030 agenda sustainable development goals in Spain. *Sustainability (Switzerland)*, 12(6), p. 2546. https://doi.org/10.3390/su12062546

[17] Gaham, M., Brahim, B., and Achour, N., 2015. Human-in-the-loop cyber-physical production systems control (HiLCP2sC): A multi-objective interactive framework proposal. *Studies in Computational Intelligence*, 594, pp. 315–325. https://doi.org/10.1007/978-3-319-15159-5_29

[18] Hu, L., Xie, N., Kuang, Z., and Zhao, K., 2012. Review of cyber-physical system architecture. *15th International Symposium on Object/Component/Service-Oriented Real-Time Distributed Computing Workshops*, pp. 25–30.

[19] Dumitrache, I., Sacala, I., Moisescu, M., and Caramihai, S., 2012. A conceptual framework for modeling and design of cyber-physical systems. *Studies in Informatics and Control*, 26(3), pp. 325–334. https://doi.org/10.24846/v26i3y201708

[20] Dey, N., Ashour, A. S., Shi, F., Fong, S. J., and Tavares, J. M. R. S., 2018. Medical cyber-physical systems: A survey. *Journal of Medical Systems*, 42(4). https://doi.org/10.1007/S10916-018-0921-X

[21] NIST-National Institute of Standards and Technology, 2017. Framework for Cyber-Physical Systems: Volume 2, Working Group Reports. https://doi.org/10.6028/NIST.SP.1500-202
[22] NIST-National Institute of Standards and Technology, 2017. Framework for Cyber-Physical Systems: Volume 1, Overview. https://doi.org/10.6028/NIST.SP.1500-201
[23] Cruz, E., Carrillo, L., and Salazar, L., 2023. Structuring cyber-physical systems for distributed control with IEC 61499 standard. *IEEE Latin America Transactions*, 21(2), pp. 251–259.
[24] Hamzah, M., Islam, M., Hassan, S., and Akhtar, M., 2023. Distributed control of cyber physical system on various domains: A critical review. *Systems*, 11(4), p. 208.
[25] Feist, M., Pacher, M., and Brinkschulte, U., 2023. Evaluating the comprehensive adaptive chameleon middleware for mixed-critical cyber-physical networks. *Lecture Notes in Computer Science (Including Subseries Lecture Notes in Artificial Intelligence and Lecture Notes in Bioinformatics), 13949 LNCS*, pp. 200–214. https://doi.org/10.1007/978-3-031-42785-5_14
[26] Cook, M., Marnerides, A., Johnson, C., and Pezaros, D., 2023. A survey on industrial control system digital forensics: Challenges, advances and future directions. *IEEE Communications Surveys & Tutorials*, 25(3), pp. 1705–1747.
[27] Pothuganti, K., Haile, A., and Pothuganti, S., 2016. A comparative study of real time operating systems for embedded systems. *International Journal of Innovative Research in Computer and Communication Engineering*, 4(6), p. 12008.
[28] Stankovic, J. A., 1996. Real-time and embedded systems. *ACM Computing Surveys*, 28(1), pp. 205–208. https://doi.org/10.1145/234313.234400
[29] Laskey, K., 2009. Service oriented architecture. *Wiley Interdisciplinary Reviews: Computational Statistics*, 1(1), pp. 101–105. https://doi.org/10.1002/wics.8
[30] Niknejad, N., Ismail, W., Ghani, I., Nazari, B., and Bahari, M., 2020. Understanding service-oriented architecture (SOA): A systematic literature review and directions for further investigation. *Information Systems*, 91, p. 101491.
[31] Wan, Y. L., Zhu, H. P., Mu, Y. P., and Yu, H. C., 2014. Research on IOT-based material delivery system of the mixed-model assembly workshop. *Proceedings of 2013 4th International Asia Conference on Industrial Engineering and Management Innovation, IEMI 2013*, pp. 581–593. https://doi.org/10.1007/978-3-642-40060-5_56
[32] Cai, H., Xu, B., Jiang, L., and Vasilakos, A. V., 2017. IoT-based big data storage systems in cloud computing: Perspectives and challenges. *IEEE Internet of Things Journal*, 4(1), pp. 75–87. https://doi.org/10.1109/JIOT.2016.2619369
[33] Chung, K., Yoo, H., Choe, D., and Jung, H., 2019. Blockchain network based topic mining process for cognitive manufacturing. *Wireless Personal Communications*, 105, pp. 583–597. https://doi.org/10.1007/s11277-018-5979-8
[34] Qi, Q., and Tao, F., 2019. A smart manufacturing service system based on edge computing, fog, computing, and cloud computing. *IEEE Access*, 7, pp. 86769–86777.
[35] Khan, W. Z., Ahmed, E., Hakak, S., Yaqoob, I., and Ahmed, A., 2019. Edge computing: A survey. *Future Generation Computer Systems*, 97, pp. 219–235. https://doi.org/10.1016/j.future.2019.02.050

[36] Boschert, S., and Rosen, R., 2016. Digital twin—The simulation aspect. In *Mechatronic Futures* (pp. 59–74). Springer International Publishing. https://doi.org/10.1007/978-3-319-32156-1_5
[37] Jacob, E., Astorga, J., Unzilla, J. J., Huarte, M., García, D., and López-De-Lacalle, L. N., 2018. Towards a 5G compliant and flexible connected manufacturing facility. *Dyna (Spain)*, 93(6), pp. 656–662. https://doi.org/10.6036/8831
[38] David, I., Archambault, P., Wolak, Q., Vinh Vu, C., Lalonde, T., Riaz, K., Syriani, E., and Sahraoui, H., 2023. Digital twins for cyber-biophysical systems: Challenges and lessons learned. *ACM/IEEE 26th International Conference on Model-Driven Engineering Languages and Systems (MODELS)*.
[39] Dusadeerungsikul, P., Nof, S., and Bechar, A., 2019. Collaborative control protocol for agricultural cyber-physical system. *Procedia Manufacturing*, 39, pp. 235–242. https://doi.org/10.1016/j.promfg.2020.01.330.
[40] Rad, C., Hancu, O., Takacs, I., and Olteanu, G., 2015. Smart monitoring of potato crop: A cyber-physical system architecture model in the field of precision agriculture. *Agriculture and Agricultural Science Procedia*, 6, pp. 73–79.
[41] Wang, T., Wang, X., and Jiang, Y., 2022. Hybrid machine learning approach for evapotranspiration estimation of fruit tree in agricultural cyber-physical systems. *IEEE Transactions on Cybernetics*, 53(9), pp. 5677–5691. https://doi.org/ 10.1109/TCYB.2022.3164542.
[42] Wang, B., Zheng, P., Yin, Y., Shih, A., and Wang, L., 2022. Toward human-centric smart manufacturing: A human-cyber-physical systems (HCPS) perspective. *Journal of Manufacturing Systems*, 63, pp. 471–490.
[43] Romero, D., Bernus, P., Noran, O., Stahre, J., and Fast-Berglund, A., 2016. The operator 4.0: Human cyber-physical systems & adaptive automation towards human-automation symbiosis work systems. *Advances in Production Management Systems. Initiatives for a Sustainable World: IFIP WG 5.7 International Conference, APMS*, pp. 677–686.
[44] Flores, E., Xu, X., and Lu, Y., 2020. Human cyber-physical systems: A skill-based correlation between humans and machines. *IEEE 16th International Conference on Automation Science and Engineering (CASE)*, pp. 1313–1318.
[45] Nikolakis, N., Maratos, V., and Makris, S., 2019. A cyber physical system (CPS) approach for safe human-robot collaboration in a shared workplace. *Robotics and Computer-Integrated Manufacturing*, 56, pp. 233–243. https://doi.org/10.1016/j.rcim.2018.10.003
[46] Zhou, J., Zhou, Y., Wang, B., and Zang, J., 2019. Human–Cyber–Physical Systems (HCPSs) in the context of new-generation intelligent manufacturing. Engineering, 5(4), pp. 624–636. https://doi.org/10.1016/j.eng.2019.07.015
[47] Luiz Da Silva, V., Kovaleski, J., Pagani, R. N., Corsi, A., Kovaleski, J. L., Augusto, M., and Gomes, S., 2020. Human factor in smart industry: A literature review. *Future Studies Research Journal*, 12(1), pp. 31–53. https://doi.org/10.24023/FutureJournal/2175-5825/2020.v12i1.473
[48] Darwish, A., and Hassanien, A. E., 2018. Cyber physical systems design, methodology, and integration: The current status and future outlook. *Journal of Ambient Intelligence and Humanized Computing*, 9(5), pp. 1541–1556. https://doi.org/10.1007/S12652-017-0575-4
[49] Haque, S. A., Aziz, S. M., and Rahman, M., 2014. Review of cyber-physical system in healthcare. *International Journal of Distributed Sensor Network*. https://doi.org/10.1155/2014/217415

[50] Schmittner, C., Ma, Z., Schoitsch, E., and Gruber, T., 2015. A case study of FMVEA and CHASSIS as safety and security co-analysis method for automotive cyber-physical systems. *Proceedings of the 1st ACM Workshop on Cyber-Physical System Security*, pp. 69–80. https://doi.org/10.1145/2732198.2732204
[51] Deka, L., Khan, S., and Chowdhury, M., 2018. Transportation cyber-physical system and its importance for future mobility. *Transportation Cyber-Physical Systems*, pp. 1–20.
[52] Mishra, A., Jha, A. V., Appasani, B., Ray, A. K., Gupta, D. K., and Ghazali, A. N., 2022. Emerging technologies and design aspects of next generation cyber physical system with a smart city application perspective. *International Journal of Systems Assurance Engineering and Management*. https://doi.org/10.1007/S13198-021-01523-Y
[53] Tushar, W., Yuen, C., Saha, T., Nizami, S., Alam, M., Smith, D., and Poor, H., 2023. A survey of cyber-physical systems from a game-theoretic perspective. *IEEE Access*, pp. 9799–9834.
[54] Lee, H., Kim, W., Han, J., and Han, C., 2012. The technical trend of the exoskeleton robot system for human power assistance. *International Journal of Precision Engineering and Manufacturing*, 13(8), pp. 1491–1497. https://doi.org/10.1007/s12541-012-0197-x
[55] Bogue, R., 2018. Exoskeletons – A review of industrial applications. *Industrial Robot*, 45(5), pp. 585–590. https://doi.org/10.1108/IR-05-2018-0109/FULL/HTML
[56] Codreanu, A., 2016. A VUCA action framework for a VUCA environment. Leadership challenges and solutions. *Journal of Defense Resources Management*, 7(2), pp. 31–38.
[57] Millar, C. C. J. M., Groth, O., and Mahon, J. F., 2018. Management innovation in a VUCA world: Challenges and recommendations. *California Management Review*, 61(1), pp. 5–14. https://doi.org/10.1177/0008125618805111
[58] Campill, M. A., and von Fircks, E., 2023. Biocenosis of the self: The dynamic of relationships. *Re-Inventing Organic Metaphors for the Social Sciences*, pp. 197–214. https://doi.org/10.1007/978-3-031-26677-5_11/COVER
[59] Alicea, B., Gordon, R., and Parent, J., 2023. Embodied cognitive morphogenesis as a route to intelligent systems. *Interface Focus*, 13(3). https://doi.org/10.1098/RSFS.2022.0067
[60] Ashby, W. R., 1991. Requisite variety and its implications for the control of complex systems. *Facets of Systems Science*, pp. 405–417. https://doi.org/10.1007/978-1-4899-0718-9_28
[61] Ávila-Gutiérrez, M. J., Martín-Gómez, A., Aguayo-González, F., & Lama-Ruiz, J. R., 2020. Eco-holonic 4.0 circular business model to conceptualize sustainable value chain towards digital transition. *Sustainability*, 12(5), p. 1889. https://doi.org/10.3390/SU12051889

Chapter 8

AI model generation methodology and software architecture for CPS 2.0 Manufacturing systems

Ander García, Telmo Fernández De Barrena, and Juan Luis Ferrando

LIST OF ABBREVIATIONS

AE	acoustic emission
AI	artificial intelligence
BiGRU	bidirectional gated recurrent unit
BiLSTM	bidirectional long short-term memory network
CNN	convolutional neural network
CPS	cyber-physical system
CV	cross-validation
DL	deep learning
GRU	gated recurrent unit
HF	high-frequency
HMI	human–machine interface
IIoT	industrial IoT
IoT	Internet of Things
IT	information technology
LF	low-frequency
LSTM	long short-term memory network
ML	machine learning
MLP	multilayer perceptron
MQTT	message queuing telemetry transport
MSE	mean square error
OPC UA	Open Communication Protocol Unified Architecture
OT	operation technology
PLC	programmable logic controller
ReLu	rectified linear unit
RF	random forest
RF-RFE	random forest–reinforcement feature extraction
RMS	root mean square
RMSE	root mean square error
RUL	remaining useful life
SCADA	supervisory control and data acquisition
SME	small and medium enterprises
TANH	hyperbolic tangent
WT	wavelet transform

 DOI: 10.1201/9781003559993-8

8.1 INTRODUCTION

Traditionally, programmable logic controller (PLC) and supervisory control and data acquisition (SCADA) systems have been in charge of the automation and control of manufacturing machines and lines. However, with the Industry 4.0 paradigm, the cyber-physical system (CPS) arises. First-generation industrial CPSs were focused on capturing industrial data to be sent to cloud services to be visualized and analyzed. As related technologies have matured, new opportunities have been opened for CPSs, leading to second-generation CPS.

However, CPS 2.0 requires new software platforms to be deployed. The technological knowledge required for their development covers the domains of operation technology (OT), information technology (IT), and artificial intelligence (AI). This chapter tackles this complexity, describing a methodology to generate AI models based on high-frequency (HF) industrial data [1] and a reference CPS 2.0 architecture [2], applying them to a relevant industrial requirement: real-time AI monitoring of manufacturing processes.

The chapter is structured as follows. After an introduction and a revision of the state of the art, Section 8.3 focuses on the methodology to generate HF data AI models. Section 8.4 describes a microservices-based software architecture to integrate OT, IT, and AI. Section 8.5 presents the validation of the methodology and the software architecture to develop CPS 2.0 for two industrial use cases. While the first use case targets the generation of a HF data AI model to estimate the remaining useful life (RUL) of a tooling machine, the second one integrates OT, IT, and AI to analyze in real time the vacuum generation process of a leak test machine.

8.2 STATE OF THE ART

The Internet of Things (IoT) has become increasingly significant, impacting various aspects of modern society in recent years. This influence has led to the creation of smart environments, presenting new opportunities for innovative applications and advancements, particularly in fields like manufacturing [3].

The CPS [4] combines electronic and electric systems (known as "cyber") with real-world objects ("physical"). This union enables physical things like machines to connect with and affect the real world by making a digital version of it. This digital version includes the physical parts of the CPS (called a cyber representation) by turning data and information into digital formats [5].

Typically, a CPS comprises two key parts: (1) a sophisticated connection system that gathers real-time data from the real world and shares information within the digital space, and (2) smart data handling, analysis, and computing abilities that build and manage the digital environment [6].

The integration of CPS into smart manufacturing systems is believed to offer competitive edges within the manufacturing sectors of leading nations [7]. Hence, the industry is actively seeking to incorporate smart connectivity through numerous sensors and devices, alongside adopting cloud computing platforms and software-defined network control and management approaches [8]. Consequently, manufacturing firms are progressing toward the creation of intelligent machinery [9] to optimize product quality, increase production efficiency, and lower expenses [10, 11].

Traditional industrial automation systems typically operate in isolated units, receiving external orders with limited internal data sharing. However, the advent of the Industry 4.0 paradigm has imposed fresh demands on these conventional PLC and SCADA automation systems to actively exchange and utilize data. The quantity of data collected from manufacturing lines is steadily expanding. To gain deeper insights into manufacturing processes, a greater array of data variables is being monitored and captured at higher frequencies—shifting from a few key variables per batch to a time series encompassing multiple variables recorded at seconds or even faster intervals. To address these evolving requisites, novel architectures are necessary to seamlessly blend IT and OT domains. This integration entails a diverse array of IT and OT technologies, standards, and specifications aligned with the principles of Industry 4.0.

The intricate nature of integrating these systems creates a knowledge barrier, stemming from the fundamentally different approaches between IT technologies and the conventional tools employed by OT engineers and maintenance teams. Small and medium-sized enterprises (SMEs), often lacking diverse teams with the requisite expertise in both IT and OT, encounter substantial challenges in capturing, monitoring, and visualizing data from manufacturing processes.

Data capturing, monitoring, and visualization are just the initial steps within the Industry 4.0 paradigm. Once data becomes actionable, AI models are trained and employed to derive insights, forecast manufacturing line behaviors, autonomously identify production issues, and potentially oversee industrial processes. However, integrating this subsequent phase amplifies the hurdles for SMEs, necessitating the integration of AI engineers into teams and adapting IT architecture to accommodate AI algorithms fed with real-time data. Additionally, integrating industrial AI capabilities into conventional industrial setups poses significant complexities.

Cloud computing is reshaping data processing methodologies. Utilizing cloud computing models, resources and capabilities are virtualized and delivered as services through the cloud, facilitating shared access to computing and storage resources on demand [12]. Nonetheless, data exchange between machinery/sensors and remote cloud locations might result in latency, heightened bandwidth usage, and increased energy consumption.

Moreover, this approach raises concerns regarding the security and reliability of external networks [13].

To address these challenges, developers are crafting edge computing solutions. These emerging infrastructures position computing, storage, and network resources closer to data origins, enabling the processing of time-sensitive data at the network edge, where it is generated [14]. This approach fulfills requirements for low latency, real-time processing, minimized network traffic, and enhanced security while optimizing resource utilization [15]. Edge computing involves analyzing and storing data in proximity to the devices producing and utilizing it, mitigating drawbacks from cloud computing and proving advantageous in manufacturing scenarios [16].

Edge computing devices now boast robust computational capabilities, capable of executing demanding industrial AI applications [17]. However, implementing these solutions necessitates expertise in OT, IT, and AI. Despite proposed architectures to streamline the integration of Industry 4.0 data collection and monitoring solutions, key challenges persist in Industry 4.0 and the industrial Internet of Things (IIoT). These challenges encompass security concerns and standardized data exchange between devices, machines, and services (across industries, not just within one), posing ongoing challenges [18]. A comprehensive review in [19] delves into the application of edge computing paradigms in manufacturing scenarios, outlining architectures, advancements, and lingering challenges.

For instance, in recent publications, authors [20] introduced an architectural framework designed for capturing and monitoring time series data. Earlier work by [21] introduced an MQTT-based (or message queuing telemetry transport) IoT cloud platform utilizing Node-RED, adaptable for edge environments, sharing similarities with the MING stack. Additionally, [22] presented an affordable, highly adaptable modular SCADA method grounded in Node-RED. Recently, [23] proposed a system for Proton Exchange Membrane (PEM) hydrogen generators leveraging Grafana and Node-RED.

However, to the best of the authors' knowledge, no existing architectures integrate industrial AI models, neither based on LF or HF data, into machinery and processes with a primary focus on reducing the IT, OT, and AI expertise barrier for the deployment of CPS 2.0. Consequently, this chapter concentrates on introducing a user-friendly architecture to help users integrate IT, OT, and AI technologies.

Regarding the role of AI within industrial applications, it plays a pivotal role in industry, particularly in predictive maintenance—a swiftly advancing domain within manufacturing. Its primary objective is to optimize maintenance protocols by forecasting and preventing equipment failures. Through the utilization of data analytics, sensor technology, and machine learning (ML) algorithms, predictive maintenance empowers manufacturers to continually monitor equipment health in real time and foresee maintenance requirements, thereby diminishing downtime, enhancing productivity, and curbing expenses.

Within predictive maintenance, RUL is a prominent concept, representing the time until a component ceases functioning. Estimating RUL, also known as time-to-failure, constitutes a prognostic method. Prognostics, increasingly valued over diagnosis in the machinery industry, assist in devising optimal maintenance strategies and resource allocation. Various prognostic prediction methods categorically analyze subsystems or components' RUL. These methods fall into three classes: physics-based (constructing mathematical models grounded in failure mechanisms or fundamental principles of damage), data-based (applying statistical and computational intelligence), and hybrid (mixing the other two classes) [24].

With advancements in computational infrastructure, deep learning (DL) has emerged as a key focus in prognostics research. As a subset of ML, DL possesses the capability to comprehend intricate hierarchical relationships inherent in deep structures [25]. Recent literature growth in this domain underscores a rising interest, suggesting a promising role for DL in RUL prediction. DL RUL prediction methods rely solely on data-driven approaches, necessitating extensive databases of run-to-fail trajectories for the development of reliable models [24].

To facilitate data-based ML and AI models, diverse HF signals must be captured, typically including force, acoustic emission (AE), and electric current signals within the industrial machining environment [26]. HF data commonly undergoes preprocessing, often segmented into time windows and subjected to various techniques, such as the wavelet transform (WT). Feature extraction from this processed data is customary. In the machining industry's production environment, unlike laboratory settings, where numerous signals are measurable, limitations exist on the number of machine-captured signals. Hence, a clear understanding of the optimal signals for capture and their required preprocessing is crucial for constructing confident predictive models.

The resultant HF data models and preprocessing techniques hold potential in predicting manufacturing line outcomes, autonomously identifying production issues, and potentially controlling industrial processes. However, selecting and correctly applying the proper HF data preprocessing techniques and the generation of HF data are still a complex issue. Thus, this chapter proposes a methodology to lead users on these HF data tasks.

8.3 HIGH-FREQUENCY AI MODEL GENERATION METHODOLOGY

This section presents a methodology to generate industrial AI models based on HF data. The methodology, depicted in Figure 8.1, comprises two primary phases. The initial phase, illustrated within the light-green box, focuses

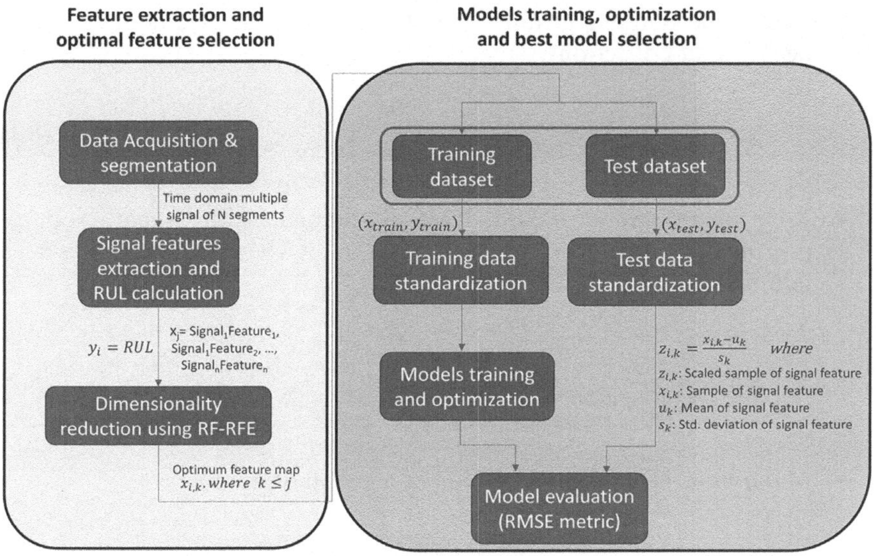

Figure 8.1 Graphical description of the followed methodology.

on exploring the best features for predicting tool RUL. Meanwhile, the subsequent phase, represented by the dark-green box, concentrates on identifying the optimal ML model using these selected features.

Initially, data from 12 distinct tools was gathered and divided into in N segments of 1 s duration (where $0 < i \leq N$). From each segment, various features (x_j) were derived, and the RUL value (y) for each tool was computed. Subsequently, the random forest–recursive feature elimination (RF-RFE) technique was applied to identify the most effective k features for predicting tool RUL.

In the subsequent phase, the dataset was divided into two sets: a training set encompassing eight tools, and a test set containing four tools, chosen randomly. This division, adhering to the common practice of a 2/3 training and 1/3 testing split, was followed [27]. Both sets underwent standardization. The training set was then utilized to train and fine-tune the models. These optimized models were subsequently assessed using the test set, and their performance was compared using the root mean square error (RMSE) metric. Further elaboration on these steps is provided in subsequent sections.

8.3.1 Signal features extraction

Twenty-six distinct signals were gathered to forecast tool RUL. These signals were divided into N segments, each lasting 1 s. From these segments, five unique features were derived across all the signals:

- *Mean.* The sum of a group of numbers divided by the amount of numbers of that collection:

$$\mu = \frac{1}{N}\sum_{i=1}^{N} x_i \tag{8.1}$$

- *Root mean square (RMS).* RMS is defined as the square root of the mean square (the arithmetic mean of the squares of a set of numbers):

$$RMS = \sqrt{\frac{1}{N}\sum_{i=1}^{N} x_i^2} \tag{8.2}$$

- *Maximum.* The maximum value of a group of numbers:

$$MAX = \max(x_i) \tag{8.3}$$

- *Skewness.* A measure of symmetry in a distribution:

$$Skewness = \frac{\sum_{i=1}^{N}(x_i - \mu)^3}{(N-1)\sigma^3} \tag{8.4}$$

- *Kurtosis.* A measure of the "tailedness" of the probability distribution of a variable:

$$Kurtosis = \frac{\sum_{i=1}^{N}(x_i - \mu)^4}{(N-1)\sigma^4} \tag{8.5}$$

After these features were derived from the signals, a set of j variables was generated, where x_j denotes each feature extracted from the signals. Each $x_{i,j}$ stands for a sample representing a feature of a signal. To maintain consistency with industrial standards, a maximum allowable wear of 250 µm for a tool was defined, derived from the maximum flank wear commonly accepted in the industry [28]. Consequently, signals recorded at wear values beyond this threshold were removed from the dataset. The RUL value, the target for prediction (y), was computed by assessing the duration between the time the tool reaches a wear of 250 µm and the present time.

8.3.2 Dimensionality reduction

Due to the multitude of captured signals and the extraction process outlined in the previous section, a total of 130 predictors was acquired. It is expected that not all these predictors will equally forecast the RUL, and some may exhibit correlation, providing redundant information. To discern which features better articulate the RUL progression of tools, the RF-RFE algorithm was employed. Consequently, correlated and repetitive features were pruned, diminishing the number of predictors to k. This algorithm aids in decreasing the computational load during training for ML algorithms and minimizes prediction errors.

8.3.3 Data split and standardization

Following the identification of the top k variables, the data was divided randomly into two subsets: a training set encompassing eight tools and a test set with four tools. Subsequently, the data underwent standardization. This step is commonplace for numerous ML estimators; when data does not adhere to a standard normal distribution, their performance might be suboptimal. The applied operation standardizes features by removing the mean and scaling to unit variance:

$$z = \frac{x - u}{s} \tag{8.6}$$

where u and s are the mean and standard deviation of the training set, respectively.

8.3.4 Models training and optimization

Upon the partitioning and standardization of the data, bidirectional long short-term memory networks (BiLSTM) and bidirectional gated recurrent units (BiGRU) models were employed for training and testing using features obtained from the RF-RFE algorithm. Additionally, to assess their performance against other conventional models, the same process was replicated with LSTM, GRU, convolutional neural network (CNN), multilayer perceptron (MLP), and RF employing 500 trees. CNN, MLP, and RF models are supplied with a two-dimensional array, as illustrated in Figure 8.2, where:

- Rows: a single record of the features.
- Columns: different features used to train the models.

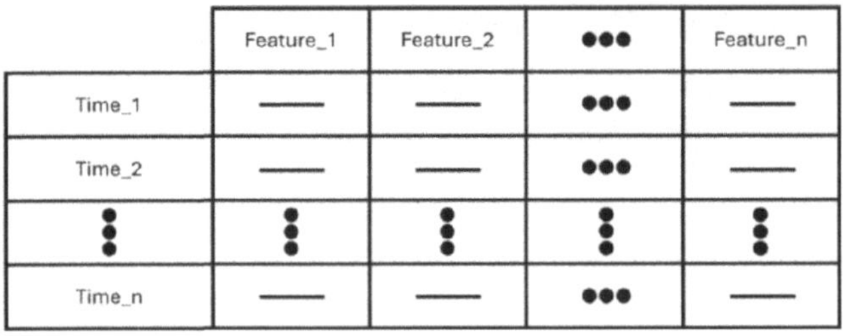

Figure 8.2 Data structure to feed traditional ML models.

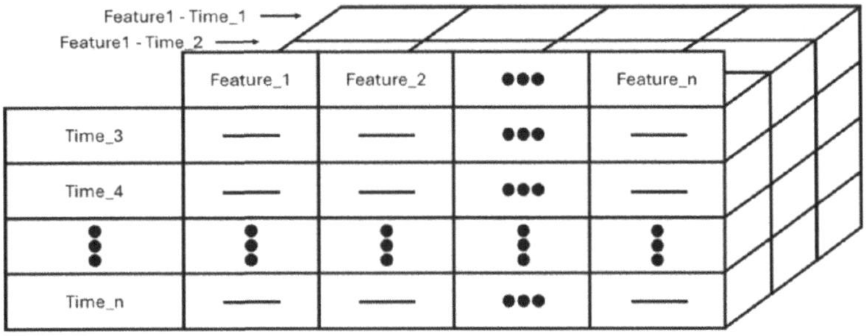

Figure 8.3 Data structure to feed LSTM and GRU models.

However, for the LSTM, BiLSTM, GRU, and BiGRU models, the structure of the input data must be modified. They are fed with a three-dimensional array, represented in Figure 8.3, where:

- Rows: a single record of the features.
- Columns: different features used to train the models.
- Depth: number of time steps used. The time step value indicates the number of previous samples added to the actual one.

When training these type of models, three parameters must be indicated: the batch size (the number of training examples, the rows, in one forward/backward pass), the number of features, and the time steps used.

To prevent overlap between the end and start of successive tools during model training, each tool's data structure was individually adjusted before merging them. Similarly, for RUL prediction, the strategy was maintained to prevent overlap between the final and initial phases of consecutive tools within the same batch size, enabling the individual prediction of each tool's RUL.

8.3.5 Models' optimization

To optimize the RUL prediction, different techniques were applied to the DL models:

- *Early stopping.* During model training, the primary aim is to reduce the loss function, specifically the MSE, in this context. Early stopping intervenes by halting the training when the monitored metric ceases to enhance. This metric is assessed after each epoch. The patience parameter denotes the count of successive epochs where loss reduction is absent, determining when to conclude the training process. Here, it was configured to 6.
- *Training process hyperparameters optimization.* In the training phase, cross-validation (CV) is employed to evaluate various combinations of predetermined hyperparameters within the chosen model (refer to Table 8.1). These combinations are assessed, and the one yielding the most favorable outcomes is selected. In this study, the dataset was divided into training (eight tools) and test sets (four tools). During model training, the CV was configured with four folds, optimizing the following hyperparameters.

  Once the models were fine-tuned, their comparison involved assessing execution time and the RUL prediction error, measured by the RMSE metric. Given the stochastic nature of the process, involving data splitting and RUL prediction, each model underwent 100 iterations, utilizing varying tool selections for training and testing in each instance. The conclusive execution time and prediction error outcomes were determined as the average across these 100 iterations.
- *Evaluation metric.* Furthermore, for a more distinct assessment of the predictive capabilities of each individual signal concerning tool RUL, every signal was separately employed as input for each model. This procedure was replicated 100 times, and the mean along with the standard deviation of the results were computed. The metric used to

Table 8.1 Optimized hyperparameters for DL models

Hyperparameter	*Values*
Number of layers	From 1 to 4
Number of neurons	From 1 to 5
Batch size [29]	8, 16, 32, 64
Activation function	Rectified linear unit (ReLu); linear, sigmoid and hyperbolic tangent (TANH)
Learning rate [30, 31]	0.01, 0.05, 0.1

evaluate the performance of the models is the RMSE, which is defined as follows:

$$RMSE = \sqrt{\frac{1}{n}\sum_{i=1}^{N}(\hat{y}_i - y_i)^2} \quad (7)$$

where $\hat{y}_i$ is the predicted value, y_i the observed values, and n the observed sample size.

8.4 SOFTWARE ARCHITECTURE TO DEPLOY CPS 2.0

This section presents a software architecture to ease the integration of the IT, OT, and AI technologies required to develop and deploy a CPS 2.0. The proposed architecture (Figure 8.4) is based on microservices. Excluding the industrial machine/process, these services are deployed as docker containers to ease the deployment and improve its scalability. The following figure summarizes the main components of the architecture.

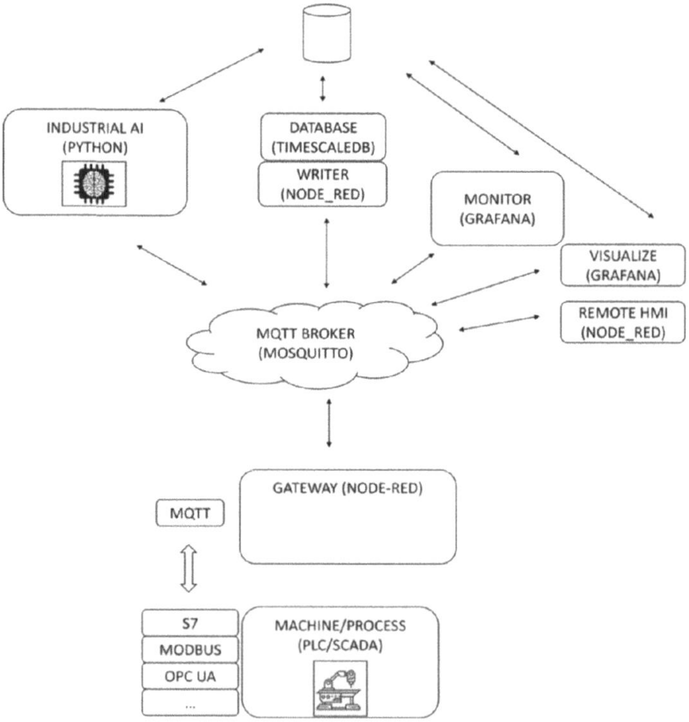

Figure 8.4 Main components of the architecture proposed to generate CPS 2.0.

- *Machine*. It represents the machine, process, or station of the shop floor. Its automation program is out of the cope of this chapter, and it is a responsibility of automation engineers for safety and low-level control. The automation engineer exposes the machine as a black box with defined inputs and outputs. Inputs and outputs link to automation program variables/methods, accessible via OT communication protocols like Modbus, Siemens S7, or OPC UA (Open Platform Communications Unified Architecture).
- *Gateway*. The gateway translates the OT protocols of the machine. It offers an interface converting OT communication protocols to MQTT. MQTT, a robust protocol demanding minimal computation, presents a wide support across hardware and software platforms. MQTT incorporates a broker system receiving messages transmitted by clients. Each message contains a subject (topic) and potential data (payload). Clients can subscribe to specific topics, and when the broker receives messages under those topics, it forwards them to subscribed clients.

The gateway maps each defined input/output of the machine to MQTT topics and message payloads. Additionally, the gateway may integrate some basic program logic to process inputs and/or outputs and transform them to new variables or methods. Several programming languages can be used to develop the gateway, such as Python, Java, or NodeJS. However, Node-RED is the suggested program to develop the gateway due to its code-free approach, simplifying gateway programming and maintenance.

- *Broker*. The MQTT broker lies between components of the architecture to ease communication between them following the messages with topics and payloads approach.
- *Writer*. It is responsible for persistence of information. It subscribes to relevant MQTT topics, decoding the payload from messages and generating the corresponding queries to write data to the database. As before, although Node-RED is the preferred option to develop it, users may choose between several programming language for its development.
- *Database*. Traditional relational databases are not encouraged for storing time series data produced by machines, as it can grow beyond the capacities of these databases easily. Within time series databases, TimescaleDB, an extension of PostgreSQL, is proposed for storing relational, JSON, and time series data within a single database, akin to InfluxDB functionalities.
- *Remote human–machine interface (HMI)*. A simplified web-based machine HMI. Node-RED is recommended, but any web technology stack suffices.
- *Visualization and monitoring*. Uses Grafana to connect to the database, creating customizable dashboards and alerts for machine monitoring.

Grafana is an open-source analytics and visualization web app adaptable to various databases, featuring an intuitive interface for customization and alert generation.

- *Industrial AI algorithms.* Deployed as MQTT-subscribed containers, these soft AI models process real-time data to generate control commands for the machine. Python is suggested for programming due to its robust AI capabilities. The gateway translates commands to the machine's OT protocol, enabling responses to AI-generated instructions.

Components described work together to generate an advanced cyber-physical system following this workflow to control the machine based on the execution of industrial AI algorithms.

- Step 1: After the process is started, the machine generates data. This data is read by the gateway using OT protocols, continuously reading machine data. Then, the gateway processes this data and sends it to the MQTT broker in real time.
- Step 2: The MQTT broker receives machine data from the gateway and publishes it, sending to the rest of the components. Then, these actions happen in parallel:
  - The writer receives data and stores it in the database.
  - The HMI is updated in real time.
  - Grafana visualizes and monitors data in real time.
  - AI algorithms receive data and use it to control the process. The algorithms are executed, and when results of these algorithms identify that some action is required, algorithms generate a command. This command is sent to the MQTT broker.
- Step 3: The command is received by the gateway, which translates it to the corresponding OT protocol compatible with the machine.
- Step 4: The machine executes the received control order.

8.5 VALIDATION

This section presents the validation of the methodology and the software architecture to develop CPS 2.0 for two industrial use cases: RUL estimation of a tooling machine based on HF data, and real-time analysis of the vacuum generation process of a leak test machine.

8.5.1 HF data model generation methodology

HF data model generation methodology has been validated analyzing HF data from a tooling machine to generate a RUL prediction model. To ensure robust predictive models considering the substantial variability in wear during machining, 12 repetitions of trial were performed during the validation. These trials were conducted under consistent cutting conditions: a cutting speed (*Vc*) of 200 m/min, a feed rate (*fv*) of 0.1 mm/rev, and a depth of cut (*ap*) of 2 mm. This approach aimed to establish models with a high level of confidence.

The material employed for the tests was a 19NiMoCr6 steel, which has a totally bainitic microstructure. Concerning the cutting tools, P25-grade uncoated inserts were employed, reference Widia TPUN160308TTM. These were clamped to a Widia CTGPL2020K16 toolholder, which gives an effective rake and clearance angles of 5° and 6°, respectively, with a positioning angle of 90°.

The cutting procedure was as follows:

- Machining of a predefined length of the workpiece, commonly 1/3 of the available length (70 mm).
- Cleaning of the tool insert to remove adhered material and to enable a correct measurement of tool wear.
- Tool wear measurement using an Alicona Infinite Focus G4 profilometer. This profilometer permits the 3D measurement of the wear in the flank (flank wear-Vb) and rake faces (crater wear Kt).
- Restart processes 1–3 until wear in the flank face (*Vb*) exceeds a value of 250 µm, being a reference of maximum flank wear employed in the industry.

The cutting forces were captured by clamping the toolholder onto a Kistler 9121 dynamometer. Data was recorded at a frequency of 50 kHz using a National Instrument cDAQ-9178 equipped with an analog input module NI-9239.

To monitor vibrations during turning, two triaxial accelerometers (PCB356A16) were utilized. One was affixed near the lathe's spindle (accelerometer 1), while the other was attached to the toolholder (accelerometer 2), as shown in Figure 8.5. Data was collected at 50 kHz using a National Instrument cDAQ-9178 with an analog input module NI-9234.

Sound signals were captured using a B&K 4189-A-021 microphone, recorded at a frequency of 50 kHz via a National Instrument cDAQ-9178 with an analog input module NI-9234.

AE was registered using a Kistler 8152B sensor coupled with a Type 5125B conditioning system, magnetically positioned on the toolholder (as depicted in Figure 8.5). The data was acquired at a frequency of 1 MHz utilizing a National Instrument cDAQ-9178 with an analog input module NI-9223.

Apart from the tools summarized in Table 8.2, measurements of current and voltage signals during cutting operations were conducted. These measurements encompassed both the motor of the y-axis drive and the spindle motor. The spindle motor currents were assessed utilizing LEM ITB 300-S

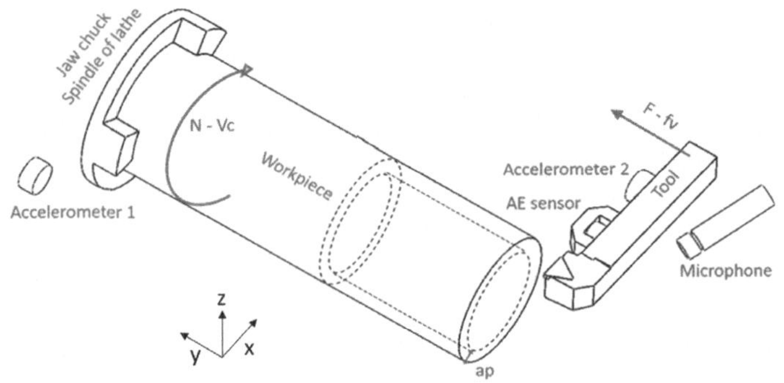

Figure 8.5 Schematic representation of the turning process and location of sensors.

Table 8.2 Data capture tool summary

Name	*Description*
Kistler 9121	Dynamometer
NI cDAQ-9178	HF data acquisition unit from National Instrument
PCB356A16 Accelerometers	Two axial accelerometers
B&K 4189-A-021	Sound acquisition sensor
Kistler 8152B	AE signal acquisition sensor
LEM ITB 300-S	Current monitor

transducers. Data acquisition occurred at a frequency of 50 kHz, employing a National Instrument cDAQ-9178 with analog input modules NI-9225 and NI-9244 for voltages, while NI-9227 was used for current measurements.

The RF-RFE technique was employed to determine the most effective input features for the models, derived from the signals. Analysis using RF-RFE revealed that among all the extracted features, the F_{y_RMS}, representing the RMS value of the feed force, stands as the most reliable predictor of tool RUL. This single feature demonstrates the capability to predict tool RUL, significantly simplifying the complexity and costs associated with the data acquisition system to just one sensor. Additionally, this reduction in the complexity of signal pre-processing, as well as in the ML and DL models, results in accurate and expedited model development.

Various regressive ML and DL models, inclusive of RF, MLP, LSTM, GRU, BiLSTM, BiGRU, and CNN, were trained and optimized using the F_{y_RMS} feature, identified by the RF-RFE algorithm as the optimal predictor. Consistent with existing literature on tool wear [32] and RUL [33] prediction, BiGRU and BiLSTM models exhibit superior performance compared to LSTM, GRU, MLP, RF, and CNN. Introducing bidirectionality to LSTM and GRU models improves their RUL prediction RMSE by 3.18 s (9.66%) and 1.26 s (4.04%), respectively. Furthermore, the BiGRU model, a simpler variant of BiLSTM, trains 8.95% faster than BiLSTM. In scenarios prioritizing computational optimization, the BiGRU model might be preferred, whereas if maximizing predictive performance is paramount, the BiLSTM model could be more suitable.

8.5.2 Architecture to integrate AI, OT, and AI to generate a CPS 2.0

This section describes the process of generating and deploying a CPS 2.0 of a vacuum generation procedure of a leak test machine. This process encompasses a vacuum chamber, three valves, a regular pump, a specialized pump, and an apparatus for vacuum measurement. A Siemens 1500 PLC manages this process, connecting to a Pfeiffer ASM 340 vacuum generator and measurement device. It presents a conventional HMI, enabling actions like commencing or halting the vacuum process, controlling valve states and chamber access, initiating or stopping the pumps, and monitoring vacuum levels. This vacuum generation process constitutes the initial phase of a developing leak detection apparatus.

In conventional settings, the PLC necessitates pre-configuration of the total process time and the moment to activate the specialized pump. These durations are manually set by operators, often involving trial and error until the desired vacuum level (e.g., 10e-1, 10e-2, 10e-3 . . .) is achieved. To exert control over these durations through soft AI models, the process has been initially conceptualized as a black box, characterizing its inputs and outputs (as detailed in Table 8.3).

Instead of altering the initial PLC program, a fresh data block was crafted, housing variables associated with specified inputs and outputs. Subsequently, leveraging SIOME software, an OPC UA data structure was devised, and each data block variable was linked to a variable within the OPC UA structure. This correlation was then integrated into the TIA portal to establish an OPC UA server.

To streamline process control, two new directives for initiating and halting the vacuum process were introduced. The gateway interprets these directives,

Table 8.3 Inputs and outputs of the process

Inputs	*Outputs*
Commands to open/close the valves	State of the valves
Commands to start/stop the pumps	State of the pumps
	Real-time vacuum level

converting them into suitable open/close commands for the pumps and valves anticipated by the PLC. The various inputs and outputs were aligned with these specific topics:

- /machine_name/start: Input command to start a vacuum generation process.
- /machine_name/created: Output to acknowledge that a new process has been created at the database and the vacuum process has been started. At this step, only the regular pump is running. The payload includes the identifier of the new process.
- /nachine_name/superpump: Input command to start the special pump, with the objective of reaching lower vacuum levels.
- /nachine_name/data: Output the share real-time data about the vacuum level.
- /nachine_name/stop: Input command to stop the vacuum generation process.
- /nachine_name/humidity: Input command to notify humidity has been detected during the process. In this case, the user is notified and the process is stopped.

The Node-RED user interface (Figure 8.6) has been designed to respond to these messages, creating the HMI. The interface exhibits various controls: three gauges visually display the current vacuum level in real time, each with a distinct range to portray different vacuum levels—high (0–1,000 mbar), medium (0–10 mbar), and low (0–1 mbar). Multiple data fields showcase the process identifier, real-time vacuum value, process status (starting/started/stopped), and elapsed time. Additionally, three buttons facilitate the initiation of a vacuum generation process, manual termination, or activation of the special pump. Two switches visually represent the pump states within the interface.

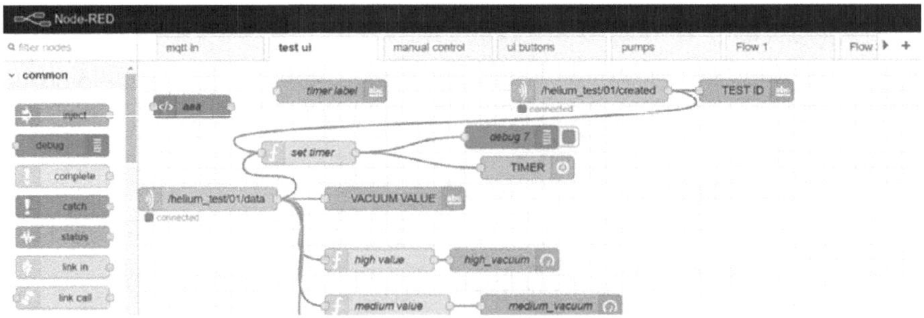

Figure 8.6 User interface of the HMI.

Grafana (Figure 8.7a) has been incorporated to produce a dashboard enabling visualization of both current and past data. The user can choose single or multiple vacuum generation processes for display on the dashboard. The displayed processes depict the vacuum generation operations executed using the same part within the chamber.

(a)

(b)

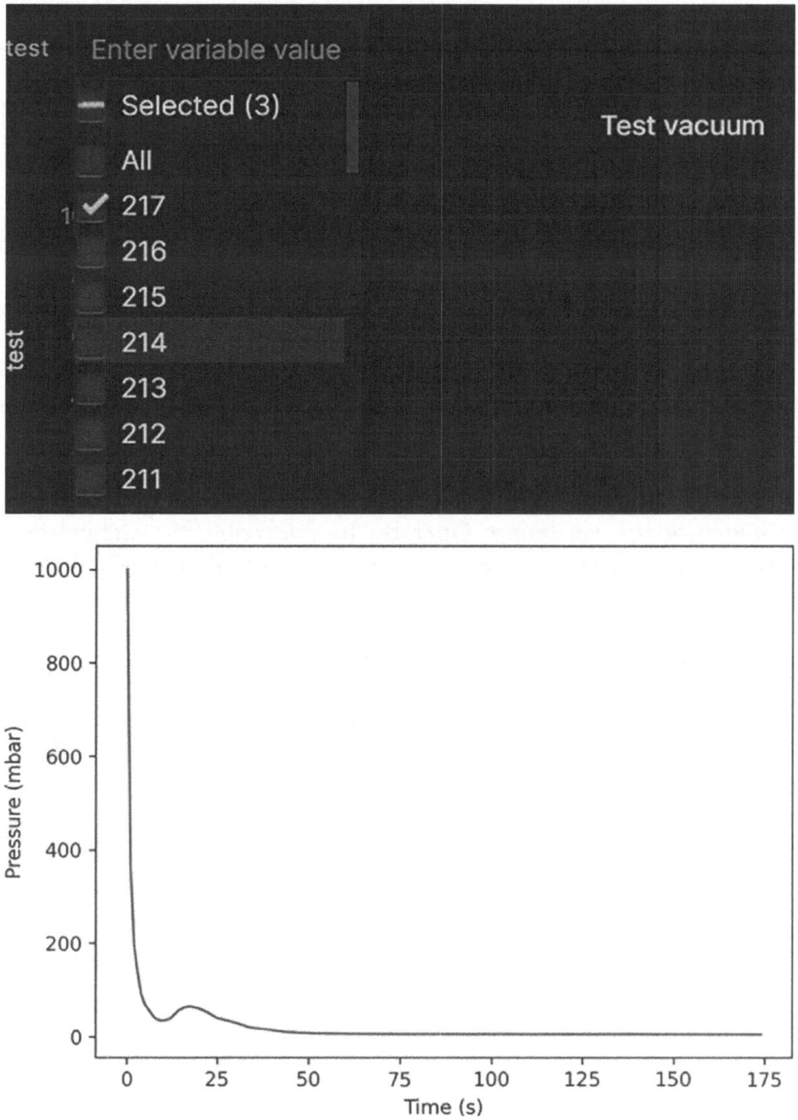

Figure 8.7 (a) Dashboard to show vacuum measurements at Grafana. (b) Humidity effect.

Finally, industrial AI, employing soft models, examines real-time vacuum measurements. The resulting analysis determines the process state for control purposes, deciding on activating the special pump, identifying process completion, or detecting humidity to stop the process.

These models were trained using a dataset derived from the vacuum machine's operation under traditional PLC control, storing vacuum measurements in the database. The dataset encompasses multiple manual runs conducted with two types of parts: aluminum and 3D printed. These runs varied in duration and activation times for the special pump, each lasting between 40 and 180 s. Due to the nature and evolution of the physical measurements of the process (temperatures and pressures), HF was not required for this validation, and initially, data was captured at a frequency of 3 Hz in the database but was subsequently down-sampled to average values per second for model training. However, internal testing has validated the capacity of the architecture to manage data sources of KHZ, and even a few MHz, without changes. Additionally, some parts were immersed in water before entering the chamber in certain processes. The final training dataset comprised over 120 labeled individual vacuum processes. This dataset was utilized to create three distinct models. Following an analysis of various methods, ranging from traditional digital signal processing techniques to neural networks, these models were developed and deployed within Docker containers:

- *Switch-on special pump.* Operates by analyzing the vacuum generation process. Initially, the first pump is activated, achieving a vacuum level near 1, and it relies on the second pump to reach lower levels. Activation of the special pump is crucial once a specific vacuum level is attained, and the rate of vacuum reduction begins to plateau over time. This model works on a 20-point window of data. Initially, it fits this window to an exponential function and calculates the gradient of the fitted curve. When the gradient values are significantly negative, it signifies a steep decay in vacuum levels. Conversely, smaller negative values indicate a flattening curve of vacuum decay. By comparing the maximum gradient value against a set threshold, the model identifies the point where the vacuum decay trend begins to flatten. This point denotes the optimal time to trigger the special pump for further reduction of the vacuum level.
- *Stop the process.* The "stop the process" model functions when a specific vacuum level is attained and both pumps fail to further reduce it. To determine the optimal timing for halting the process, the model follows a series of steps. First, it establishes the target vacuum value for precision purposes. Subsequently, if the objective vacuum value is reached, the process concludes. However, there are scenarios where reaching the desired vacuum level may not be feasible due to various

factors like chamber contamination or mechanical issues. In such cases, extending the process time becomes unproductive. This model operates by analyzing 15-point data windows, where it computes upper and lower bounds. The algorithm then evaluates whether the remaining window points fall within these calculated bounds. If they do, indicating that the vacuum value is no longer decreasing, the model decides to terminate the process.

- *Humidity detection.* Identifies the impact of humidity on the decay of vacuum values over time, causing a slight increase in pressure evolution. This phenomenon is visually depicted in Figure 8.7b. To detect humidity-related issues during the process, the algorithm utilizes a 20-point window to calculate the differences between adjacent data points within the window. If any of these differences exceed a predetermined threshold, it signifies the presence of humidity. In response, a command is issued to signal and cease the ongoing process.

The software architecture has been successfully deployed to generate a CPS 2.0 of the process connected to the machine, capturing and visualizing data and applying AI algorithms to control the real manufacturing process of the CPS 2.0.

8.6 CONCLUSIONS

CPS 2.0 leverages robust computational abilities, AI, and big data processing to conduct near-real-time data analysis. The development of CPS 2.0 necessitates expertise across OT, IT, and AI domains. This chapter (1) introduces a methodology for creating AI models using HF industrial data and (2) outlines a reference architecture to deploy CPS 2.0 integrating IT, OT, and AI technologies, specifically addressing real-time AI monitoring within manufacturing processes.

Successful validation of this methodology and the architecture provides a crucial step toward further real-world validation in industrial settings. This validation process aims to streamline the integration of CPS 2.0 into industrial operations.

8.6.1 Acknowledgment

The research was partially supported by the Centre for the Development of Industrial Technology (CDTI) and the Spanish Minister of Science and Innovation (IDI-20210506) and by the Economic Development, Sustainability, and Environment Department of the Basque Government (KK-2022/00119, KK-2020/00103).

REFERENCES

[1] T. F. De Barrena, J. L. Ferrando, A. García, X. Badiola, M. S. de Buruaga, and J. Vicente, "Tool remaining useful life prediction using bidirectional recurrent neural networks (BRNN)," *The International Journal of Advanced Manufacturing Technology*, vol. 125, no. 9–10, pp. 4027–4045, Apr. 2023, doi: 10.1007/s00170-023-10811-9.

[2] A. Garcia, X. Oregui, J. Franco, U. Arrieta, J. Ferreres, and J. A. Valencia, "Time series manufacturing data edge monitoring and visualization to support industrial maintenance teams," *SN Computer Science*, vol. 5, no. 1, p. 131, Dec. 2023, doi: 10.1007/s42979-023-02442-4.

[3] E. Ahmed, I. Yaqoob, A. Gani, M. Imran, and M. Guizani, "Internet-of-things-based smart environments: State of the art, taxonomy, and open research challenges," *IEEE Wireless Communications*, vol. 23, no. 5, pp. 10–16, Oct. 2016, doi: 10.1109/MWC.2016.7721736.

[4] X. Fei *et al.*, "CPS data streams analytics based on machine learning for Cloud and Fog Computing: A survey," *Future Generation Computer Systems*, vol. 90, pp. 435–450, Jan. 2019, doi: 10.1016/j.future.2018.06.042.

[5] V. Alcácer and V. Cruz-Machado, "Scanning the Industry 4.0: A literature review on technologies for manufacturing systems," *Engineering Science and Technology, an International Journal*, vol. 22, no. 3, pp. 899–919, Jun. 2019, doi: 10.1016/j.jestch.2019.01.006.

[6] J. Lee, B. Bagheri, and H.-A. Kao, "A cyber-physical systems architecture for Industry 4.0-based manufacturing systems," *Manufacturing Letters*, vol. 3, pp. 18–23, Jan. 2015, doi: 10.1016/j.mfglet.2014.12.001.

[7] L. Shabtay, P. Fournier-Viger, R. Yaari, and I. Dattner, "A guided FP-growth algorithm for mining multitude-targeted item-sets and class association rules in imbalanced data," *Information Sciences (New York)*, vol. 553, pp. 353–375, Apr. 2021, doi: 10.1016/j.ins.2020.10.020.

[8] T. Taleb, I. Afolabi, and M. Bagaa, "Orchestrating 5G network slices to support industrial internet and to shape next-generation smart factories," *IEEE Network*, vol. 33, no. 4, pp. 146–154, Jul. 2019, doi: 10.1109/MNET.2018.1800129.

[9] C. Zhang, G. Zhou, J. Li, F. Chang, K. Ding, and D. Ma, "A multi-access edge computing enabled framework for the construction of a knowledge-sharing intelligent machine tool swarm in Industry 4.0," *Journal of Manufacturing Systems*, vol. 66, pp. 56–70, Feb. 2023, doi: 10.1016/j.jmsy.2022.11.015.

[10] C. Liu, X. Xu, Q. Peng, and Z. Zhou, "MTConnect-based cyber-physical machine tool: A case study," *Procedia CIRP*, vol. 72, pp. 492–497, 2018, doi: 10.1016/j.procir.2018.03.059.

[11] C. Zhang, G. Zhou, J. Li, T. Qin, K. Ding, and F. Chang, "KAiPP: An interaction recommendation approach for knowledge aided intelligent process planning with reinforcement learning," *Knowledge-Based Systems*, vol. 258, p. 110009, Dec. 2022, doi: 10.1016/j.knosys.2022.110009.

[12] F. Tao, L. Zhang, Y. Liu, Y. Cheng, L. Wang, and X. Xu, "Manufacturing service management in cloud manufacturing: Overview and future research directions," *Journal of Manufacturing Science and Engineering*, vol. 137, no. 4, Aug. 2015, doi: 10.1115/1.4030510.

[13] D. Borsatti, G. Davoli, W. Cerroni, and C. Raffaelli, "Enabling industrial IoT as a service with multi-access edge computing," *IEEE Communications Magazine*, vol. 59, no. 8, pp. 21–27, Aug. 2021, doi: 10.1109/MCOM.001.2100006.
[14] M. Nikravan and M. Haghi Kashani, "A review on trust management in fog/edge computing: Techniques, trends, and challenges," *Journal of Network and Computer Applications*, vol. 204, p. 103402, Aug. 2022, doi: 10.1016/j.jnca.2022.103402.
[15] J. Leng, Z. Chen, W. Sha, S. Ye, Q. Liu, and X. Chen, "Cloud-edge orchestration-based bi-level autonomous process control for mass individualization of rapid printed circuit boards prototyping services," *Journal of Manufacturing Systems*, vol. 63, pp. 143–161, Apr. 2022, doi: 10.1016/j.jmsy.2022.03.008.
[16] M. Alam, J. Rufino, J. Ferreira, S. H. Ahmed, N. Shah, and Y. Chen, "Orchestration of microservices for IoT using Docker and edge computing," *IEEE Communications Magazine*, vol. 56, no. 9, pp. 118–123, Sep. 2018, doi: 10.1109/MCOM.2018.1701233.
[17] R. Singh and S. S. Gill, "Edge AI: A survey," *Internet of Things and Cyber-Physical Systems*, vol. 3, pp. 71–92, 2023, doi: 10.1016/j.iotcps.2023.02.004.
[18] Y. Lu, "Industry 4.0: A survey on technologies, applications and open research issues," *Journal of Industrial Information Integration*, vol. 6, pp. 1–10, Jun. 2017, doi: 10.1016/j.jii.2017.04.005.
[19] T. Qiu, J. Chi, X. Zhou, Z. Ning, M. Atiquzzaman, and D. O. Wu, "Edge computing in industrial Internet of Things: Architecture, advances and challenges," *IEEE Communications Surveys & Tutorials*, vol. 22, no. 4, pp. 2462–2488, 2020.
[20] A. Garcia, X. Oregui, J. Franco, and U. Arrieta, "Edge Containerized Architecture for Manufacturing Process Time Series Data Monitoring and Visualization," in *Proceedings of the 3rd International Conference on Innovative Intelligent Industrial Production and Logistics*, SCITEPRESS—Science and Technology Publications, 2022, pp. 145–152. doi: 10.5220/0011574500003329.
[21] C. Rattanapoka, S. Chanthakit, A. Chimchai, and A. Sookkeaw, "An MQTT-based IoT cloud platform with flow design by node-RED," in *2019 Research, Invention, and Innovation Congress (RI2C)*, IEEE, 2019, pp. 1–6. doi: 10.1109/RI2C48728.2019.8999942.
[22] I.-V. Niţulescu and A. Korodi, "Supervisory control and data acquisition approach in node-RED: Application and discussions," *IoT*, vol. 1, no. 1, pp. 76–91, Aug. 2020, doi: 10.3390/iot1010005.
[23] F. J. Folgado, I. González, and A. J. Calderón, "Data acquisition and monitoring system framed in industrial Internet of Things for PEM hydrogen generators," *Internet of Things*, vol. 22, p. 100795, Jul. 2023, doi: 10.1016/j.iot.2023.100795.
[24] Y. Wang, Y. Zhao, and S. Addepalli, "Remaining useful life prediction using deep learning approaches: A review," *Procedia Manufacturing*, vol. 49, pp. 81–88, 2020, doi: 10.1016/j.promfg.2020.06.015.
[25] M. Ma, C. Sun, and X. Chen, "Discriminative deep belief networks with ant colony optimization for health status assessment of machine," *IEEE Transactions on Instrumentation and Measurement*, vol. 66, no. 12, pp. 3115–3125, Dec. 2017, doi: 10.1109/TIM.2017.2735661.

[26] G. Byrne, D. Dornfeld, I. Inasaki, G. Ketteler, W. König, and R. Teti, "Tool Condition Monitoring (TCM)—The status of research and industrial application," *CIRP Annals*, vol. 44, no. 2, pp. 541–567, 1995, doi: 10.1016/S0007-8506(07)60503-4.
[27] Q. H. Nguyen *et al.*, "Influence of data splitting on performance of machine learning models in prediction of shear strength of soil," *Mathematical Problems in Engineering*, vol. 2021, pp. 1–15, Feb. 2021, doi: 10.1155/2021/4832864.
[28] ISO 3685:1993, "Tool-life testing with single-point turning tools," 1993.
[29] X. Zhang, X. Lu, W. Li, and S. Wang, "Prediction of the remaining useful life of cutting tool using the Hurst exponent and CNN-LSTM", doi: 10.1007/s00170-020-06447-8/Published.
[30] J. Zhang, Y. Zeng, and B. Starly, "Recurrent neural networks with long term temporal dependencies in machine tool wear diagnosis and prognosis," *SN Applied Sciences*, vol. 3, no. 4, Apr. 2021, doi: 10.1007/s42452-021-04427-5.
[31] J. Yao, B. Lu, and J. Zhang, "Tool remaining useful life prediction using deep transfer reinforcement learning based on long short term memory networks," *International Journal of Advanced Manufacturing Technology*, vol. 118, pp. 1077–1086, 2022, doi: 10.1007/s00170-021-07950-2.
[32] X. Wu, J. Li, Y. Jin, and S. Zheng, "Modeling and analysis of tool wear prediction based on SVD and BiLSTM," *The International Journal of Advanced Manufacturing Technology*, vol. 106, no. 9–10, pp. 4391–4399, Feb. 2020, doi: 10.1007/s00170-019-04916-3.
[33] Q. An, Z. Tao, X. Xu, M. El Mansori, and M. Chen, "A data-driven model for milling tool remaining useful life prediction with convolutional and stacked LSTM network," *Measurement*, vol. 154, p. 107461, Mar. 2020, doi: 10.1016/j.measurement.2019.107461.

Chapter 9

Cyber-physical system formulations for health monitoring–informed bridge infrastructure assets

Ekin Ozer, Maria Q. Feng, Serdar Soyoz, and Paul Fanning

ABBREVIATION LIST

B	bridge
CPS	cyber-physical systems
H	hospital
SHM	structural health monitoring
T	test
WN	white noise

9.1 INTRODUCTION

CPS theory emerged as a response to complex infrastructure–computer interactions attempting to bundle sensory data and advanced mathematical models for improved control and decision-making of industrial assets. Taking the form of next-generation embedded systems in the late 2000s [1], distributed computing and control elements needed a collective/integrated form rather than their individual examination within isolated layers, phrased by Wolf as "control-computing codesign" [2]. In line with the United States National Science Foundation's ambitions, Baheti et al. [3] narrated CPS aims as (1) system description and modulation at a variety of abstraction levels, (2) formulating the interactions among subsystems, and (3) maximizing operational gains while minimizing the risks [3]. Sanislav et al. identified key CPS properties incorporating functionality, performance, security, and cost, as well as the core features, including physical input/feedback, distributed control, simultaneous analytics, and spatially distributed large-scale control environments [4]. Ashibani et al. defined three layers for a typical CPS: (1) perception layer (e.g., sensors), (2) transmission layer (e.g., communication), and (3) application layer (e.g., smart infrastructure), where the physical object is linked to a sensor and model-engaged digital knowledge [5]. Digital knowledge herein is a predecessor for the distributed control environment and facilitates actions compatible with the CPS protocols and architecture.

DOI: 10.1201/9781003559993-9

Among a variety of application domains, such as vehicular control, critical infrastructure, health care, and more [6], this study presents a civil engineering perspective to define the context of CPSs and their interpretation within SHM-enabled decision support systems, particularly for bridges. Civil infrastructure is subject to ageing and deterioration, and efficient civil infrastructure stock management relies on accurate evaluation of their conditions. Transportation systems, for instance, depend on the performance of their weak links, which are commonly inferred as bridges. Failure of civil infrastructure, for example, bridges, results in numerous negative consequences that disrupt functionality and initiate losses and casualties and, indirectly, carbon emission. Therefore, timely and effective condition assessment has been a long-lasting effort in civil infrastructure maintenance.

Within the CPS context, there is a growing need for maximizing the physical performance of civil infrastructure via the perception, transmission, and application layers proposed by [5]. On the other hand, traditionally, bridge condition assessment used to depend on visual inspection with subjective human judgment [7], way behind the digital age and far from real-time operations. The state of the structure is assessed via expert visits and reports, and accordingly, visual inspection can feed into the bridge management process, where maintenance-related decisions can be effectively taken [8, 9]. While the conventional visual inspection approaches necessitate human participation, there is a rising trend toward more automated methods through emerging technologies, for example, uninhabited aerial vehicles [10]. Such vehicular systems can become pre-eminent in case of regular or post-disaster bridge evaluation efforts [11].

An alternative technology-engaged effort to diagnose bridge conditions is through vibration-based SHM, where sensors instrumented on a bridge can capture damage indicators and inform the authorities in an automated and objective manner [12]. SHM discipline studies the changes observed in vibration characteristics of mechanical systems, for example, civil infrastructures, through sensor deployment, data acquisition, signal processing, and damage identification [13–15]. The traditional instrumentation used to be centralized and cable-connected, whereas the last two decades have experienced a radical improvement in the usage of wireless SHM networks [16]. With the synergic merge of advanced statistical learning techniques [17], there has been significant improvement in the detection accuracy and computational efficiency of SHM platforms. And with prognosis, SHM systems are becoming a core component of infrastructure-related decision-making processes through estimations of future conditions [18]. However, environmental and operational variability is the key outstanding challenge that has not completely been removed from SHM findings despite all the progress [19].

As expressed earlier, algorithmic developments, physics-informed learning, and sensory decision support systems opened a new avenue for SHM research. Malekloo et al. defined *digital twins* as one of the four breakthroughs linking modern SHM with machine learning paradigms [20]. According to Malekloo et al., model-driven SHM is one roadmap toward infrastructure condition assessment and beyond. Model-driven SHM stems from the notion that models are not fully representative of reality unless they are calibrated with actual data from physical infrastructure [21–23]. However, physical infrastructure, as it reaches large scales and is of non-homogeneous material behavior, is subject to numerous uncertainties, similar to the loads imposed on it. Therefore, uncertainty is a key component of model updating [24] and numerous sources of failure-induced risks [25]. Jointly handling the model and hazard-related uncertainties in an associated manner is possible via probabilistic approaches, for example, Monte Carlo simulation, where each realization per random variable can be sampled from the non-updated vs. updated distributions [26].

CPS theory was born as a superset of the aforementioned effort within a more generalized context, computerized and real assets, in some cases, jointly interacting with humans [27–29]. CPS and bridge SHM crossed roads in the 2010s [30]. Legatiuk et al. identified that structural control and health monitoring are the core CPS application fields in civil engineering [31]. The idea arose from the integration of physical and digital knowledge pertaining to structures, for example, more particularly, bridge finite element models and their calibration via mobile sensors for SHM-integrated reliability assessment [32]. Similar examples include mobile sensors monitoring the hoisting phenomenon of bridge girders [33], model-engaged decision support for various infrastructures [34, 35], and load assessment based on machine vision [36].

Ozer et al. defined the cyclic character of cyber-physical SHM systems connecting the physical asset with a model calibration framework resulting in reliability-driven bridge infrastructure decisions, which eventually reflect on the physical asset through maintenance and retrofit processes [37–39]. The framework stemmed from a model-updating-based SHM philosophy, where infrastructure physical features are represented via measurements, modeled through theoretical parameters, and risk estimations can be made based on calibrated model behavior. These estimations feed into the decision support systems, which eventually can redefine the physical features. The proposed procedure is also sensitive to structural damage and, therefore, can extend the lifetime of the structure through accurate evaluation of the structural conditions.

In line with these, this study presents a collection of CPS protocols in line with bridge infrastructure assets being monitored and draws a roadmap toward model-driven decision support systems empowered by SHM principles. In the following sections, the authors introduce a brief overview of the civil infrastructure CPS idea incorporating SHM information together with model updating, reliability analysis, decision-making, and three demonstrative examples at different implementation stages.

To discover the steps represented earlier in further detail, Section 9.2 sets the CPS analogy with the model-driven SHM processes. Section 9.3 and Section 9.4 discuss the model updating process, reliability analysis, and decision analysis, respectively. Finally, Section 9.5 presents three case studies to provide examples of the proposed procedure.

9.2 OVERVIEW OF THE CYBER-PHYSICAL SYSTEM NOTION FROM BRIDGE SHM PERSPECTIVE

A CPS process incorporating SHM consists of a series of processes linking the physical infrastructure to the digital representations and their associated decisions. Figure 9.1 presents a summary of the workflow of the relation between SHM-support infrastructure decisions, physical changes made to the structure, and associated updates in the digital domain as a result of those changes. The initiation of the process starts at the current state of the structure, where vibration data is collected via sensor measurements. These measurements are processed to retrieve modal characteristics of the structure, which are damage-sensitive features. The original mathematical (finite element) model developed based on structural information, for example, dimensions and material properties, is calibrated to best fit the identified modal characteristics. Later on, the calibrated model, which is more representative of the actual structural state, is used to perform reliability estimation. Structural reliability indicates the probabilistic means where the structure remains functional, and when combined with failure consequences, it can correspond to the risks with and without structural interventions. Decision analysis is then performed to maximize the expected utility or minimize expected losses, which will eventually result in an infrastructure-related action, for example, retrofit. With the new norms in CPS and model-driven SHM, the majority of the proposed processes can be performed without human intervention in an automated manner, except those related to the intervention operations. Therefore, a CPS approach can speed up the analytics starting from sensory data reaching onto automated decision recommendations.

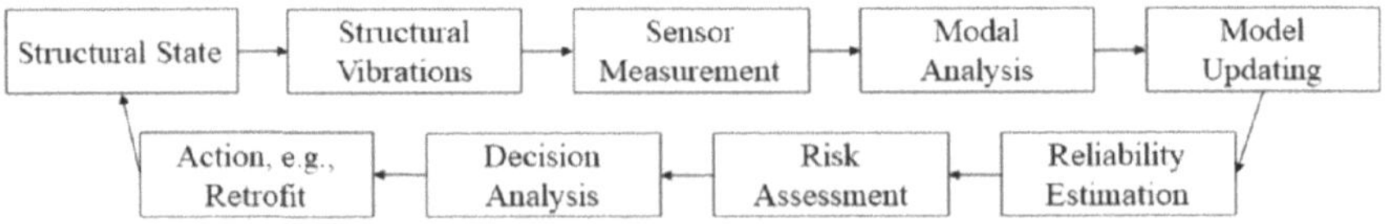

Figure 9.1 CPS framework constituting model-driven SHM processes.

9.3 MODEL-DRIVEN STRUCTURAL HEALTH MONITORING

This section presents the integration of modal identification results with finite element model calibration or updating procedure. Model updating relies on a population of models with varying physical parameters, where the damage-sensitive feature is directly or indirectly represented. For example, studies have shown that structural stiffness is sensitive to structural damage, and changes in the stiffness values are reflected in the stiffness matrix of the structure, hence eigenvalues of the mass–stiffness matrix couple of the system. To optimize the model based on modal parameter outputs, an objective function O is defined in equation 1:

$$O(k_1,\ldots,k_i,\ldots,k_n) = \sum_{m=1}^{m=m_{max}} \alpha_m \left[\frac{\left(f_m^* - f_m\right)}{f_m^*}\right]^2 + \beta_m \left[\frac{\left(\xi_m^* - \xi_m\right)}{\xi_m^*}\right]^2 + \gamma_m \left[1 - MAC_m\right]^2 \quad (9.1)$$

where k represents the member stiffness of the damage-sensitive component of the system, that is, plastic hinge locations of a bridge pier subject to seismic excitation. The subscript i corresponds to the ith member influencing the stiffness matrix among n members contributing to the objective function. The index m corresponds to the vibration mode taken into consideration to quantify the disagreement between the identified and model modal parameters. f, ξ, and MAC correspond to the modal frequency, modal damping ratio, and modal assurance criteria, respectively. Each model parameter is assigned to a coefficient determining their importance on the objective function, for example, α, β, and γ denoting the weights of frequencies, damping ratios, and mode shapes. Finally, the asterisk (*) denotes the reference modal parameter value obtained from identification results (e.g., those obtained from frequency domain decomposition or stochastic subspace identification; see [40]), whereas the counterpart term without asterisk corresponds to a single-model realization corresponding to the combination of stiffness values influencing the objective function. An example objective function with a similar structure can be found in [38].

A baseline state is achieved if vibration measurements are collected and processed prior to damage. Then, the post-disaster damage state can be quantified in terms of reduction in stiffness values, such as in nonlinear moment–curvature behavior of a bridge pier represented with the effective stiffness EI_{eff} and disaster-induced ductility demand μ in terms of the original stiffness EI, formulated in equation 2:

$$EI_{eff} = \frac{EI}{\mu} \tag{9.2}$$

Similar relationships can be defined for other terms, for example, damping ratio, with further examples provided in [38]. Through the proposed updating procedure, vibration data collected before, during, and after a damaging event can be utilized to update the model, which fits the modal characteristics better compared with the original non-updated model. The better-fit model can then serve for reliability analysis, which will provide the probabilistic means toward quantifying risks and prioritizing decisions.

9.4 HAZARD ANALYSIS, STRUCTURAL RELIABILITY, RISK, AND DECISION-MAKING

Structural performance, given an expected hazard level, depends on the fragility curve of the structure, as well as the intensity measure. The fragility curve F can be directly constructed through a maximum likelihood estimation process based on cumulative distribution function fitting in the logarithmic domain (see equation 3):

$$F(a) = \phi(\ln(a / c) / \sigma) \tag{9.3}$$

where a, c, and σ correspond to the ground motion intensity measure, mean of the fragility curve, and standard deviation of the fragility curve, respectively. The curve fitting process is based on a series of nonlinear time history analyses where a limit state threshold is whether exceeded or not, therefore has a binary outcome. As formulated in equation 4, the fragility curve feeds into the maximum likelihood L, such as:

$$L = \prod_{r=1}^{r=r_{max}} [F(a_r)]^{x_r} [1 - F(a_r)]^{1-x_r} \tag{9.4}$$

Herein, r is the number of nonlinear time history analyses existing in the database, and x is the binary term corresponding to failure (i.e., 1) or no failure (i.e., 0). Finally, the c and σ terms are determined based on the likelihood function's derivative approaching 0, such as in equation 5:

$$\frac{d \ln L}{dc} = \frac{d \ln L}{d\sigma} = 0 \tag{9.5}$$

A more detailed narrative is presented in [38], and the procedure can be repeated for different damage limit states, each representing a different fragility curve (e.g., minor, moderate, major). Once the fragility curve is constructed, probabilistic hazard analysis is conducted to quantify the intensity measure a_b (e.g., peak ground acceleration) of the bth bridge, and the failure probability is determined as $F_b(a = a_b)$, whereas structural reliability R is equal to the value in equation 6:

$$R_b(a = a_b) = 1 - F_b(a = a_b) \tag{9.6}$$

The aforementioned expression R stands for the reliability of a single bridge, whereas in the case of a population size b_{max} of bridges, the systemic reliability—denoted in equations 7 and 8—of the bridge network can take various forms, such as:

$$R_{sys}\left(\left[a_1, a_2, \ldots, a_b, \ldots, a_{b_{max}}\right]\right) = \prod_{b=1}^{b=b_{max}} R_b(a = a_b) \tag{9.7}$$

For bridge populations in series:

$$R_{sys}\left(\left[a_1, a_2, \ldots, a_b, \ldots, a_{b_{max}}\right]\right) = 1 - \prod_{b=1}^{b=b_{max}} 1 - R_b(a = a_b) \tag{9.8}$$

For bridge populations in parallel, and for combinations of those for each alternative route. Cascaded reliability effects of different routes will result in an array of decision outcomes when combined with their associated consequences. Finally, equation 9 and the utility expression U define the gain/loss from making a bridge or bridge network decision which will determine its comparative performance with respect to the opportunity costs. In other words, concerning civil infrastructure systems, utility takes the form of the inverse of costs C associated with a particular decision:

$$U = -R * C \tag{9.9}$$

Otherwise, benefits B (i.e., equation 10) combined with the reliability associated with the civil infrastructure conditions paired with the decision:

$$U = R * B \tag{9.10}$$

Therefore, expected utility maximization $E(U)$ seeks the minimal value of C or the maximal value of B, depending on the nature of the decision problem. R takes the form of R_b for individual infrastructure problems, whereas R_{sys} looks into the systemic performance of an infrastructure network. Given a decision d among d_{max} possible decisions, one can define the optimal action as based on the comparative U values per d, that is, U_d. Equation 11 suggests that the decision d becomes the optimal d_{opt}, where the utility U is maximized:

$$U_{d_{opt}} = \max\left(\left[U_{d=1} : U_{d=d_{max}}\right]\right) \tag{9.11}$$

An SHM-informed reliability estimation and decision analysis process can support the proposed CPS formulation in two streams. These can be summarized as (1) removal of estimation bias coming at various levels, that is, hazard as well as structural system's representation, and (2) reduction in uncertainties associated with hazards and modeling. Modeling input in hazards herein corresponds to the characterization of structural demand, whereas modeling input in structures corresponds to the physical parameters borne from material and section characteristics—in other words, structural capacity. And with reduced bias and reduced uncertainties, a decision is likely to be closer to the "actual" optimal decision rather than a "simulated" one. It should be mentioned that there can be a variety of civil infrastructure decisions that can take different forms; a few examples include infrastructure retrofit, traffic management, complete closure or service disruption during inspection, as well as transportation network problems, such as finding the optimal route in case of a disastrous event.

In the forthcoming sections, the authors present an SHM-integrated decision analysis example empowered by modern mobile sensing technology via smartphone accelerometers.

9.5 APPLICATIONS

In the previous sections, a bridge infrastructure decision analysis approach incorporating SHM knowledge has been introduced with the combined notions of model updating, reliability assessment, and expected utility theorem. In this section, the authors provide examples of these steps to demonstrate how infrastructure conditions are sensitive to damage, and their impacts on reliability, novel forms of SHM data collection under limited modeling input, and region-scale infrastructure decisions, where conventional systems cannot cope with the monitoring needs in a timely manner and, instead, a ubiquitous sensing and computing approach is needed. These examples are (1) damage–reliability relationship via large-scale experimentation, (2) mobile data–integrated reliability estimation with limited modelling data, and (3) upscaling transportation infrastructure decisions via network-level model-driven SHM, respectively. The first and the second studies only extend into the reliability assessment range, whereas there is an additional layer in the third study, which is decision analysis, again based on bridge reliability outcomes. All studies presented in what follows incorporate a model-driven

SHM approach through a calibration protocol followed by reliability assessment. See Table 9.1 for the summary of these cases.

9.5.1 Bridge reliability under varying damage (case 1)

This case study presents experimental evidence on the bridge condition–residual reliability relationship through a series of seismic shake table tests conducted at the University of Nevada, Reno, under the leadership of Prof. Saiid Saidii and Prof. David Sanders, with open-access data available by [41], as introduced as case no 1. A two-span three-bent reinforced concrete bridge structure was subjected to ground motion excitation with increasing intensities. As the larger-intensity-measure earthquakes are imposed on the bridge, incremental damages concentrated on column plastic hinge locations were observed (e.g., cracks and spalling). The bridge was instrumented with 11 accelerometers monitoring the vibration response of the bridge during the earthquake exposure. Figure 9.2 shows the experimental setup and the

Table 9.1 Case studies incorporating SHM-oriented CPS framework

Case no.	*Context of the study*	*Scope*
1	Experimental	Quantifying the reliability effects of identified damage on bridge piers
2	Field, individual	Calibrating a bridge model with limited knowledge through smartphone data
3	Field, network	Post-disaster mitigation decisions for a remote vicinity with bridge populations

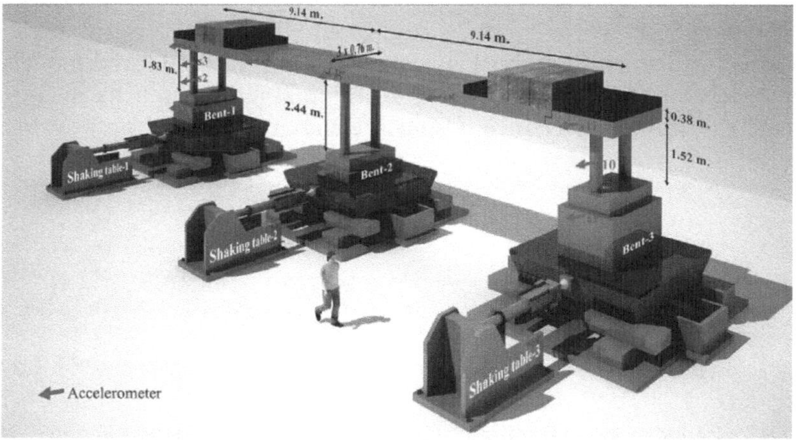

Figure 9.2 Bridge shake table test setup and instrumentation.

Source: Retrieved from [41].

accelerometer configurations, together with the shake tables located at the base of each bridge pier. Johnson et al. present a more detailed summary of the tests conducted in this experimental campaign [42].

Before and after each seismic excitation, vibration measurements are collected from the bridge through the accelerometric network. In further detail, the bridge was exposed to low-amplitude (non-damaging) white noise excitations to represent the vibration character of the structure during those stages. In the experimental campaign, these tests are denoted as white noise (WN), whereas the earthquake tests (T) represent the damaging events. Figure 9.3 presents the entire ground motion history experienced by the bridge specimen, with *WN* and *T* experiments marked with respect to the ground motion sequence. It can be observed that the earlier tests correspond to very low-intensity measures (e.g., less than 0.1 g peak ground acceleration), whereas the latter tests approach very high values, such as 1.7 g. As a result, the structural damage is subject to variation throughout the tests; for instance, the structure is intact at WN1, whereas Bent 1 and Bent 3 experience damage prior to WN2, and Bent 2 prior to WN3. All piers are subject to significant damage as WN4 takes place, and some bridge piers even experience buckling toward this state. Nevertheless, with a vibration-based model updating scheme, this gradual decline in structural stiffness can be tracked, and its impacts on residual reliability estimation can be quantified. For instance, Figure 9.4 shows the identified modal frequencies reported in [38], reducing from 2.93 Hz to 1.56 Hz throughout the shake table tests, which can also be represented with stiffness reduction in bridge piers from a model updating perspective.

Modal identification results are used to calibrate the bridge model, and an objective function parallel with section 3 is developed, accounting for modal frequencies and mode shapes. As a result, one can observe the bridge pier stiffness reduction as the model is calibrated with respect to the identification results, such as shown in Figure 9.5. Finally, the calibrated models are used to develop the fragility curves directly linking the intensity measure to damage threshold exceedance. Using the maximum likelihood estimation

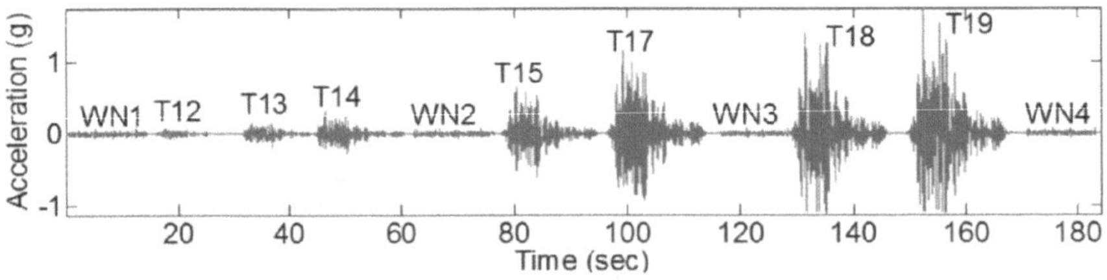

Figure 9.3 Ground motion record showing the consecutive earthquakes imposed on the bridge.

Source: [38].

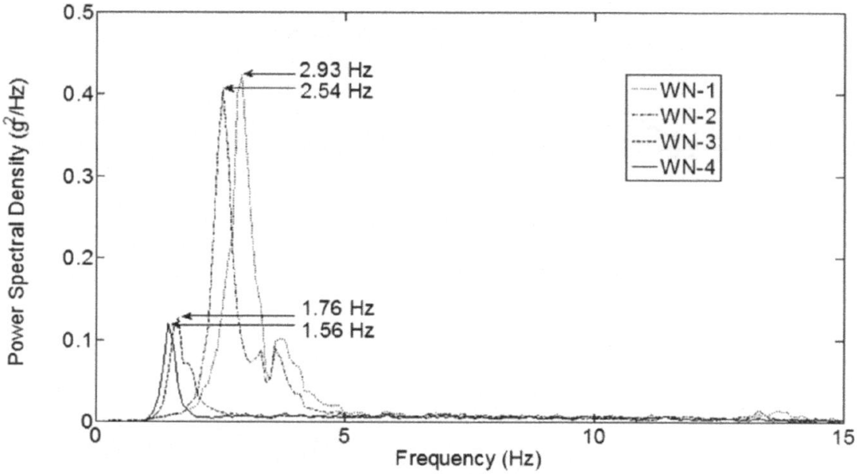

Figure 9.4 Frequency decay observed during different WN states.

Source: [38].

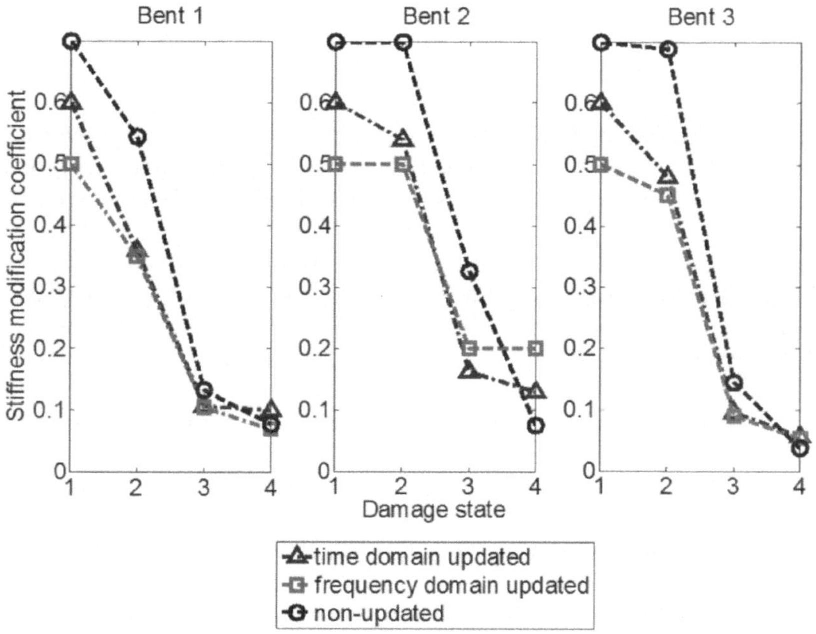

Figure 9.5 Stiffness reduction is observed on different bridge piers as different WNs take place.

Source: [38].

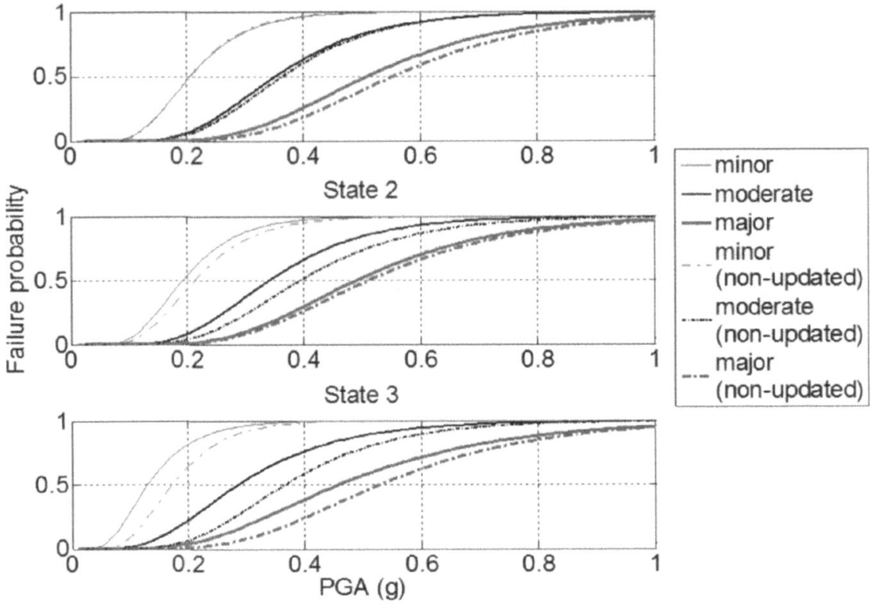

Figure 9.6 Fragility curves at different damage states, with/without updating.

Source: [38].

approach [43], Figure 9.6 presents fragility curves during different stages of the test, which imply increasing failure probabilities for given intensity measures, as the damage progresses throughout the tests. In summary, if the fragility curves show a leftward shift due to imposed damage, it corresponds to reduced residual reliability under a forthcoming earthquake.

The aforementioned process demonstrates that vibration-based SHM can reflect the actual condition of the bridge via sensors and instrumentation, and the actual condition can be used to better interpret the bridge reliability subject to natural hazards. If the bridge reliability can be better estimated together with appropriate consequence models, bridge infrastructure decisions can be based on a quantitative and objective framework, one core notion of CPSs. Moreover, changes made to the structure as a result of those decisions can also be validated/verified, or the model can be recalibrated to refer to the latest status of the structure, completing the cyber-physical loop. The power of model-driven SHM and the relevant CPS framework is that models can extrapolate the existing sensory knowledge through physics-informed strategies and engineering knowhow. For instance, the aforementioned case study is valid under the earthquake scenarios of uniform excitation, whereas, in reality, different bridge supports can experience a variation in ground motion (also called multi-support excitation).

An alternative model calibration and reliability estimation procedure can be restated while parts of the framework remain the same, showing the adaptiveness of the CPS protocol from a bridge infrastructure perspective. Ozer et al., for instance, reshape the model-driven SHM procedure to account for ground motion incoherence observed at different foundations of the bridge bents, through a Monte Carlo simulation approach [44]. In this case study, the structure was instrumented with a high-fidelity measurement platform, and the baseline models were developed using detailed design drawings and material property knowledge. Under these circumstances, the main variation in updated model parameters arose from seismic damage imposed on the bridge piers, which can be quantified through the model updating process. However, in real life, many bridges have neither advanced instrumentation nor detailed modeling dimensions. Under those circumstances, can one still develop a model-driven SHM and, therefore, a CPS strategy? To answer this question, the next case study is introduced.

9.5.2 Reliability assessment with modelling and instrumentation deficiencies (case 2)

Revisiting case study 2 introduced earlier, Ozer et al. were among the first to connect the two parallel themes, CPSs and model-driven SHM, with a novel approach for collecting bridge vibration data [37]. The traditional vibration monitoring platforms are expensive, are time-consuming, and require advanced labor, whereas novel sensor networks are empowered by everyday technologies, for example, smartphones [32]. Such technologies enable ubiquitous data resources at minimal costs to administrators, especially if collected with an administrator free strategy, that is, crowdsourcing [45]. Developing a mobile and cloud-connected application software, the authors delegated student crowdsourcers to collect vibration data from an on-campus pedestrian bridge at Columbia University, the Mudd-Schapiro Bridge, shown in Figure 9.7.

Considering a scenario in which there is an insufficient source of information to generate a model-driven SHM framework, the authors proposed a reverse engineering paradigm to retrieve bridge properties from community contributions, for example, on-site observations and smartphone-based vibration measurements. The proposed framework overlaps with the CPS motivations, where the digital and physical infrastructure environment is connected to each other via sensing and engineering computations. They intend to interact with each other through reliability-based decisions, such as retrofit intervention and associated changes imposed on the structure. According to this idealization, the cyber components are based on the advanced engineering modeling prospects (e.g., batch finite element analysis and their statistical distributions), and the physical components derive from the structure itself, measured via sensor technologies, as well as interventions/deteriorations imposing changes on the structure. These two domains are linked to each other via a calibration protocol, which complements the cyber-physical process (see Figure 9.8 demonstrating the looped behavior of the process which ideally links the sensory data with future structural states).

Figure 9.7 Mudd-Schapiro bridge testbed used for the cyber-physical SHM system use case [37].

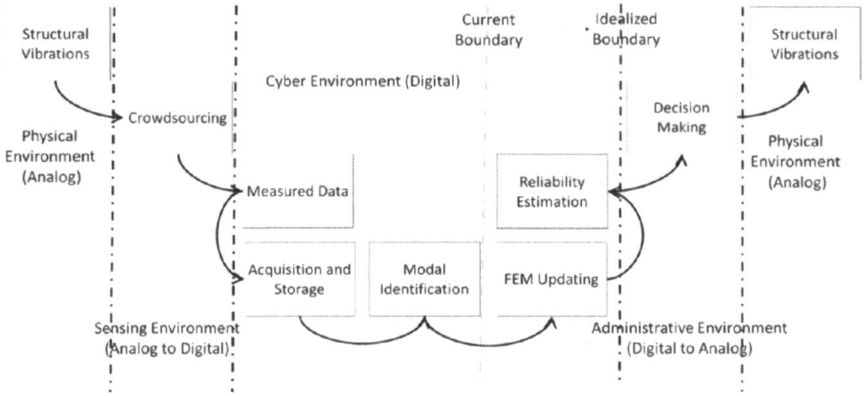

Figure 9.8 CPS process reflecting community-engaged, model-driven SHM.

Source: [37].

Ozer et al. introduced a calibration protocol to develop a baseline finite element model for the testbed bridge and considered mass stiffness features and boundary conditions as uncertain parameters influencing the bridge model [37]. Based on the parameterization of these uncertain features, they introduced an objective function incorporating modal frequencies as error indicators between the model and the reality retrieved from the measurements. Figure 9.9 presents the objective function surfaces corresponding to alternative boundary conditions and the variation of the error per mass–stiffness term combinations. While they did not attempt to compare the

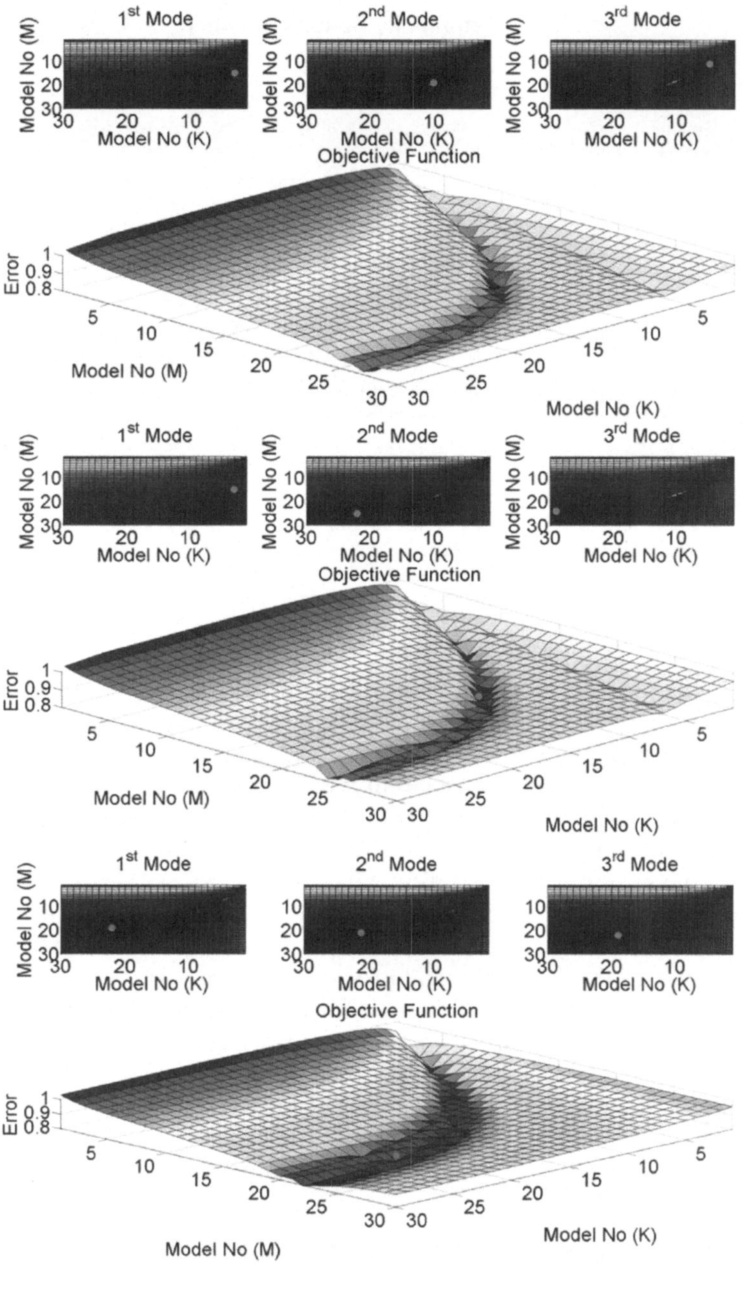

Figure 9.9 Optimizing the unknown model features based on vibration data and identified modal parameters from an administration-free perspective.

Source: [37].

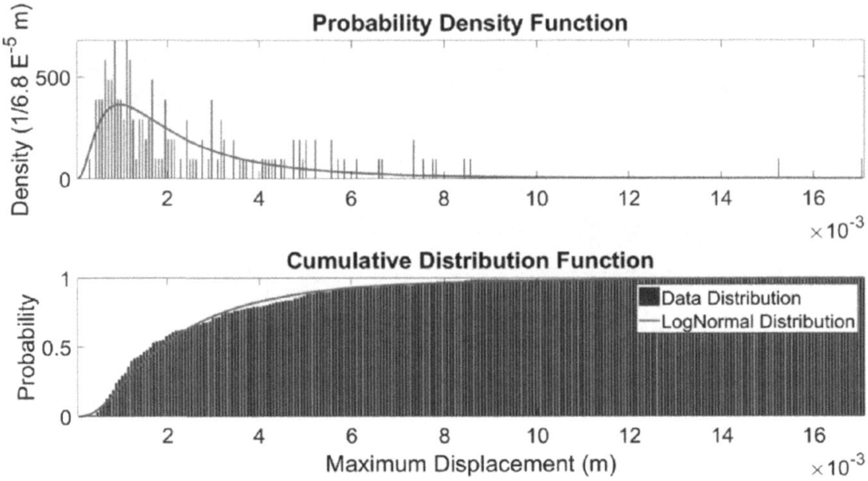

Figure 9.10 Maximum displacement demand pertaining to facade safety under a given ground motion distribution.

Source: [37].

findings with ground truth in terms of identified parameters, they demonstrated that real ground motion datasets imposed on the bridge model show a log-normal distribution for the engineering demand parameter shown in Figure 9.10, and a simplistic reliability analysis can account for the probability of glass facades being broken under a given intensity measure.

Considering suppositional deflection limits such as 0.01, 0.005, and 0.002 m, the exemplified reliability values for the bridge facade were computed as 0.99, 0.87, and 0.58 under the given ground motion dataset (indicating that exceeding a high deflection limit is less likely to occur). The authors envisage the cyber-physical process such that maintenance decisions can be taken based on the collective information gathered from the digital assets and the physical content arising from measurements. The proposed framework was an important step not only because it is a potential decision support tool completing the cyber-physical loop but also because it is scalable in terms of handling spatially distributed infrastructure, such as bridge networks, discussed in the forthcoming case study.

9.5.3 Post-earthquake transportation network decisions (case 3)

This subsection relates to the final case study introduced previously, case study 3. Model-free SHM initiatives—as opposed to model-driven SHM initiatives—are primarily data-driven, for example, anomaly detection based on a given reference healthy dataset. While model-free approaches can simplify the information workflow and expedite the detection process, they can have limitations due to the lack of physical or consequential meaning of the damage. The matter is discussed from a bridge damage detection perspective [46]. Whether model-driven or data-driven, Rytter was a pioneer in unifying the understanding of a typical SHM process through a modulated framework description through the following questions: (1) Is there any damage? (2) If yes, where is the damage? (3) And what is the extent of the damage? (4) And finally, what is the remaining useful life or, alternatively, the consequence of the damage? [47]. The ultimate step was among the least addressed, perhaps due to the differing notion of the technical challenges [48].

To expand on the least-addressed notion of the SHM paradigm, Ozer et al. introduced the reliability-driven SHM framework as a decision support tool that can monitor the conditions of distributed bridge infrastructure assets while being aware of the consequences of bridge failure [39]. For instance, a remote vicinity distantly connected to a hospital network is expected to efficiently use the transportation infrastructure following a disastrous event. In other words, the post-disaster performance of the transportation system relies on its nodes, that is, bridges that will ensure the accessibility of health services to the vicinities with emergency needs.

According to the proposed scheme, the conditions of the bridges have to be known, and such knowledge can be interpreted from a CPS perspective through model-driven SHM. If bridge vibration data is collected from the population of bridges and baseline models are developed based on reconnaissance rounds, one can better estimate how the bridges perform in case of a seismic event. In other words, with the calibrated bridge models feeding into the transportation demand-and-supply models, one can generate regional evacuation routes in the form of a decision support system. The decision support system would rely on a utility function, which is characterized by the (1) serviceability of the hospitals and their locations as sink nodes, (2) the transportation length from source to sink, and (3) bridge configurations as weak links and their systemic reliabilities constituting the hazard-prone transportation network decision metrics. A summary of the proposed framework is laid out in Figure 9.11.

To specify further, the utility term was determined based on each individual route's travel time, hospital capacity/demand per each possible destination, and bridge-impacted route reliability (to be calibrated with mobile sensor data). The route with maximum utility would govern the decision, and suppositional exercises were carried out to show different causes of decision changes. Figure 9.12 presents the transportation network modeled to perform the decision operations accounting for different seismicity scenarios. Parallel with the previous case studies, the bridge reliability dataset was generated accounting for the vibration data collected from each of the 20 bridges and model updating based on those measurements.

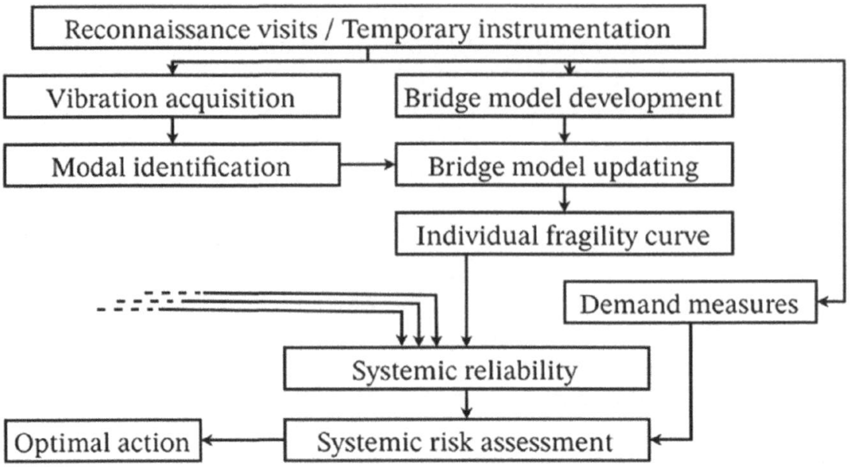

Figure 9.11 Methodological framework for the mobile SHM-integrated post-disaster transportation network decision support system.

Source: [39].

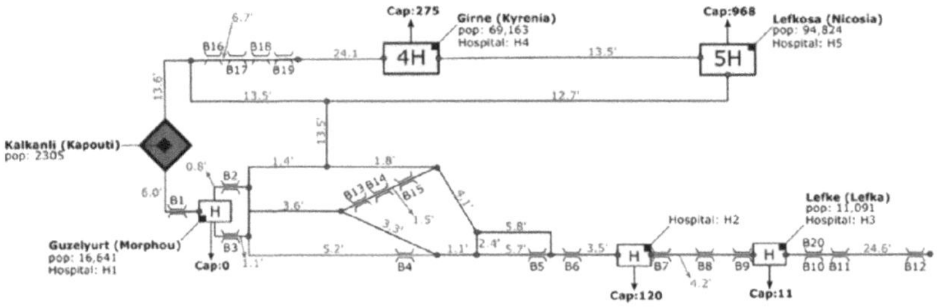

Figure 9.12 The abstract bridge network (B1–20), hospital (*H*) location and capacities, expected travel times, and the reference source node (Kalkanli) define the post-disaster route optimization problem.

Source: [39].

The reference testbed area corresponds to the western part of Northern Cyprus, and an example seismic scenario is demonstrated in Figure 9.13. In other words, seismic demand combined with the bridge fragility curves (capacity) will determine the bridge's reliability under a given seismicity scenario. Figure 9.14 presents one of the fragility curves belonging to the bridge network (Bridge 2), linking the intensity measure directly to bridge failure. Per a given intensity measure, a fragility curve (a log-normal cumulative distribution function with certain mean and standard deviation values) indicates a particular failure probability for the designated damage state.

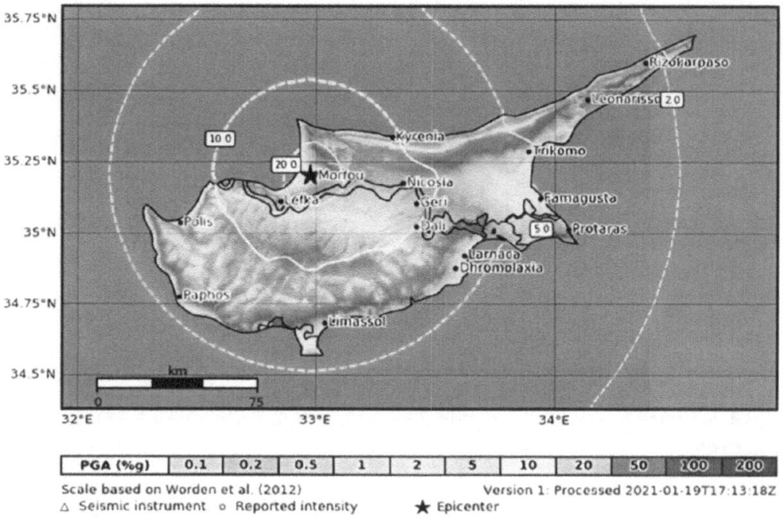

Figure 9.13 An example seismic intensity measure distribution out of 18 different scenarios; contours represent % g values, reducing with the increasing distance from the star-denoted epicenter.

Source: [39].

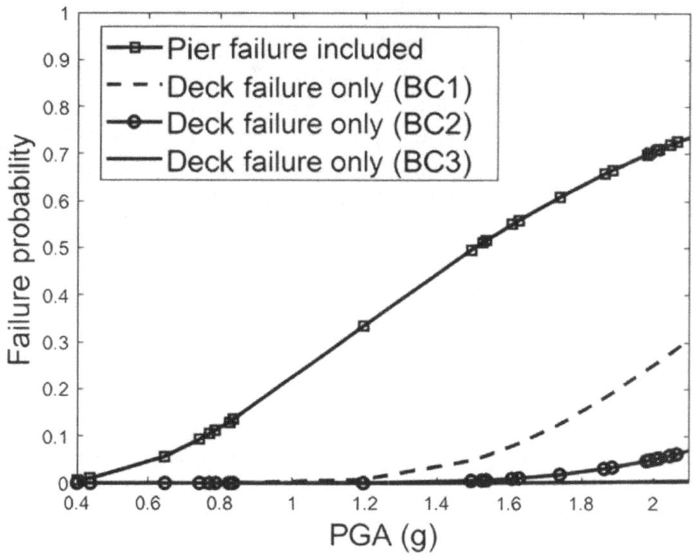

Figure 9.14 An example bridge (Bridge 2 in the 20-bridge network) fragility curve accounting for alternative boundary conditions.

Source: [39].

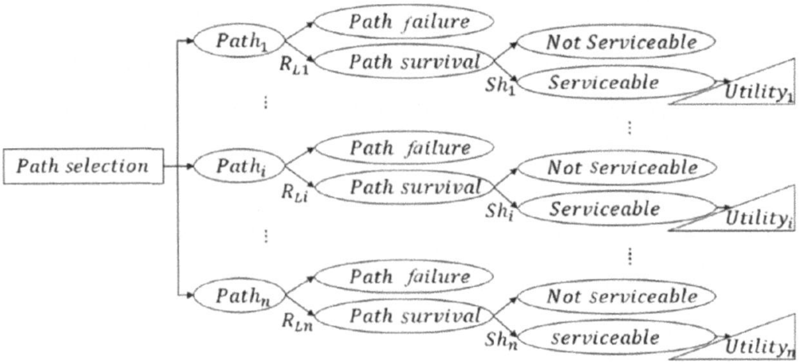

Figure 9.15 Optimal route selection and decision-making via decision trees.
Source: [39].

As the final step of a cyber-physical SHM process is depicted as decision-making, the post-disaster route selection problem can be expressed as a decision tree, where the maximum utility corresponds to the inverse of travel time to a hospital combined with the probability multipliers related to health services and bridge survival. Figure 9.15 conceptualizes the decision tree. It should be expressed that the decision herein is not expected to impact the structural state; however, network problems associated with retrofit prioritization can indicate otherwise [49].

It should be noted that CPS research consists of a broad range of technical challenges and conceptualization efforts. This chapter attempts to convey the concept from a civil infrastructure point of view with a model-driven SHM roadmap. There are various aspects of CPSs that are not addressed herein and can be found in the remainder of the book, that is, security. For further discussion on the model-driven SHM context and its relation to CPSs, the readers are referred to the reference list in [37].

9.6 CONCLUSIONS AND FUTURE WORK

In this chapter, the authors presented an interpretation of CPSs from a civil infrastructure perspective, particularly bridges. A model-driven SHM framework is integrated with a reliability estimation and decision analysis process to fulfill the evolving nature of the bridge infrastructure models linked to their physical assets via sensory data and identification processes. Structural reliability herein is a key feature that backbones the risk quantification and, accordingly, the decision optimization phases of monitored infrastructures. Combined with quantitative consequence models, such a framework can minimize human intervention in the process of infrastructure-related decisions, or at least support the process with supplementary knowledge. For demonstration purposes, three case studies utilizing a model-driven approach are taken into consideration to formulate the analogy between CPSs and the health monitoring of transportation infrastructure.

In summary, model-driven SHM is just one synergistic tool for civil infrastructure CPSs. The observations presented in this book chapter attempt to clarify the relationship between the two notions and visualize a roadmap toward further development of the proposed concept. Nevertheless, CPSs are expected to have a radical impact on future civil infrastructures through the integration of complex engineering knowledge, condition awareness, and most importantly, infrastructural intelligence in the form of automated maintenance decisions. More research in developing alternative SHM-integrated decision support systems is highly encouraged since there is still a clear knowledge gap on how stakeholder interests can better benefit from such automation initiatives.

REFERENCES

[1] Wolf, W., 2007. The good news and the bad news. IEEE Computer Magazine, 40(11), pp. 104–105.
[2] Wolf, W., 2009. Cyber-physical systems. Computer, 42(3), pp. 88–89.
[3] Baheti, R. and Gill, H., 2011. Cyber-physical systems. The Impact of Control Technology, 12(1), pp. 161–166.
[4] Sanislav, T. and Miclea, L., 2012. Cyber-physical systems-concept, challenges and research areas. Journal of Control Engineering and Applied Informatics, 14(2), pp. 28–33.
[5] Ashibani, Y. and Mahmoud, Q. H., 2017. Cyber physical systems security: Analysis, challenges and solutions. Computers & Security, 68, pp. 81–97.
[6] Zanero, S., 2017. Cyber-physical systems. Computer, 50(4), pp. 14–16.
[7] Graybeal, B. A., Phares, B. M., Rolander, D. D., Moore, M. and Washer, G., 2002. Visual inspection of highway bridges. Journal of Nondestructive Evaluation, 21(3), pp. 67–83.
[8] Gattulli, V. and Chiaramonte, L., 2005. Condition assessment by visual inspection for a bridge management system. Computer-Aided Civil and Infrastructure Engineering, 20(2), pp. 95–107.
[9] Quirk, L., Matos, J., Murphy, J. and Pakrashi, V., 2018. Visual inspection and bridge management. Structure and Infrastructure Engineering, 14(3), pp. 320–332.
[10] Seo, J., Duque, L. and Wacker, J., 2018. Drone-enabled bridge inspection methodology and application. Automation in Construction, 94, pp. 112–126.
[11] Mandirola, M., Casarotti, C., Peloso, S., Lanese, I., Brunesi, E. and Senaldi, I., 2022. Use of UAS for damage inspection and assessment of bridge infrastructures. International Journal of Disaster Risk Reduction, 72, p. 102824.

[12] Carden, E. P. and Fanning, P., 2004. Vibration based condition monitoring: A review. Structural Health Monitoring, 3(4), pp. 355–377.
[13] Farrar, C. R. and Worden, K., 2007. An introduction to structural health monitoring. Philosophical Transactions of the Royal Society A: Mathematical, Physical and Engineering Sciences, 365(1851), pp. 303–315.
[14] Brownjohn, J. M., 2007. Structural health monitoring of civil infrastructure. Philosophical Transactions of the Royal Society A: Mathematical, Physical and Engineering Sciences, 365(1851), pp. 589–622.
[15] Worden, K., Farrar, C. R., Manson, G. and Park, G., 2007. The fundamental axioms of structural health monitoring. Proceedings of the Royal Society A: Mathematical, Physical and Engineering Sciences, 463(2082), pp. 1639–1664.
[16] Lynch, J. P., 2007. An overview of wireless structural health monitoring for civil structures. Philosophical Transactions of the Royal Society A: Mathematical, Physical and Engineering Sciences, 365(1851), pp. 345–372.
[17] Worden, K. and Manson, G., 2007. The application of machine learning to structural health monitoring. Philosophical Transactions of the Royal Society A: Mathematical, Physical and Engineering Sciences, 365(1851), pp. 515–537.
[18] Farrar, C. R. and Lieven, N. A., 2007. Damage prognosis: The future of structural health monitoring. Philosophical Transactions of the Royal Society A: Mathematical, Physical and Engineering Sciences, 365(1851), pp. 623–632.
[19] Sohn, H., 2007. Effects of environmental and operational variability on structural health monitoring. Philosophical Transactions of the Royal Society A: Mathematical, Physical and Engineering Sciences, 365(1851), pp. 539–560.
[20] Malekloo, A., Ozer, E., AlHamaydeh, M. and Girolami, M., 2022. Machine learning and structural health monitoring overview with emerging technology and high-dimensional data source highlights. Structural Health Monitoring, 21(4), pp. 1906–1955.
[21] Zárate, B. A. and Caicedo, J. M., 2008. Finite element model updating: Multiple alternatives. Engineering Structures, 30(12), pp. 3724–3730.
[22] Mottershead, J. E., Link, M. and Friswell, M. I., 2011. The sensitivity method in finite element model updating: A tutorial. Mechanical Systems and Signal Processing, 25(7), pp. 2275–2296.
[23] Ereiz, S., Duvnjak, I. and Jiménez-Alonso, J. F., 2022. Review of finite element model updating methods for structural applications. Structures, 41, pp. 684–723.
[24] Steenackers, G. and Guillaume, P., 2006. Finite element model updating taking into account the uncertainty on the modal parameters estimates. Journal of Sound and Vibration, 296(4–5), pp. 919–934.
[25] Hughes, A. J., Barthorpe, R. J., Dervilis, N., Farrar, C. R. and Worden, K., 2021. A probabilistic risk-based decision framework for structural health monitoring. Mechanical Systems and Signal Processing, 150, p. 107339.
[26] Ozer, E., Feng, M. Q. and Soyoz, S., 2015. SHM-integrated bridge reliability estimation using multivariate stochastic processes. Earthquake Engineering & Structural Dynamics, 44(4), pp. 601–618.
[27] Lee, E. A., 2015. The past, present and future of cyber-physical systems: A focus on models. Sensors, 15(3), pp. 4837–4869.
[28] Mishra, A., Jha, A. V., Appasani, B., Ray, A. K., Gupta, D. K. and Ghazali, A. N., 2023. Emerging technologies and design aspects of next generation cyber physical system with a smart city application perspective. International Journal of System Assurance Engineering and Management, 14(3), pp. 699–721.

[29] Jha, A. V., Appasani, B., Ghazali, A. N., Pattanayak, P., Gurjar, D. S., Kabalci, E. and Mohanta, D. K., 2021. Smart grid cyber-physical systems: Communication technologies, standards and challenges. Wireless Network, 27, pp. 2595–2613. https://doi.org/10.1007/s11276-021-02579-1
[30] Park, K. J., Zheng, R. and Liu, X., 2012. Cyber-physical systems: Milestones and research challenges. Computer Communications, 36(1), pp. 1–7.
[31] Legatiuk, D., Dragos, K. and Smarsly, K., 2017. Modeling and evaluation of cyber-physical systems in civil engineering. Proceedings in Applied Mathematics and Mechanics, 17(1), pp. 807–808.
[32] Ozer, E. (2016). Multisensory smartphone applications in vibration-based structural health monitoring. Columbia University.
[33] Han, R., Zhao, X., Yu, Y., Guan, Q., Hu, W. and Li, M., 2016. A cyber-physical system for girder hoisting monitoring based on smartphones. Sensors, 16(7), p. 1048.
[34] Bhuiyan, M. Z. A., Wu, J., Wang, G. and Cao, J., 2016. Sensing and decision making in cyber-physical systems: The case of structural event monitoring. IEEE Transactions on Industrial Informatics, 12(6), pp. 2103–2114.
[35] Chen, X., Eder, M. A., Shihavuddin, A. S. M. and Zheng, D., 2021. A human-cyber-physical system toward intelligent wind turbine operation and maintenance. Sustainability, 13(2), p. 561.
[36] Hou, R., Jeong, S., Lynch, J. P. and Law, K. H., 2020. Cyber-physical system architecture for automating the mapping of truck loads to bridge behavior using computer vision in connected highway corridors. Transportation Research Part C: Emerging Technologies, 111, pp. 547–571.
[37] Ozer, E. and Feng, M. Q., 2019. Structural reliability estimation with participatory sensing and mobile cyber-physical structural health monitoring systems. Applied Sciences, 9(14), p. 2840.
[38] Ozer, E. and Soyoz, S., 2015. Vibration-based damage detection and seismic performance assessment of bridges. Earthquake Spectra, 31(1), pp. 137–157.
[39] Ozer, E., Malekloo, A., Ramadan, W., Tran, T. T. and Di, X., 2023. Systemic reliability of bridge networks with mobile sensing-based model updating for postevent transportation decisions. Computer-Aided Civil and Infrastructure Engineering, 38(8), pp. 975–999.
[40] Tran, T. T. and Ozer, E., 2021. Synergistic bridge modal analysis using frequency domain decomposition, observer Kalman filter identification, stochastic subspace identification, system realization using information matrix, and autoregressive exogenous model. Mechanical Systems and Signal Processing, 160, p. 107818.
[41] Saiidi, M., 2005. Large-Scale Experimental Seismic Studies of a Two-Span Reinforced Concrete Bridge System. (accessed on 15 August 2020). Available online: www.designsafe-ci.org/data/browser/public/nees.public/NEES-2005-0032.groups/
[42] Johnson, N., Ranf, R. T., Saiidi, M. S., Sanders, D. and Eberhard, M., 2008. Seismic testing of a two-span reinforced concrete bridge. Journal of Bridge Engineering, 13(2), pp. 173–182.
[43] Shinozuka, M., Feng, M. Q., Lee, J. and Naganuma, T., 2000. Statistical analysis of fragility curves. Journal of Engineering Mechanics, 126(12), pp. 1224–1231.
[44] Ozer, E., Feng, M. Q. and Soyoz, S. 2015. SHM-integrated bridge reliability estimation using multivariate stochastic processes. Earthquake Engineering & Structural Dynamics, 44(4), pp. 601–618.

[45] Ozer, E., Feng, M. Q. and Feng, D., 2015. Citizen sensors for SHM: Towards a crowdsourcing platform. Sensors, 15(6), pp. 14591–14614.
[46] Tran, T. T. and Ozer, E., 2020. Automated and model-free bridge damage indicators with simultaneous multiparameter modal anomaly detection. Sensors, 20(17), p. 4752.
[47] Rytter, A., 1993. Vibrational based inspection of civil engineering structures. Doctoral Dissertation, Aalborg University.
[48] Doebling, S. W., Farrar, C. R. and Prime, M. B., 1998. A summary review of vibration-based damage identification methods. Shock and Vibration Digest, 30(2), pp. 91–105.
[49] Malekloo, A., Ozer, E. and Ramadan, W., 2022. Bridge network seismic risk assessment using ShakeMap/HAZUS with dynamic traffic modeling. Infrastructures, 7(10), p. 131.

Chapter 10

Security and privacy in industrial cyber-physical systems

Concerns, challenges, and countermeasures

Yuchen Jiang, Jilun Tian, Shimeng Wu, Hao Luo, and Tianyi Gao

10.1 INTRODUCTION

The safe and reliable operation of industrial cyber-physical systems (ICPSs), including industrial equipment, facilities, and processes, is not only the basis for ensuring industrial production capacity but is also closely related to social safety, security, and stability. In various pillar industries such as the chemical industry, the metallurgy industry, the power and energy industry, as well as key infrastructure that involves dynamic processes, there are many new technical demands, and the advancements in basic theories and effective algorithms are the cornerstone of innovation and the indispensable prerequisite of large-scale research and development (R&D) activities. As the subjects of concern become increasingly complex and require much more scalability, scientists and engineers are devoting huge efforts to bridge the physical and digital worlds, such that both worlds/spaces are deeply intertwined, forming the backbone of Industry 4.0, the ICPSs.

In such new industrial context, safe and reliable operation is under the threat of brand-new types of cyberattacks, particularly those that lead to security and privacy concerns [1]. It is possible and is forming a large loophole. For example, attackers can tamper with the sensor data and the control commands which are transmitted over the network, resulting in performance degradation or system out-of-control. The essence of such new threat is the emergence of operational safety problems induced by information security problems. These need to be studied at the bottom system monitoring and control layer [2, 3].

10.1.1 Background of industrial cyber-physical systems (ICPSs)

The birth and evolution of ICPSs should be traced back to the convergence of several technological advancements and historical developments. It is undeniable that significant historical events and milestones shape the development of ICPSs.

DOI: 10.1201/9781003559993-10

In the late 18th to early 20th century, James Watt's invention of the steam engine (1775) revolutionized industrial processes by introducing mechanization and automation of manual labor. In this era, the foundation for automated manufacturing processes was also laid. Visionary Henry Ford is a typical example who steered the wheel of realizing assembly lines and mass production.

In the mid-20th century, the development of computing and information technology laid the necessary foundation for the later emergence of ICPSs. At this stage, computers were invented in the 1940s. In the 1960s, industrial control systems were constructed for centralized monitoring and control of industrial processes, such as supervisory control and data acquisition (SCADA). After that, programmable logic controllers (PLCs) have been widely deployed, greatly enhancing the programmability.

Near the end of the 20th century, Internet of Things (IoT) and advanced communication technologies grew rapidly. In the 1990s, the rapid development of the Internet and wireless communication technology enabled various devices and systems to interconnect. The concept of IoT lays the foundation for ICPS by seamlessly integrating physical objects, sensors, and actuators with computing systems.

The term "Industry 4.0" was coined in 2011 to describe the Fourth Industrial Revolution. The terminology emphasizes the integration of CPSs with industrial processes. Especially, it considers the mutual interactions of the latest advancements in technologies, such as artificial intelligence (AI) and cloud computing. The technological advancements have further enhanced the capabilities of ICPS, enabling real-time data analytics, predictive maintenance, and autonomous decision-making in various industries.

Although the aforementioned historical events are not comprehensive enough to elaborate the complexity in the development and evolvement of ICPS, they showcase the roles and trend of pillar technologies converging in the emergence of ICPSs. To be more specific, it can be learned from the historical development stages that the physical systems and the information/networking technologies are gradually converged to bring novel values, improving productivity, transparency, reliability, and many other favorable features [4].

10.1.2 Significance of security and privacy in ICPS

Due to the unique new features in ICPSs, security and privacy are assigned with novel meanings and challenges in the new industrial context, especially where there might be impacts induced by malicious breaches and unidentified vulnerabilities. As such, we summarize the core differences of ICPS-contexed privacy and security issues from traditional industrial scenarios or cyberattack scenarios.

First, ICPS brings together the physical and digital realms [5]. Second, ICPS implements much higher level of interconnectivity between various components and systems, which amplifies the potential impact of a security breach or privacy violation. Third, ICPSs generate and process vast amounts of data in real time, which often include sensitive information about industrial processes, production lines, and potentially, personal data of customers. Protecting the data from misuse is needed to respect the privacy and confidentiality requirements. Fourth, there is a larger attack surface for ICPSs to be exposed to external attacks. Fifth, the impact and consequences of security and privacy breaches are long-lasting in ICPSs.

In a nutshell, there are vital considerations in ICPSs to prevent attacks on infrastructure, to safeguard sensitive data, to protect user profiles, and to maintain normal operations.

10.2 FUNDAMENTALS OF PRIVACY AND SECURITY IN ICPSs

10.2.1 Definitions and concepts

Industrial cyber-physical systems (ICPSs) are integrated systems that combine physical components with computational elements and networking elements. As they are integrated systems, the core value lies in the ability to realize novel functionalities that would otherwise be very expensive or infeasible when limited to a single industry or a single discipline of study. For example, production scheduling involves complex optimization problems. An oil refining company can produce several types of final products. Each production process corresponds to certain costs and rewards. It is not hard to imagine that it is a complex problem to determine how to allocate the production power of each final product, because there are numerous factors forming the constraints of the optimization problem. We take raw material as an example: the cost of raw material is fluctuating; some rare raw materials are unable to guarantee sufficient supply, and the constituents are mixtures with different proportions of impurities. These three factors alone lead to large uncertainty and, thereby, largely complicate the production scheduling problem. How to seek solutions from ICPSs?

In terms of the supply and cost of raw material, we can establish information systems to monitor and predict the raw material supply chain and market. Traditionally, such workload belongs to the responsibility of the OT (operational technology) sectors. Now we need to leverage the power of the IT (information technology) sector to informationize and digitalize the market status so that the downstream sectors get informed, especially the production scheduling sector in this case. In terms of the difference in raw material constituents, it seems a more challenging problem, requiring online adaptive adjustment of the downstream processing technologies. In a nutshell, the cyberspace in ICPSs mainly plays the role of describing, quantizing, planning, reasoning, deducting what is happening in the real world, or the events in the past and future. On the other hand, the physical entities converge all possibilities and constitute values in real economy. From here, we can understand more about why they say "cyber-physical systems (CPSs) are the backbone of Industry 4.0." ICPSs are enabling real-time perception, dynamic control, and information services for complex industrial processes and scenarios. The increasing demands on safe and reliable operation, as well as scalable design of large-scale industrial plants against cyberattacks, have facilitated the advancement of novel approaches to anomaly detection and diagnosis by interdisciplinary and transdisciplinary collaborations [6, 7].

In ICPSs, *security* refers to the protection of critical assets, infrastructure, and operations from various threats, both physical and digital. It involves the design of strategies, adoption of measures, and taking actions to ensure the confidentiality, integrity, and availability of systems, data, and processes. For easily understanding the terms "confidentiality," "integrity," "and availability," we imagine a scenario that Company A takes the cloud storage service provided by Company B. From Company A's perspective, the cloud storage service is not secure unless (1) the stored data are not leaked to a third party, (2) the stored data are not lost, and (3) the stored data can be accessed and transferred whenever needed. From a macroscopic view, security strategies require multifaceted approaches to protect critical assets, maintain operational continuity, and mitigate potential risks. The relevant/dominant (1) *physical factors*, (2) *digital factors*, and (3) *human factors* all need to be addressed.

10.2.1.1 Physical factors

Physical factors involve the physical infrastructure of industrial systems, including manufacturing plants, production lines, warehouses, and equipment. Physical security focuses on the protection of the aforementioned entity assets. It can be realized by deploying security devices, such as video surveillance; borderline-based security systems, such as access control system; as well as advanced monitoring systems, such as intrusion detection systems [8]. On many occasions, security personnels are necessary, particularly in scenarios where making complicated decisions and being able to take responsibility are needed. As typical examples of physical security, one may consider risks related to the environment, such as temperature specifications for databases and cloud servers. That can explain why site selection for many is so special: inside the mountains (Huawei) or at the bottom of the sea (Microsoft).

10.2.1.2 Digital factors

Cybersecurity involves protecting the digitalized, networked, interconnected systems that operate on "bits." The space of bits suffer from unauthorized access, cyberattacks, data breaches, and disruptions [9]. These types of vulnerabilities can be mirrored to and associated with similar counterparts of physical attacks. The correspondences of whether these misbehaviors have impact on the three key security elements are shown in the following.

In recent years, the governmental and academic sectors have devoted much effort in helping maintain a secure environment in the cyberspace. The network security monitoring system can detect and trace abnormal traffic and suspicious transactions. This needs close collaboration with telecom operators and service providers of online platforms so that they can leverage the power of big data while, in the meantime, realize precise tracing and localization. On the other hand, as for the ICPS-based smart factories and industrial plants equipped with networked intelligent devices, they are newly connected and share information with front-end departments, such as purchasing departments, marketing departments, and planning and finance departments. In other words, OT sectors are bridged to the IT sectors. As a result, the cyberspaces of the enterprises are expanded, introducing novel security issues. It is worth noting that, still the best practice for many scenarios, physical isolation, which cuts off every possibility of external unauthorized parties from internal industrial production sites, is still the primary strategy to ensure security. However, in such context, security is achieved by sacrificing connectivity and transparency.

Now, considering the fact that an elevated degree of connectivity is a prerequisite for ICPSs to realize most of their functions and is really the foundation of interoperability, real-time perception, distributed computing, and high-fidelity simulation (e.g., digital twin), one needs to get prepared for the long-term defense and mediation with malicious parties (attackers) using some countermeasures [10]. At the network communication layer, some popular solutions include implementing network firewalls, deploying intrusion detection systems, and regularly updating security patches. At the monitoring and control layer, apart from built-in security design for industrial control

Table 10.1 Influence of digital misbehaviors on key security elements

Elements / *Misbehaviors*	*Confidentiality*	*Integrity*	*Availability*
Unauthorized access	O		
Cyberattacks	O	O	O
Data breaches	O		
Disruptions		O	O

system (ICS), such as supervisory control and data acquisition (SCADA) and distributed control systems (DCS), it also realizes functions of access controls, authentication, encryption, and network monitoring using intelligent algorithms [11].

10.2.1.3 Human factors and other factors

Human factors can never be a negligible factor in industrial system security. Employee awareness and training shall constitute the first lesson before one takes up confidential tasks or gets in touch with sensitive information. Training employees on security best practices, recognizing social engineering attempts, and promoting a security-conscious culture are vital. This includes educating employees about password security, phishing awareness, physical security protocols, and the responsible use of company resources.

Compared with external remote attacks, internal attacks may lead to fatal failures through the core control units, causing a major destruction to the facilities and processes. The Maroochy water service (Australia) went through a dark time in 2000. A disaffected employee of the SCADA system installation service provider attacked the wastewater treatment control system and gave unauthorized instructions. Around 800,000 L of sewage leaked to local parks and rivers. We recommend interested readers to refer to the article by Jill Slay and Michael Miller, "Lessons Learned from the Maroochy Water Breach."

The concepts of privacy and security are closely intertwined and interdependent. They are both concerned with the protection of sensitive information, assets, and industrial process operations. Among these tasks, *privacy* shows a distinction regarding the protection of sensitive information to maintain confidentiality, data anonymization, and secure data storage. As such, the focus is on the appropriate use, collection, and handling of these information, safeguarding them from unauthorized parties [12]. On the other hand, privacy and security are complementary and mutually supportive. Privacy often relies on security practices to protect sensitive data from unauthorized access or breaches. In turn, reliable security practices contribute to privacy preservation.

10.2.2 New security and privacy challenges in ICPSs

The security of ICPSs faces multiple challenges. Firstly, the increased connectivity of devices to the network/Internet has introduced new vulnerabilities. The networked systems become potential targets for network attacks, where the attack surface is expanded and the possibility of unauthorized access or manipulation is increased. Secondly, many components of ICPS were implemented before security became a major issue. Legacy systems may lack dedicated design and fail to implement strong security mechanisms. As a result, there are vulnerabilities that can be exploited by malicious parties. Thirdly, the integration of information technology (IT) and operational technology (OT) breaks the traditional "isolation principle," where the OT systems must be physically isolated from the IT systems. Such integration blurs the boundaries between previously separated areas. Fourthly, malicious attackers today tend to make better preparation and get involved in long-term attacks [13, 14]. They may be equipped with good knowledge about the underlying mechanisms of the process operation and utilize the knowledge to launch concealed attacks that would not trigger alarms. In other words, a new target of the attackers is at the monitoring and anomaly detection systems [3].

At the same time, new challenges related to privacy have also emerged in ICPS. On one hand, ICPS conducts extensive data collection, leading to concerns about the types of data collected and their use, leading to privacy issues. On the other hand, when data is shared and integrated among various stakeholders, the lack of appropriate controls and protocols may bring privacy risks. In addition, the data collected by ICPS is very detailed and may disclose sensitive information about individuals, processes, or operations. Therefore, a balance must be struck between data value, protecting personal privacy, and preventing unauthorized access or inference of confidential information.

10.3 THREATS IN INDUSTRIAL CYBER-PHYSICAL SYSTEMS

10.3.1 Common threats, vulnerabilities, and potential risks

Figure 10.1 illustrates the connections between common threats and vulnerabilities and the corresponding sources and consequences. It can be learned that the potential consequences of a malicious party manipulating physical equipment in an ICPS environment are serious.

For example, unauthorized access can pose risks to critical infrastructure, service disruption, and physical harm. Unauthorized access can be caused by supply chain risks and compromised ICPS components from different suppliers. Also, the assembly process and the distribution process may be compromised during manufacturing, when back doors could be introduced into the system. Hence, strong supply chain security practices are required to mitigate such risks. Unauthorized access can also be caused when lacking standardized security measures. When different devices and different components have inconsistent security measures, it creates weaknesses that attackers can exploit. Furthermore, interconnected networks such as IoT (Internet of Things) and cloud infrastructure can lead to unauthorized access. They bring new entry points for cyberattacks. As we mentioned earlier, a single vulnerability can threaten multiple components and systems in ICPS, which has a cascading effect.

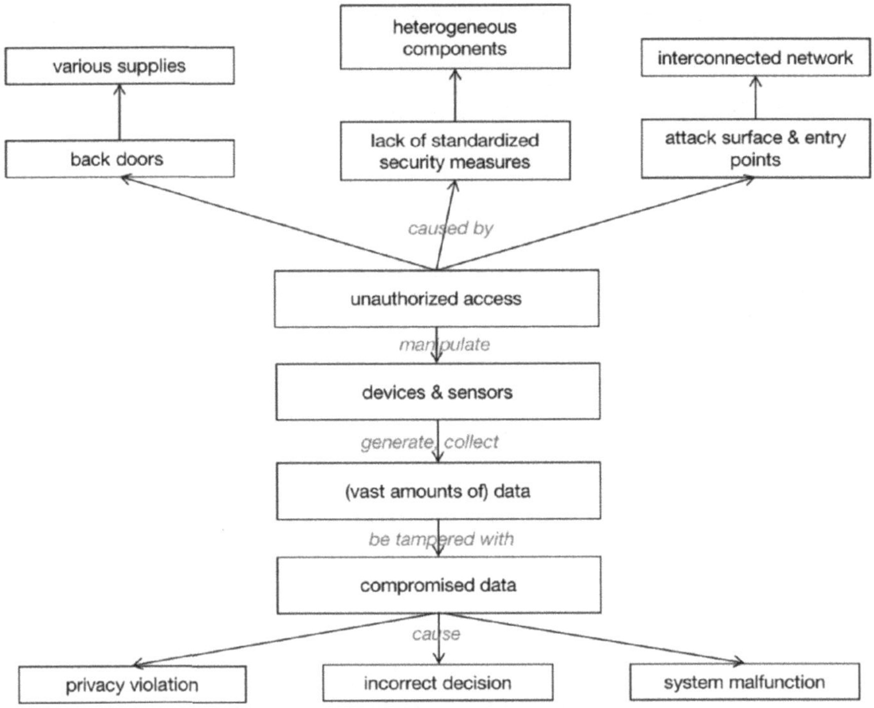

Figure 10.1 Relationship of common threats, vulnerabilities, and risks in ICPSs.

Tampering with the data collected by sensors or transmitted through the network can lead to incorrect decisions, system malfunctions, and privacy violation [15]. Data privacy issues are concerned with the vast amounts of data generated and collected by sensors and devices, which often include patterns [16]. Therefore, when the data are inadequately protected, it risks data trustworthiness.

The role of human factors in privacy and security risks within ICPS is indispensable. Human error, intentional actions, improper configurations, and inadequate training on system security all have impact on the links in the diagram [17].

10.3.2 Attack surfaces in ICPSs

The attack surfaces in ICPSs refer to potential vulnerabilities and entry points that can be exploited by malicious parties to compromise the security and functionality of these systems. The following summarizes some key attack surfaces (Table 10.2) and the countermeasures (Table 10.3).

Table 10.2 Categories of attack surfaces and the characteristics

Category \ *Characteristics*	*Cause*	*Consequence*	*Examples*
Physical access points	Left unprotected or unmonitored	Directly connect and compromise the ICPS	USB ports; serial interfaces; maintenance ports
Network interfaces	Vulnerabilities in network protocols; inadequate access controls; weak encryption; misconfigurations	Gain unauthorized access; perform network-based attacks; intercept sensitive data	Wired and wireless connections
Human–machine interfaces (HMIs)	Weak authentication mechanisms; insecure communication channels; inadequate user training	Manipulate or compromise HMIs to disrupt or modify the behavior of the ICPS	Head-up displays; wearable glasses
Control system software	Logic flaws or insecure coding practices; untimely update and patch of the control system software	ICPS operation being manipulated or disrupted	Stuxnet virus targeting at SCADA–Siemens WinCC
Firmware and hardware	Insecure firmware updates; back doors; weaknesses in hardware design	Data breaches; system downtimes; safety risks	Stuxnet that damage Iran's nuclear program
Integration points	Inadequate access controls; insecure APIs; vulnerabilities in integrated systems	Gain unauthorized access to the ICPS	Enterprise networks; third-party applications; cloud services
Supply chain	Insecure supply chain practices, for example, counterfeit components, compromised suppliers, inadequate verification	Compromise the integrity of the data, components, or software; trust erosion	SolarWinds attack; IoT devices

10.3.3 Real-world examples of security breaches

Security breaches are not imaginary risks or something invented by the academia [18–20]. By contrast, there are ongoing attacks in real industrial plants and production sites, albeit many are not (and will not be) disclosed to the wide public due to various strategic considerations. Nevertheless, we attempted to collect several world-shocking events related to security flaws.

Table 10.3 How to identify and mitigate the risks associated with the attack surfaces

	Method	*Description*
Identification of attack risks	Regular security assessments	Systematically scheduling evaluations of cybersecurity measures to proactively identify vulnerabilities and attack risks in systems, applications, and processes
	Penetration testing	Performing a controlled, simulated cyberattack on systems and networks to identify vulnerabilities and assess susceptibility to real-world threats
	Threat intelligence	Identifying attack risks by staying informed of evolving threats and vulnerabilities, by collection, analysis, and dissemination of information about current and potential cyber threats
Mitigation of attack risks	Access controls	Restricting and managing user permissions, limiting access to sensitive data and systems
	Network segmentation	Dividing large-scale networks into smaller, isolated segments, preventing lateral movement of attackers and minimizing the secondary impact
	Strong authentication mechanisms	Adding an extra layer of security by requiring users to provide multiple forms of verification, reducing the risk of compromised credentials
	Encryption	Transforming data into an unreadable format, ensuring that even if data is stolen, it remains confidential and secure
	Regular patching	Addressing known vulnerabilities
	Security monitoring	Detecting and responding to suspicious activities in real time by continuous surveillance
	Employee training	Educating personnel about cybersecurity best practices, raising awareness and reducing the risk of social engineering attacks and human errors

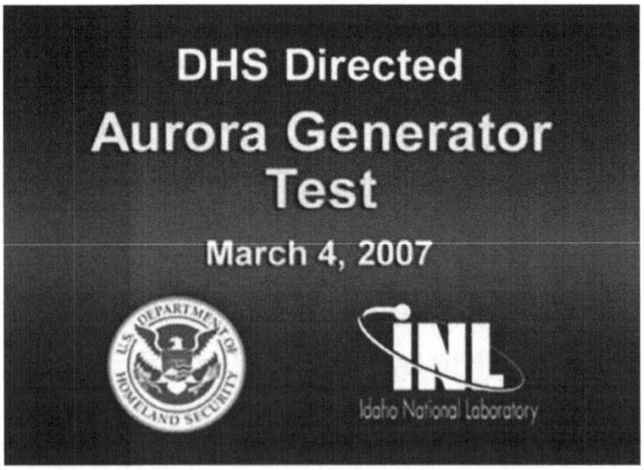

Figure 10.2 The first malicious code to make damage to devices in lab experiments.

Source: Key frames extracted from a video source from the Internet.

Figure 10.2 (Continued)

Figure 10.2 (Continued)

Figure 10.2 (Continued)

Case 1: Aurora Generator Test (2007). It was a test on a widely used generator in the industry. It was carried out by the Idaho National Laboratory in 2007. Hackers injected malicious codes to the machine, and after a while, the generator vibrated greatly. Some parts bounced off. We can see black smoke coming out of the machine. It was finally destroyed by a few lines of erroneous codes. This test is claimed as a very early malicious code that targets making damage to industrial devices in lab experiments (see Figure 10.4).

Case 2: Stuxnet (2009–2010). Stuxnet targeted Iran's nuclear program and specifically aimed at disrupting the centrifuges used for uranium enrichment. Stuxnet exploited multiple vulnerabilities, including zero-day exploits, to infect the control systems and manipulate the physical processes, causing significant damage. The event shocked the industry safety and security domain as the very first malicious code that led to real damages in physical systems. The virus quickly infected the PC in 2010. Upon February 2011, 20% of the centrifuges were infected. More than 1,000 uranium enrichment facilities in the Natanz and many more in the Bushehr nuclear power plants were affected. The centrifuges became uncontrollable and caused explosions (see Figure 10.5).

Case 3: Ukraine power grid Attack (2015 and 2016). In December 2015 and December 2016, Ukraine experienced two separate cyberattacks

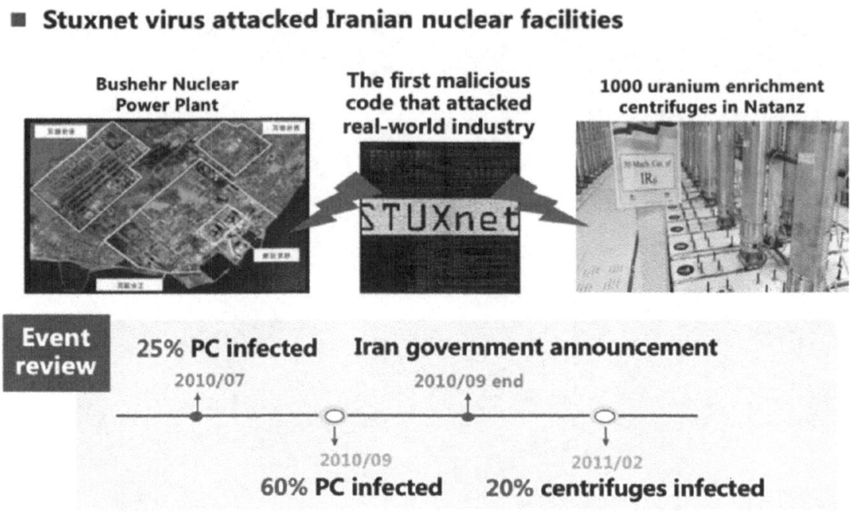

Figure 10.3 The first malicious code that targeted and made damage to real-world deployed industrial facilities.

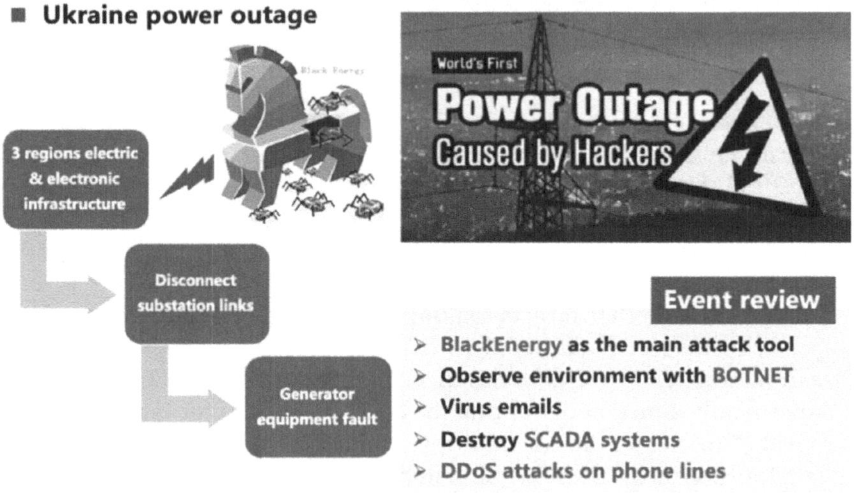

Figure 10.4 The world's first large-scale power system failure caused by cyberattacks.

that resulted in power outages. The attackers gained access to the ICPS infrastructure and remotely manipulated the control systems, leading to widespread blackouts. These incidents highlighted the vulnerability of critical infrastructure to cyberattacks. Before Christmas 2015, hackers attacked the SCADA of a power distribution company. It caused six hours of power outage that affected over 200,000 citizens. This time the attack tool (BlackEnergy) is a super Trojan that supports pluggable malicious components. Botnet was used to collect information and observe the environment. In the meantime, phone lines were also under DDoS attacks. It caused more time for engineers to get notified about the situation (see Figure 10.6).

Case 4: Triton/Trisis (2017). The Triton/Trisis attack was launched on a petrochemical plant in the Middle East. It targeted the controllers of the safety instrumented system (SIS), which are responsible for shutting down processes in the event of hazardous conditions. The malware attempted to reprogram the SIS controllers, putting workers at the scene and the surrounding environment at risk.

Case 5: Colonial Pipeline ransomware attack (2021). In May 2021, the United States declared a state of emergency due to a cyberattack on its largest fuel pipeline provider, Colonial Pipeline, which had to shut down key fuel supply networks, affecting gasoline and diesel supplies on the East Coast of the United States.

10.4 PRIVACY CONSIDERATIONS IN INDUSTRIAL CYBER-PHYSICAL SYSTEMS

In ICPSs, collecting more data than necessary, or keeping collected data longer than necessary for the intended purposes, can lead to privacy concerns. These practices increase the risk of misuse and the leakage of sensitive information. Also, during data collection processes, insufficient security measures also bring threats to the privacy-preserving demand from the customers and the enterprises. For example, in a manufacturing scenario, industrial robots and automation systems are used. If privacy measures are insufficient, it can lead to data misuse and eavesdropping by external attackers, such as illegal disclosure of sensitive and personal data to the public. The section discusses privacy protection considerations in the context of ICPSs, with an overview and analysis of the existing privacy protection methodologies and techniques for industrial systems and data. It should be noted that in practical applications, these schemes do not normally operate in isolation, but rather, a systematic combination of these schemes is realized to make use of the strengths from different aspects.

10.4.1 Data collection and privacy concerns

Data collection is a critical process to acquire data information using sensors and measuring devices [21]. Nowadays, intelligent data collection has attracted focuses on academia and industry due to the increasing needs to integrate complex industrial systems with advanced sensors and network technology. Among them, industrial Internet of Things (IIoT) technology leverages real-time data on equipment, production, and environment to add values for intelligent production and maintenance [22]. Meanwhile, the integration of AI enables multi-source, multi-mode data processing and automatic analysis to provide trustworthy data and high-quality distorted data for subsequent uses.

Data collection is the very first stage, where the physical world is perceived, thus acting as the basis of data-driven design, recalibration, synchronization, optimization, monitoring, and control, as well as the necessities to train powerful AI models [23]. Meanwhile, some related privacy issues are worthy of consideration, including personal privacy breaches, data security, and data abuse issues. In ICPSs, the involved data often contains trade secrets and personal information, such as production plans, employee salaries, etc. If these data are leaked, it will bring huge losses to enterprises and individuals [24]. Besides, because the collected data in ICPS may involve frequent and large-scale transmission and storage, it is also crucial to ensure that these critical processes are not stolen or tampered with. In the context of ICPSs, some best practices for the privacy preservation of data collection include data minimization, informed consent, anonymization and pseudonymization, encryption, secure data storage and transfer, privacy by design, user control, and so forth. More details and key ideas will be introduced in the following section.

10.4.2 Privacy-preserving techniques in ICPS

ICPSs collect and process raw industrial data through directional and bidirectional communication, dispatching commands, and sending and receiving terminal data. However, real-time industrial data contain large amounts of sensitive and private information about industrial operations, which can lead to serious impacts on physical infrastructure, operational continuity, sensitive data, and public safety in the event of leakage. Malicious attackers can conduct eavesdropping and advanced data analysis techniques, such as those combined with the spatial and temporal correlation, to peek on the operating status of industrial systems, operation, work tasks, and other confidential content.

To mitigate illegal access to private data in industrial systems, Aggrawal and Srikan introduced a new research area in 2008 [25]. During the research, privacy-preserving technology is considered as a set of methods to modify, transform, distribute, and hide information, which are used to prevent exposing the original data during processing [26], and to protect confidential data from unauthorized users while processing it on the network. The difficulties encountered in the research of this technology mainly include the complexity of industrial systems, the strong coupling, and the quantitative problem of privacy assessment [27]. The main solutions to this problem today include encryption-based [28, 29], perturbation-based [29, 30], differential privacy (DP) [31], authentication-based [32], and blockchain-based ones.

Table 10.4 Pros and cons of the main solutions for privacy preservation

	Pros	*Cons*
Encryption-based	Wide applicability on algorithms	Possibility of data leakage and attacks, high computing resources, and limited computing functions
Perturbation-based	Wide applicability on algorithms	Reduction of data quality
Differential-based	Strong privacy protection, good flexibility, and wide applicability of algorithms	Reduction of data quality, high cost of privacy protection
Authentication-based	High efficiency, controllability, and flexibility	Complex authentication, higher maintenance costs, and possibility of verification vulnerabilities
Blockchain-based	Decentralized, secure, reliable, and tamperproof	Difficulty in privacy protection, high consumption of storage space and computing resources, scalability issues

10.4.2.1 Encryption-based privacy preservation

Encryption-based privacy preservation is a method of protecting data privacy by encrypting data to protect its confidentiality and prevent unauthorized access and leakage [28]. It is typically used to preserve data privacy during data sharing and processing. Encryption methods can be divided into encrypted data and computation for processing objects. Encrypted data can be stored on some untrusted storage devices, and only authorized users can access and decrypt relevant data. Encrypted computing allows data to be encrypted and transmitted to untrusted computing devices for processing, and only authorized users can access and decrypt the calculation results. Therefore, this method can help organizations and individuals protect privacy when sharing and processing sensitive data, thereby improving data security and protecting personal privacy.

Homomorphic encryption tools play an important role in data analysis scenarios. They allow customers to use data for analysis while ensuring the security of their private data. In this process, customers encrypt their private data using homomorphic encryption algorithms. The encrypted data is then sent to the third party, who can only analyze the encrypted data and return the encrypted analysis results. This means that the third party cannot access the customer's original private data.

On the customer side, the encrypted analysis results can only be decrypted using a private key, thus obtaining the final analysis results. This ensures that the analysis results can only be decrypted by the customer, maintaining control over their own data.

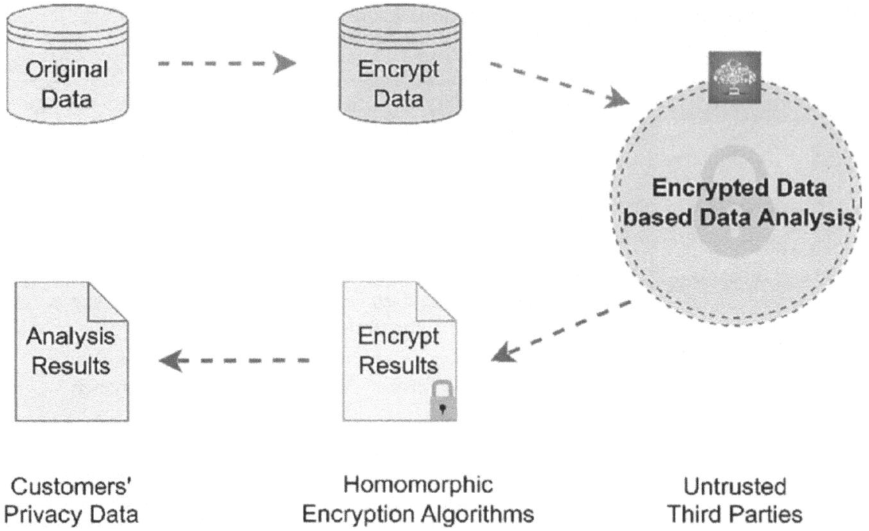

Figure 10.5 The procedure of encryption-based privacy-preserving method.

10.4.2.2 Perturbation-based privacy preservation

Perturbation-based privacy preservation is a commonly used privacy protection technique that protects data privacy by making small perturbations to the original data [29]. By utilizing perturbations such as noise, random signals, and encryption, the original data is transformed into a new format, and while maintaining the statistical characteristics of the data, attackers cannot infer the original information from the perturbed data.

By protecting data privacy through perturbation and keeping statistical features unchanged, the perturbed data can still be analyzed and applied. Therefore, this method can be flexibly used in various scenarios. However, data quality can be affected by disturbances, which can reduce the performance of data analysis. The main challenge of these methods is to balance privacy protection and data utility, that is, to balance data privacy and quality. However, perfect privacy protection cannot coexist with perfect data utility, and attackers can still query disturbed data through multiple visits, inferring deep statistical information of the data, thus limiting protection. Nevertheless, due to its flexibility, it still has certain research value and application scenarios.

10.4.2.3 Differential privacy preservation

A series of privacy protection strategies has been introduced to overcome privacy threats in various situations. However, due to the limitations of the computing power of sensors, real-time encryption is difficult to apply in practice [33]. In public key cryptography (also known as asymmetric cryptography), the generation and distribution of public and private keys are a computationally complex task that is not easily accomplished on small devices with limited resources. It also requires the interconnection of network nodes. In the event of an interconnection failure, the lack of keys also makes data collection more difficult. Anonymization protection strategies cannot fully protect data privacy, and these weaknesses have also been demonstrated in many applications. The researchers study three months of credit card records for 1.1 million people and show that four spatiotemporal points are enough to uniquely reidentify 90% of individuals [34]. We show that knowing the price of a transaction increases the risk of reidentification by 22%, on average. Finally, we show that even datasets that provide coarse information at any of or all the dimensions provide little anonymity, and that women are more re-identifiable than men in credit card metadata.

The main focus of privacy protection is on databases, and relying on anonymization technology to protect complete data is often incomplete. In contrast, differential privacy has become a feasible solution by developing a sound theoretical framework based on privacy requirements [35]. The strategy of differential privacy usually involves differential identifiability and membership privacy. Differential identification enables the databases to obtain the maximum difference between any two data for a certain query, and membership privacy specifies the number of data queries. These contents, as important elements of differential privacy, make it difficult for attackers to determine whether an individual has participated in the database. Therefore, it can ensure that any query result does not display any sufficient information that could lead to its identification.

In differential privacy, based on the addition of random perturbations within noisy and especially processed data, it is difficult to determine which data belongs to which individuals, and it cannot confidently infer sensitive information of any dataset, thus protecting individual privacy. Meanwhile, it ensures that the perturbation calculation of specific data will not undergo substantial changes after the original data is updated. The differential privacy method uses effective statistical methods, such as Gaussian and Laplace mechanisms, to prevent inference and data poisoning attacks, thus ensuring complete privacy.

A typical industrial application of differential privacy is gradient encryption in federated learning. Since the original samples can be reconstructed through gradient data and model parameters, the gradient data transmitted between federated clients and federated servers is also part of private data. In federated learning algorithms, the differential encryption module is commonly used to project and encrypt the gradient data of a local federated learning worker. After the federated learning master performs fusion on the encrypted gradient data, the reconstructed gradient data is obtained using the differential encryption module and used for local model parameter

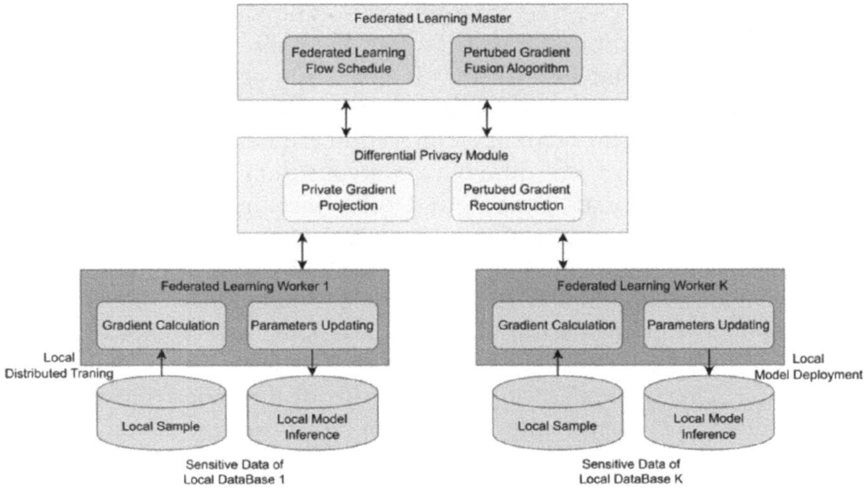

Figure 10.6 A gradient encryption method in federated learning for differential privacy preservation.

updating. The core metric for evaluating the differential encryption module is accuracy, which measures the amount of loss in accuracy after encryption and reconstruction.

10.4.2.4 Authentication-based privacy preservation

The authentication-based methods rely on system authentication mechanisms to protect data and system privacy, such as single sign on, federated identity, and key management [32]. It protects data privacy by initially setting up authentication to authorize users to access data, making it inaccessible to unauthorized visitors.

Due to the characteristics of identity verification, it has high efficiency, controllability, and flexibility. It can quickly achieve privacy protection and precise control based on authorization registration and has a wide range of application scenarios. However, identity verification itself is complex and requires the system to pay more costs to maintain, while also having drawbacks, such as verification vulnerabilities. While implementing this method, it is also necessary to ensure the personal information of the visitor.

In federated learning, authentication-based privacy preservation also has important applications. In a federated learning architecture, thousands of clients typically request to participate in federated computations, but there may be untrusted clients among them. In this case, clients need to register with the server through an authentication mechanism before participating in federated computations.

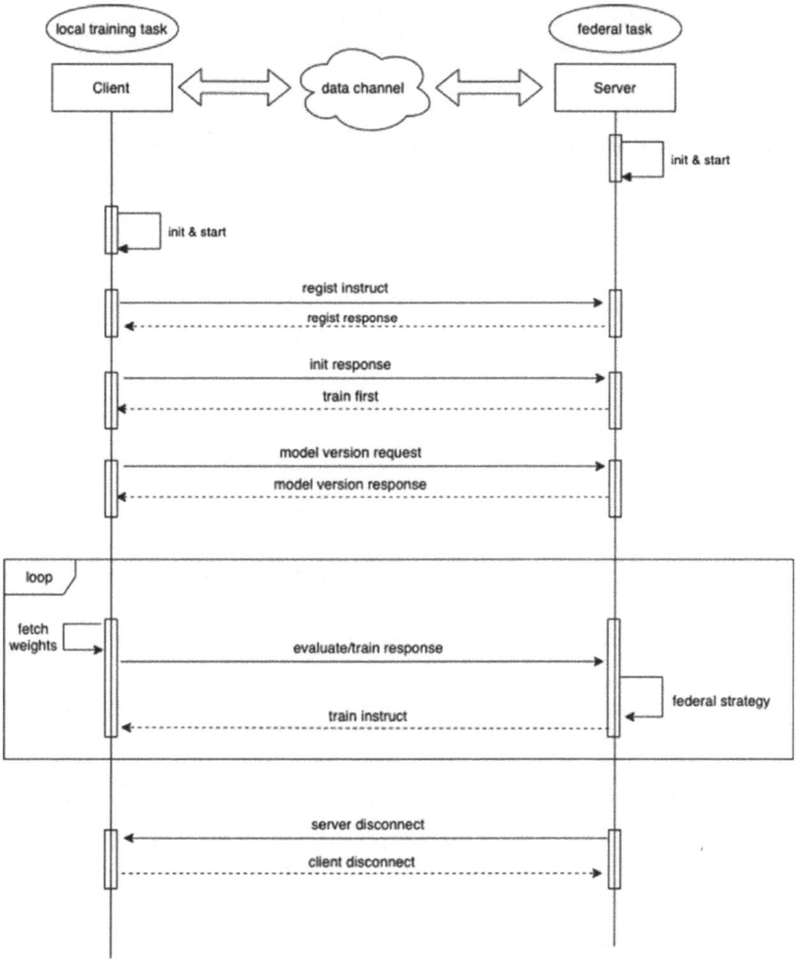

Figure 10.7 A detailed framework in federated learning of authentication-based methods.

The authentication mechanism ensures that only authorized clients are allowed to join the federated learning process. It helps prevent unauthorized or malicious clients from accessing sensitive data or interfering with the federated computation.

10.4.2.5 Blockchain-based privacy preservation

Blockchain is a decentralized, secure, reliable, and tamperproof distributed database technology. Unlike traditional centralized databases, blockchains do not require a centralized institution to manage the data, which eliminates the need to assume the security and reliability of the data. The decentralization of data is achieved through blockchain technology, which stores information across multiple nodes. This eliminates the possibility of single points of failure and attacks. Cryptographic algorithms are also employed to ensure data security and comparability. The fundamental concept of blockchain technology involves recording data in a block and linking these blocks together to create a chain, hence the name "blockchain." Each block comprises transaction records and a pointer to the previous block, forming an immutable chain of transaction records. Initially used for recording transactions in digital currencies like Bitcoin, blockchain technology is now widely implemented in industries such as finance, logistics, health care, and energy.

The blockchain-based privacy protection approach applies the concept of blockchain, which is a peer-to-peer encrypted connection for protecting network nodes or data transactions [36]. Peers come from a distributed network in which each peer operates as a node of the network and can contribute to the computation of a solution to a hash-based puzzle problem to confirm the integrity of the transaction. Each transaction record is compressed into the existing blockchain as a block. The recorded block content is considered a ledger. The entire block is synchronized and updated across the network so that each peer retains a record of the same ledger [37].

Two popular mining techniques, proof of work (PoW) and proof of stake (PoS), have been applied to Bitcoin and Ether, respectively, to verify the legitimacy of transactions within a block and add new blocks [38]. To solve this challenge, PoW miners rely on computational power, while PoS uses deterministic algorithms and sometimes applies hard forks to lose some blocks [39]. However, both techniques can be violated if a malicious miner's computer has more than 51% of the network's computational power; this is known as a 51% attack [39].

10.5 SECURITY MEASURES FOR INDUSTRIAL CYBER-PHYSICAL SYSTEMS

10.5.1 Access control and authentication

Access control is a security technique that regulates access to resources. *Authentication* is a critical security measure that involves verifying the identity of a user, device, or process [40]. Together, these techniques grant access to resources within an information system, ensuring confidentiality and restricting access to data sources, which form the foundation of a secure information system, safeguarding against unauthorized access and ensuring data privacy [41].

10.5.1.1 Access control

Access control mechanisms come in various forms, including discretionary access control (DAC), mandatory access control (MAC), role-based access control (RBAC), and attribute-based access control (ABAC). DAC allows the object owner to set rules, which can be done through an access control list (ACL) or access control matrix [42]. MAC, on the other hand, controls access to objects and subjects based on their levels, making it useful in highly restricted environments, where a central authority makes access decisions [42]. RBAC regulates subject access by their designated role and the rules outlining permissible access for each role. This simplifies the process of monitoring user permissions and granted access, since a few roles can represent numerous users. Lastly, ABAC evaluates control rules against attributes such as subject, object, action, and environment, making it ideal for fine-grained access control [43].

10.5.1.2 Authentication

Various studies have been conducted to determine the most effective methods of identity authentication for industrial cyber-physical systems. Based on these studies, four main categories of authentication have been identified [44]. The first is knowledge-based authentication, which entails the use of user IDs and passwords to verify an individual's identity. The second is possession-based authentication, whereby authentication is based on the user's credentials, RFID, or other unique identifiers. The third is inherence-based authentication, or biometric-based authentication, which utilizes biometric features, such as fingerprints or iris data, to authenticate an individual's identity [45]. Lastly, multifactor authentication, which combines two or more of these methods for added security, has been identified as the most comprehensive form of identity authentication [46].

10.5.1.3 Issues and blockchain-based solutions

Traditional access control and authentication methods rely on a central authority for request assessment, which can be a single point of system failure. Moreover, these methods necessitate the implementation of control servers and trusted third parties, which lead to additional expenses in terms of hardware and deployment. Furthermore, these schemes only address identity authentication and access control within a specific domain. Various systems have different certificate forms and key management methods, leading to notable application isolation, which poses a challenge to achieving cross-domain connectivity.

As a new generation of security protection technology, blockchain can be used to address these issues due to its own characteristics, shown in Figure 10.10, including immutability (i.e., any confirmed transaction cannot

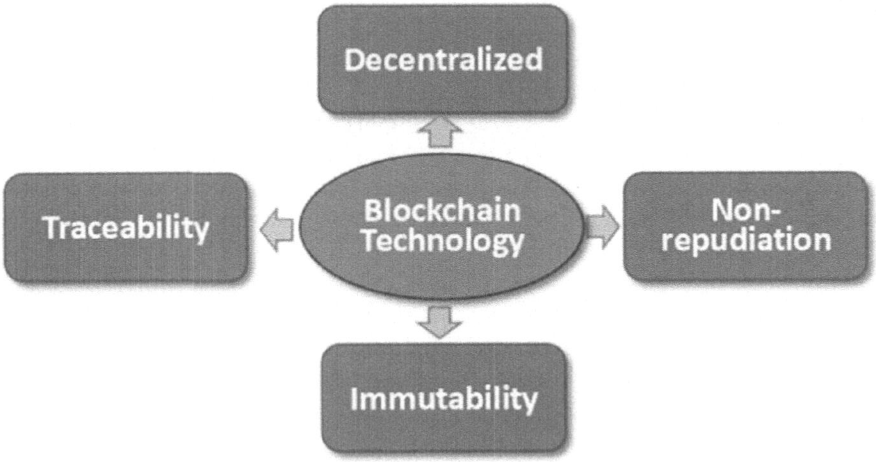

Figure 10.8 Advantages of blockchain in access control and authentication.

be altered), decentralization (i.e., no central authority to control the network), traceability (i.e., all transactions can be seen and tracked by nodes), and non-repudiation (i.e., no one can deny his action). The blockchain technology can establish a distributed trust system, which achieves distributed verification, alleviates performance bottlenecks in central institutions, meets the cross-domain identity authentication needs of CPS, and improves cross-domain communication efficiency [47].

10.5.2 Secure communication protocols

In the realm of industrial control and process systems (ICPSs), wireless communication is a commonly used method of transmitting data. However, utilizing this method also poses a security risk, as messages may not reach their intended recipient or could be tampered with. To prevent these issues, various security measures have been implemented in most protocols, such as Transport Layer Security (TLS) or Datagram Transport Layer Security (DTLS). In the context of CPS, a secure communication protocol plays a crucial role in enhancing the overall security of the system.

The scope of a secure communication protocol in CPS involves several key aspects:

Confidentiality. The protocol ensures that the information exchanged between different components of the CPS remains confidential and cannot be accessed by unauthorized entities. This involves the use of encryption techniques to protect the data during transmission and storage.

Integrity. The protocol verifies the integrity of the data to ensure that it has not been tampered with during transmission. This is achieved through the use of cryptographic mechanisms, such as digital signatures or message authentication codes, which enable the recipient to verify the authenticity and integrity of the received data.

Authentication. The protocol provides mechanisms for authenticating the entities involved in the communication. This ensures that only trusted entities can participate in the system and prevents unauthorized access or impersonation. Authentication can be achieved through various methods, such as passwords, digital certificates, or biometric authentication.

Availability. The protocol ensures that the communication channels and resources required for CPS operation are available and accessible to authorized entities. This involves implementing measures to prevent denial-of-service attacks, ensuring redundancy and fault tolerance, and managing network congestion effectively.

Resilience. The protocol incorporates resilience mechanisms to mitigate the impact of security breaches or failures. This includes techniques such as intrusion detection and prevention systems, anomaly detection, and incident response procedures to detect and respond to security incidents promptly.

Overall, a secure communication protocol for CPS aims to establish a trusted and secure environment for data exchange and interaction between different components of the system. By addressing confidentiality, integrity, authentication, availability, and resilience, it enhances the security posture of CPS and safeguards against potential cyber threats. Some of the typical secure communication protocols that are commonly used in ICPSs are reviewed in [48–50].

10.5.3 Intrusion detection and prevention systems

In the event that attackers manage to bypass border defense technologies, like network firewalls, and security defense technologies, such as software access control, it becomes imperative to carry out intrusion defense and monitoring at the bottom control layers of the ICPSs. Unlike the analysis of network throughput and access frequency, defense and monitoring at the bottom layer can leverage the relationship between measurement data and the principles of operation and control of physical processes, making it possible to thwart deliberate and organized intrusion attempts. Defenders adopt intrusion defense schemes like intrusion prevention (to prevent intrusion), intrusion detection (to detect intrusion during the act), and reduction of losses (during and after intrusion). In this section, we primarily delve into the principle and state of the art of intrusion detection and prevention systems.

10.5.3.1 Intrusion detection system

When it comes to detecting potential attacks on a system, it is crucial to have a reliable mechanism in place that can monitor any unusual activity and alert the appropriate parties in a timely manner. This is where intrusion detection comes into play. By implementing various methods and approaches, it is possible to detect and limit any intrusion to a tolerable level.

One popular method is the bad data detector. This approach involves installing detectors that can provide an alarm operation if any malicious activity is detected. Control system observers are also utilized to measure unmeasurable state variables, which can then be compared to healthy and abnormal cases. However, this method requires further refinement in the context of intrusion detection due to the increasing sophistication of attackers.

Another approach is the moving target approach. This involves introducing a time-varying dynamic auxiliary system that is difficult for attackers to identify. This system remains sensitive to external attacks, and any differences between actual operation and the nominal system are checked by residual generation and evaluation systems [51]. This method can detect covert intrusions since attackers have limited prior knowledge. However, creating moving target modules requires a detailed understanding of the physical plant [52].

Watermarking approaches are also commonly used for intrusion detection. This involves adding a set of auxiliary signals, known as watermarking signals, to the original transmission signal. The aim is to detect any malicious attacks by analyzing whether the characteristics of the auxiliary signals have been altered. This technique provides an added layer of security to ensure the integrity of the original signal. It is frequently used to detect replay attacks and requires little system knowledge [53].

Lastly, machine learning–based approaches have been gaining traction in recent years. The method involves examining the high-dimensional features of the transmitted data and analyzing the relationship between multiple variables and the controlled system or other auxiliary systems [54].

10.5.3.2 Intrusion prevention system

The methods in this category aim to prevent disclosure attacks. These attacks involve infiltrating a system to steal vital information, which can then be used in future attacks. For example, an insider may be involved (as in the Maroochy attack), or an unauthorized user may be given access to the system in a stealthy manner for an extended period of time (known as advanced persistent threats, or APTs). In addition to secure communication protocols, access control, and authentication on the cyber and communication layer, preventive measures can also be taken on the physical or monitoring and control layer.

One effective method for preventing disclosure attacks is cryptography. This involves developing and analyzing protocols that safeguard private messages and protect them from unauthorized access or public exposure. Cryptography-based techniques can also prevent interception of crucial messages at the physical layer. For example, chaotic systems and synchronous control theory have been introduced to thwart attackers from using intercepted data for system identification and state estimation [55].

Another useful method is randomization, which can confuse potential attackers by introducing unpredictability. This method is especially effective when attackers rely on predictable deterministic rules to gain access to critical information within a system. Generating an auxiliary masking signal that is highly correlated with the transmitted signal can facilitate data-driven encryption and decryption. This method is most suitable for fixed systems. Even if attackers are aware of the encryption mechanism, they cannot decipher the data without the auxiliary masking data, which can only be obtained during the offline training phase required to obtain the decryption matrix [2].

Redundant infrastructure elements are also an effective method for preventing intrusions. One can implement redundant infrastructure components, such as devices and network paths, to introduce randomization in communication patterns and timing behaviors. This makes it considerably more challenging and expensive for malicious actors to mount successful attacks. At the application layer, thwarting attacks necessitates an adequate number and variety of backup measurements so that attackers must compromise a greater number of devices to execute a sophisticated, covert, and consequential intrusion. In the realm of SE, this entails elevating the scope of status and analog measurements to more effectively detect and authenticate existing measurements. This can be achieved by increasing the status and analog measurements to detect and validate the existing measurements [56].

Overall, implementing such methods can significantly reduce the likelihood of disclosure attacks and prevent valuable information from being stolen by malicious actors. It is important to employ multiple layers of defense and regularly update security measures to stay ahead of potential threats in the ever-evolving field of cybersecurity.

10.6 CONCLUDING REMARKS

This chapter discusses key issues related to security and privacy in the modern industrial contexts. As we navigate the rapidly developing world of technology and automation, it is becoming increasingly evident that the security and privacy of ICPS are crucial. The vulnerabilities exposed in this chapter remind us that when enjoying the advantages of connecting the physical and the cyber worlds, we must remain alert against the newly induced threats.

The chapter briefly overviews the technical routes and the corresponding countermeasures to mitigate novel threats. Directions and efforts include privacy-preserving techniques, robust security protocols, deploying intrusion detection and prevention systems, and arising human (participants) awareness. It is believed that an essential aspect is to foster collaboration between industry, academia, and regulatory sectors to jointly create a safe and resilient ICPS economy.

In the upcoming decades, the novel forms of security and privacy issues shall not only be regarded as obstacles but also be treated as integral components of the design and operation of ICPSs. In other words, safeguarding the future needs proactive and holistic approaches.

10.6.1 Acknowledgments

This work was supported in part by the National Natural Science Foundation of China (62203143), the China Postdoctoral Science Foundation (2022M710965), the Natural Science Foundation of Heilongjiang Province (LH2022F024), the Heilongjiang Province Postdoctoral Foundation (LBH-Z22130), and the Fundamental Research Funds for the Central Universities (HIT.NSRIF202344).

REFERENCES

[1] M. Cheminod, L. Durante, and A. Valenzano, "Review of security issues in industrial networks," Proceedings of the 19th IFAC World Congress, vol. 9, no. 1, pp. 277–293, 2012.

[2] S. Dibaji, M. Pirani, D. Flamholz, A. M. Annaswamy, K. H. Johansson, and A. Chakrabortty, "A systems and control perspective of CPS security," Annual Reviews in Control, vol. 47, pp. 394–411, 2019.

[3] Y. Jiang, S. Wu, R. Ma, M. Liu, H. Luo, and O. Kaynak, "Monitoring and defense of industrial cyber-physical systems under typical attacks: From a systems and control perspective," IEEE Transactions on Industrial Cyber-Physical Systems, 2023, doi: 10.1109/TICPS.2023.3317237.

[4] A. Humayed, J. Lin, F. Li, and B. Luo, "Cyber-physical systems security—a survey," IEEE Internet of Things Journal, vol. 4, no. 6, pp. 1802–1831, 2017.

[5] F. Pasqualetti, F. Doŕfler, and F. Bullo, "Control-theoretic methods for cyber-physical security: Geometric principles for optimal cross-layer resilient control systems," IEEE Control Systems Magazine, vol. 35, no. 1, pp. 110–127, 2015.

[6] M. Kordestani and M. Saif, "Observer-based attack detection and mitigation for cyberphysical systems: A review," IEEE Systems, Man, & Cybernetics Magazine, vol. 7, no. 2, pp. 35–60, 2021.

[7] Y. Jiang, J. Dong, and S. Yin, "Improving the safety of distributed cyber-physical systems against false data injection attack by establishing interconnections," The 46th Annual Conference of the IEEE Industrial Electronics Society, Singapore, October 18–21, 2020.

[8] Y. Mo, S. Weerakkody, and B. Sinopoli, "Physical authentication of control systems: Designing watermarked control inputs to detect counterfeit sensor outputs," IEEE Control Systems Magazine, vol. 35, no. 1, pp. 93–109, 2015.
[9] M. Kordestani and M. Saif, "Observer-based attack detection and mitigation for cyberphysical systems: A review," IEEE Systems, Man, & Cybernetics Magazine, vol. 7, no. 2, pp. 35–60, 2021.
[10] A. Teixeira, D. Pérez, H. Sandberg, and K. Johansson, "Attack model and scenarios for networked control systems," International Conference on High Confidence Networked Systems, Beijing, pp. 55–64, 2012.
[11] D. Abbasinezhad, A. Ostad, S. Mazinani, and M. Nikooghadam, "Provably secure escrow-less chebyshev chaotic map-based key agreement protocol for vehicle to grid connections with privacy protection," IEEE Transactions on Industrial Informatics, vol. 16, no. 12, pp. 7287–7294, 2020.
[12] S. Karnouskos and F. Kerschbaum, "Privacy and integrity considerations in hyperconnected autonomous vehicles," Proceedings of the IEEE, vol. 106, no. 1, pp. 160–170, 2018.
[13] F. Pasqualetti, F. Dörfler, and F. Bullo, "Attack detection and identification in cyber-physical systems," IEEE Transactions on Automatic Control, vol. 58, no. 11, pp. 2715–2729, 2013.
[14] A. Barboni, J. Gallo, F. Boem, and T. Parisini, "A distributed approach for the detection of covert attacks in interconnected systems with stochastic uncertainties," IEEE 58th Conference on Decision and Control (CDC), pp. 5623–5628, 2019.
[15] Y. Jiang, S. Wu, H. Yang, et al., "Secure data transmission and trustworthiness judgement approaches against cyber-physical attacks in an integrated data-driven framework," IEEE Transactions on Systems, Man, and Cybernetics: Systems, vol. 52, no. 12, pp. 7799–7809, 2022.
[16] S. Lu, Z. Gao, Q. Xu, C. Jiang, A. Zhang, and X. Wang, "Class-imbalance privacy-preserving federated learning for decentralized fault diagnosis with biometric authentication," IEEE Transactions on Industrial Informatics, vol. 18, no. 12, pp. 9101–9111, 2022.
[17] S. Yin, J. Rodriguez, and Y. Jiang, "Real-time monitoring and control of industrial cyberphysical systems with integrated plant-wide monitoring and control framework," IEEE Industrial Electronics Magazine, vol. 13, no. 4, pp. 38–47, 2019.
[18] A. Colombo, S. Karnouskos, O. Kaynak, and Y. Shi, "Industrial cyber physical systems: A backbone of the fourth industrial revolution," IEEE Industrial Electronics Magazine, vol. 11, no. 1, pp. 6–16, 2017.
[19] M. Rahman, M. Mahmud, A. M. Than Oo, and H. R. Pota, "Multi-agent approach for enhancing security of protection schemes in cyber-physical energy systems," IEEE Transactions on Industrial Informatics, vol. 13, no. 2, pp. 436–447, 2017.
[20] A. Gehrmann and M. Gunnarsson, "A digital twin based industrial automation and control system security architecture," IEEE Transactions on Industrial Informatics, vol. 16, no. 1, pp. 669–680, 2020.
[21] J. Giraldo, E. Sarkar, A. A. Cardenas, M. Maniatakos, and M. Kantarcioglu, "Security and privacy in cyber-physical systems: A survey of surveys," IEEE Design & Test, vol. 34, no. 4, pp. 7–17, 2017, doi: 10.1109/MDAT.2017.2709310.

[22] A. Ashok, M. Govindarasu, and J. Wang, "Cyber-physical attack-resilient wide-area monitoring, protection, and control for the power grid," Proceedings of the IEEE, vol. 105, no. 7, pp. 1389–1407, 2017.
[23] D. G. S. Pivoto, L. F. F. de Almeida, R. da Rosa Righi, J. J. P. C. Rodrigues, A. B. Lugli, and A. M. Alberti, "Cyber-physical systems architectures for industrial Internet of Things applications in Industry 4.0: A literature review," Journal of Manufacturing Systems, vol. 58, pp. 176–192, 2021.
[24] J. Zhang, L. Pan, Q.-L. Han, C. Chen, S. Wen, and Y. Xiang, "Deep learning based attack detection for cyber-physical system cybersecurity: A survey," IEEE/CAA Journal of Automatica Sinica, vol. 9, no. 3, pp. 377–391, 2021.
[25] C. C. Aggarwal and S. Y. Philip, "A general survey of privacy-preserving data mining models and algorithms," in Privacy-Preserving Data Mining. Berlin: Springer, 2008, pp. 11–52.
[26] A. Fahad, Z. Tari, A. Almalawi, A. Goscinski, I. Khalil, and A. Mahmood, "PPF-SCADA: Privacy preserving framework for SCADA data publishing," Future Generation Computer Systems, vol. 37, pp. 496–511, 2014.
[27] M. Keshk, B. Turnbull, N. Moustafa, D. Vatsalan, and K.-K. Raymond Choo, "A privacy-preserving-framework-based blockchain and deep learning for protecting smart power networks," IEEE Transactions on Industrial Informatics, vol. 16, no. 8, pp. 5110–5118, 2019.
[28] M. S. Rahman, I. Khalil, A. Alabdulatif, and X. Yi, "Privacy preserving service selection using fully homomorphic encryption scheme on untrusted cloud service platform," Knowledge-Based systems, vol. 180, pp. 104–115, 2019.
[29] J. Lu and R. K. Wong, "Insider threat detection with long short-term memory," Proceedings of the Australasian Computer Science Week Multiconference, 2019, Art. no. 1
[30] R. V. Banu and N. Nagaveni, "Evaluation of a perturbation-based technique for privacy preservation in a multi-party clustering scenario," Information Sciences, vol. 232, pp. 437–448, 2013.
[31] M. Keshk, E. Sitnikova, N. Moustafa, J. Hu and I. Khalil, "An integrated framework for privacy-preserving based anomaly detection for cyber-physical systems," IEEE Transactions on Sustainable Computing, vol. 6, no. 1, pp. 66–79, 2021.
[32] T. A. Adesuyi and B. M. Kim, "A layer-wise perturbation based privacy preserving deep neural networks," International Conference on Artificial Intelligence in Information and Communication, 2019, pp. 389–394.
[33] P. Barbosa, A. Brito, and H. Almeida, "A technique to provide differential privacy for appliance usage in smart metering," Information Sciences, vols. 370–371, pp. 355–367, 2016.
[34] Y.-A. De Montjoye, L. Radaelli, V. K. Singh, and A. Pentland, "Unique in the shopping mall: On the reidentifiability of credit card metadata," Science, vol. 347, no. 6221, pp. 536–539, 2015.
[35] M. U. Hassan, M. H. Rehmani and J. Chen, "Differential privacy techniques for cyber physical systems: A survey," IEEE Communications Surveys & Tutorials, vol. 22, no. 1, pp. 746–789, 2020, doi: 10.1109/COMST.2019.2944748.
[36] G. Liang, S. R. Weller, F. Luo, J. Zhao, and Z. Y. Dong, "Distributed blockchain-based data protection framework for modern power systems against cyber attacks," IEEE Transactions on Smart Grid, vol. 10, no. 3, pp. 3162–3173, 2019.

[37] M. Shen, X. Tang, L. Zhu, X. Du, and M. Guizani, "Privacy-preserving support vector machine training over blockchain-based encrypted IoT data in smart cities," IEEE Internet of Things Journal, vol. 6, no. 5, pp. 7702–7712, 2019.
[38] J. Huang, L. Kong, G. Chen, M.-Y. Wu, X. Liu, and P. Zeng, "Towards secure industrial IoT: Blockchain system with credit-based consensus mechanism," IEEE Transactions on Industrial Informatics, vol. 15, no. 6, pp. 3680–3689, 2019.
[39] D. Puthal, N. Malik, S. P. Mohanty, E. Kougianos, and C. Yang, "The blockchain as a decentralized security framework [future directions]," IEEE Consumer Electronics Magazine, vol. 7, no. 2, pp. 18–21, 2018.
[40] F. Ghaffari, E. Bertin, J. Hatin, and N. Crespi, "Authentication and access control based on distributed ledger technology: A survey," 2020 2nd Conference on Blockchain Research & Applications for Innovative Networks and Services (BRAINS). IEEE, 2020.
[41] I. Butun and R. Sankar, "A brief survey of access control in wireless sensor networks," 2011 IEEE Consumer Communications and Networking Conference (CCNC). IEEE, 2011.
[42] E. Bertin, D. Hussein, C. Sengul, and V. Frey, "Access control in the Internet of Things: A survey of existing approaches and open research questions," Annals of Telecommunications, vol. 74, no. 7–8, pp. 375–388, 2019.
[43] V. Goyal, O. Pandey, A. Sahai, and B. Waters, "Attribute-based encryption for fine-grained access control of encrypted data," Proceedings of the 13th ACM Conference on Computer and Communications Security, pp. 89–98, 2006.
[44] J. Bonneau, C. Herley, P. C. Van Oorschot, and F. Stajano, "The quest to replace passwords: A framework for comparative evaluation of web authentication schemes," IEEE Symposium on Security and Privacy, pp. 553–567, 2012.
[45] A. K. Jain, A. Ross, and S. Prabhakar, "An introduction to biometric recognition," IEEE Transactions on Circuits and Systems for Video Technology, pp. 4–20, 2004.
[46] F. H. Pohrmen, R. K. Das, and G. Saha, "Blockchain-based security aspects in heterogeneous Internet-of-Things networks: A survey," Transactions on Emerging Telecommunications Technologies, vol. 30, no. 10, p. e3741, 2019.
[47] C. Li, F. Li, L. Yin, T. Luo, and B. Wang, "A blockchain-based IoT cross-domain delegation access control method," Security and Communication Networks, vol. 2021, pp. 1–11, 2021.
[48] D. Dragomir, L. Gheorghe, S. Costea, and A. Radovici, "A survey on secure communication protocols for IoT systems," 2016 International Workshop on Secure Internet of Things (SIoT). IEEE, 2016.
[49] J. Dizdarević, F. Carpio, A. Jukan, and X. Masip-Bruin, "A survey of communication protocols for Internet of Things and related challenges of fog and cloud computing integration," ACM Computing Surveys (CSUR), vol. 51, no. 6, pp. 1–29, 2019.
[50] K. T. Nguyen, M. Laurent, and N. Oualha, "Survey on secure communication protocols for the Internet of Things," Ad Hoc Networks, vol. 32, pp. 17–31, 2015.
[51] S. Wu, H. Luo, S. Yin, K. Li, and Y. Jiang, "A residual-driven secure transmission and detection approach against stealthy cyber-physical attacks for accident prevention," IEEE Transactions on Information Forensics & Security, vol. 18, pp. 5762–5771, 2023.

[52] P. Griffioen, S. Weerakkody, and B. Sinopoli, "A moving target defense for securing cyber-physical systems," IEEE Transactions on Automatic Control, vol. 66, no. 5, pp. 2016–2031, 2020.
[53] S. Weerakkody and B. Sinopoli, "Challenges and opportunities: Cyber-physical security in the smart grid," Smart Grid Control: Overview and Research Opportunities, pp. 257–273, 2019.
[54] S. Wu, Y. Jiang, H. Luo, J. Zhang, S. Yin, and O. Kaynak, "An integrated data-driven scheme for the defense of typical cyber–physical attacks," Reliability Engineering & System Safety, vol. 220, p. 108257, 2022.
[55] D. Abbasinezhad, A. Ostad, S. Mazinani, and M. Nikooghadam, "Provably secure escrow-less chebyshev chaotic map-based key agreement protocol for vehicle to grid connections with privacy protection," IEEE Transactions on Industrial Informatics, vol. 16, no. 12, pp. 7287–7294, 2020.
[56] A. Ashok, M. Govindarasu, and J. Wang, "Cyber-physical attack-resilient wide-area monitoring, protection, and control for the power grid," Proceedings of the IEEE, vol. 105, no. 7, pp. 1389–1407, 2017.

Chapter 11

Strategies for protecting serial (non-IP) industrial networks in cyber-physical systems 2.0

Ralf Luis de Moura, Tiago Tadeu Wirtti, Filipe Andersonn Teixeira da Silveira, Rodrigo Rosetti Binda, and Brenda Aurora Pires Moura

LIST OF ABBREVIATIONS

AES	Advanced Encryption Standard
CAN	controller area network
CPS	cyber-physical systems
DoS	denial of service
DMZs	demilitarized zones
EtherNet/IP	Industrial Protocol
IDS	intrusion detection systems
IEDs	intelligent electronic devices
IPS	intrusion prevention system
IoT	Internet of Things
IT	information technology
MFA	multifactor authentication
OT	operational technology
RSA	Rivest–Shamir–Adleman
TCP/IP	Transmission Control Protocol/Internet Protocol
TLS	Transport Layer Security
UART	universal asynchronous receiver-transmitter
VLANs	virtual local area network

11.1 INTRODUCTION

Cyber-physical systems (CPS) typically involve multiple interconnected systems to manipulate physical objects and processes. They integrate computational and physical capabilities and interact with humans in many ways. The wide adoption of physical and cyber systems is strongly linked to the concept of Industry 4.0, which, through a combination of technologies, provides autonomy, reliability, and control with a minimum of human participation [1].

CPSs are essential for critical infrastructure operations vital to society, and their unavailability or destruction impacts the economy, defense, and people's lives. Some examples of critical infrastructure are telecommunications, electrical systems, transport, and water supply [2]. This finding denotes the

DOI: 10.1201/9781003559993-11

importance of protecting physical and cyber systems to protect society or the companies in which they are used.

An important part of CPS is communication networks, including industrial networks. Industrial networks are the foundation for the operation of any CPS, and their functioning is critical, making any operation unfeasible in case of failures; that is why these networks' robustness and availability are crucial.

Industrial networks have been a reality since the 1970s, when the first networks were created. Between its creation and the advent of the Internet, many years passed, and many communication standards and networks were created. At that time, the primary concern was performance, stability, robustness, and interoperability. Cybersecurity was not the primary concern since these networks, in most cases, were isolated; that is, they worked in islands of automation without external contact [3]. For an invasion, it would be necessary for the attacker to be physically present at the plant, which was easily circumvented with physical and perimeter security.

With the advent of the Internet and the increased connection of business networks to public networks, the level of vulnerability and threats to companies increased. However, even so, industrial networks were still seen as isolated and distant components. Over the years, mainly due to Industry 4.0, the opportunity and the need to collect data from the automation layers has been seen. The movement, known as OT (operational technology) and IT (information technology) integration, has connected previously isolated industrial networks to the communication networks that potentially reach the Internet [4].

The integration generated many opportunities for consumption and advanced data analysis, which developed optimization, failure prediction, and monitoring models never seen before. However, as a consequence, threats increased sharply in an environment that was unprepared to be connected.

Many recently created industrial networks based on TCP/IP (Transmission Control Protocol/Internet Protocol) are prepared for interconnection because they are in an environment where it is necessary to worry about security [6]. However, many networks in operation are still not in this reality; for example, several non-IP (Internet Protocol) (serial) networks were created more than 40 years ago and are still present in many cyber-physical environments. The technology exchange cycle in automation environments does not happen as quickly as in information technology environments. Systems are often kept in operation for many years before it is decided to replace them.

There are many aggravating factors in the case of non-IP networks, since they are usually accompanied by devices with many limitations in terms of computational resources that manufacturers have either discontinued or no longer have support or update options from either the point of view of hardware or software.

Although market trends show that non-IP networks will be gradually replaced by networks based on TCP/IP, living with these networks for a few years will still be necessary. Mainly, low-level networks are used for connecting sensors, which will take longer to evolve due to economic constraints, since these sensors are typically low-cost, and an update of this level would significantly increase manufacturing costs.

International standardization organizations and other regulatory bodies currently focus on networks based on TCP/IP; therefore, creating additional protection mechanisms for these networks is imperative. An attacker who manages to reach serial networks via the highest layers will have no difficulty carrying out control commands, code changes, or modification of process control parameters that can ultimately cause financial losses and even loss of lives [7].

Therefore, as we still use these networks for a while, we must take special care to ensure they have maximum protection in case of any invasion attempt. In this chapter, we will discuss some strategies to minimize invasion risks and mitigate their effects if they occur.

11.2 CONCEPT OF CRITICAL INFRASTRUCTURES

Critical infrastructures refer to the systems, networks, and technologies vital for the continuous and secure operation of strategic sectors within a society. These infrastructures provide essential services and support crucial processes in various domains, such as energy, communications, transportation, health care, finance, and government [8]. The disruption or failure of these systems could have catastrophic consequences, impacting the economy, national security, and public well-being [9].

In industrial environments, critical infrastructures are often associated with cyber-physical systems (CPS) and the Internet of Things (IoT), where the interconnection between physical and digital systems is essential for efficient sector operations. Some examples of industrial critical infrastructures are the energy and utilities and health-care and medical services sectors. Another sector strongly reliant on critical infrastructure is the financial and banking industry. In this case, the scenario involves the operation of banking systems, payment networks, stock exchanges, and financial clearances. Critical infrastructures are pivotal in maintaining the regularity of services in both cases.

Critical infrastructures must possess essential characteristics, such as [10]:

- *Resilience*. Resilience is a vital trait of critical infrastructures, involving the ability to sustain functionality amidst adverse events and to restore operational state. Resilience is founded on two pillars: (1) operate under adverse conditions or stress, even if in a degraded

or debilitated state, while maintaining essential operational capabilities, and (2) recover to an effective operational posture in a time frame consistent with mission needs. These situations include cyberattacks, natural disasters, terrorist attacks, equipment failures, and fires. Resilient infrastructures are designed to withstand and continue operating despite these challenges, ensuring that essential services remain available and functional.
- *Cybersecurity*. Critical infrastructures must be highly secure against cyber threats. Robust cybersecurity mechanisms are necessary to safeguard these systems against attacks, intrusions, and theft of sensitive data. The adoption of appropriate security practices is essential to mitigate vulnerability risks.
- *Interconnection and integration*. Critical infrastructures are often interconnected and integrated, enabling the exchange of information and coordination of operations. This interdependence enhances efficiency but can also expose these systems to additional risks. Secure and well-planned integration ensures these infrastructures' continuous and safe operation.

Further, the key risk parameters that interconnection and integration observe include:

- *Cybersecurity vulnerabilities*. Interconnecting critical systems may introduce new attack surfaces and potential vulnerabilities that adversaries could exploit.
- *Data breaches*. Integration can lead to unintended data leakage or unauthorized access if not properly secured.
- *Operational dependencies*. Overreliance on interconnected systems can result in cascading failures if one component experiences an issue.
- *Compatibility challenges*. Different systems may have compatibility issues when integrated, leading to operational disruptions.
- *Single points of failure*. Interconnected systems might create single points of failure if not redundantly designed.

11.3 INDUSTRIAL NETWORKS

The principal idea behind implementing Industry 4.0 is to empower companies to enhance collaboration among departments, making the correct information available to the right people at the right time to facilitate decision-making, increasing efficiency and productivity. In Industry 4.0, CPS offers new production capabilities enabling crucial information flow from the shop floor to the corporate systems.

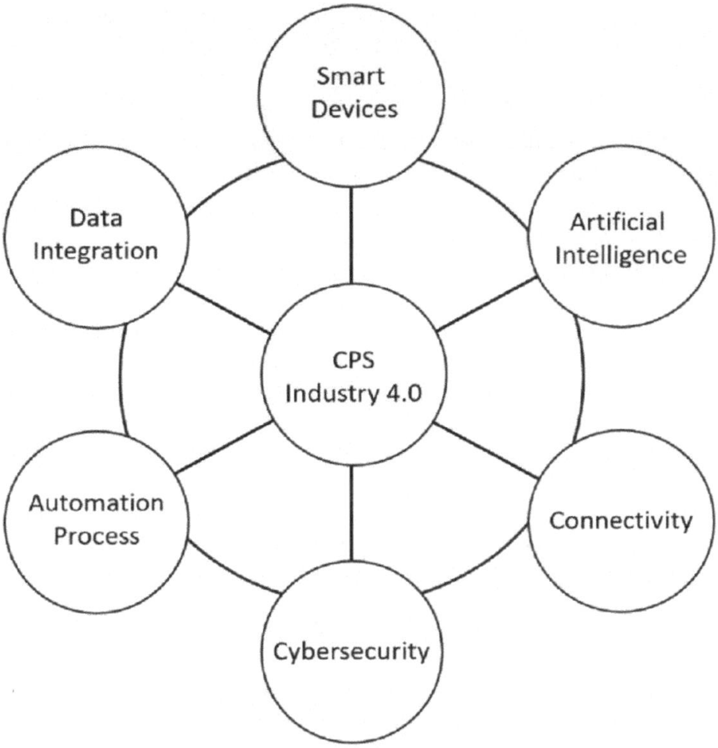

Figure 11.1 CPS in Industry 4.0.

Figure 11.1 shows non-exhaustive topics that CPS can contribute to Industry 4.0. Smart devices support capturing data from the factory floor that reaches the upper layers through data integration. Using automated processes enhanced by artificial intelligence can potentially improve production processes. This entire technology framework must be connected by industrial networks and protected through information security technologies.

Industrial networks represent a fundamental communication infrastructure within industrial production environments. They are interconnection systems that enable the efficient exchange of information among devices, machinery, and control systems across various industries. These networks are essential in CPS, ensuring integration and reliable communication throughout production.

When comparing industrial networks to conventional networks (the ones we have in our homes, for example), it becomes evident that substantial differences exist between them. While conventional networks have a broader focus on applications, industrial networks are purpose-built to address the needs of industrial automation. Therefore, industrial networks prioritize greater robustness, tolerance to adverse environments, determinism, and extremely low latencies. These networks are optimized to ensure the reliability and security of industrial operations, which can be critical for personnel safety, production efficiency, and product quality.

The relevance of industrial networks in today's world is undeniable. They play a central role in the evolution of modern industries, enabling the implementation of Industry 4.0 and advanced process automation.

Industrial networks are critical in various sectors. They facilitate efficient communication, process automation, and implementation of advanced production solutions. The increasing adoption of these networks is driving the transformation of industries, enhancing efficiency, security, and product quality. Finally, they are far more robust, fast, and secure when compared to conventional networks.

11.3.1 Types of industrial networks, their standards, and protocols

Industrial networks encompass technologies and protocols that meet specific needs within various industrial environments. These networks are designed to provide reliable and efficient communication for various applications, from process control to data collection and analysis. Here are some common types of industrial networks.

11.3.1.1 Non-IP industrial networks

Industrial networks are commonly used in industrial automation to connect field devices, such as sensors and actuators, to controllers. They enable real-time communication and control of devices within a specific localized area. Examples include PROFIBUS, CAN (controller area network), and DeviceNet.

Industrial networks are defined by several technical standards developed by international standardization organizations.

11.3.1.1.1 Profibus

PROFIBUS is an industrial communication network that utilizes the PROFIBUS communication protocol [11]. It is widely used to connect automation devices, such as sensors, actuators, and controllers, across various industries. PROFIBUS is known for its high communication speed [12, 13]. It is frequently employed in applications requiring the rapid transfer of large amounts of data.

The specific standard that defines the PROFIBUS network is the IEC 61158 standard [14]. It encompasses digital communication in industrial automation and control networks. It includes several parts detailing various PROFIBUS network types, such as PROFIBUS-DP (decentralized peripherals) and PROFIBUS-PA (process automation). In addition to IEC 61158, the PROFIBUS network is also defined by standards EN 50170 and EN 50254, along with IEC 61158–2 in the case of PROFIBUS-PA.

PROFIBUS-DP exhibits the following characteristics:

- *Topology and architecture*. PROFIBUS-DP is designed for a decentralized topology where peripherals are directly connected to the network, eliminating the need for intermediate controllers.
- *Fast and deterministic communication*. This technology offers fast and deterministic communication between devices, making it suitable for applications that require precise response times.
- *Applications in process automation and manufacturing*. PROFIBUS-DP is widely used in process automation and manufacturing, where devices such as sensors, actuators, and controllers must communicate efficiently and reliably.
- *Standards and specifications*. The PROFIBUS technology is standardized by the IEC 61158 standard. PROFIBUS-DP is a variation of PROFIBUS that focuses on decentralized peripherals.
- *Plug and play*. This technology supports the plug-and-play concept, simplifying network installation and maintenance.
- *Variety of devices*. PROFIBUS-DP is compatible with various field devices, from simple sensors to more complex actuators.
- *Robustness and reliability*. The protocol is developed to be robust and reliable, even in harsh industrial environments with electrical interference and noise.

PROFIBUS-PA possesses the following characteristics:

- *Topology and architecture*. PROFIBUS-PA is designed for decentralized network topology, similar to PROFIBUS-DP. Field devices such as sensors and actuators are directly connected to the network, eliminating the need for intermediate controllers.
- *Communication in hostile environments*. This technology is designed to operate in harsh industrial environments where electrical interference and noise may occur, ensuring data transmission reliability in adverse conditions.
- *Energy efficiency*. PROFIBUS-PA is designed to operate with low power consumption, which is essential in industrial processes requiring efficient use of available resources.

- *Device powering.* A distinctive feature of PROFIBUS-PA is its ability to provide power to connected devices through the same communication cable, reducing system complexity (eliminating the need for separate power cables, for example).
- *Intrinsic safety compliance.* PROFIBUS-PA is suitable for use in classified areas with a risk of explosion.
- *Standards and specifications.* PROFIBUS-PA technology is standardized by the IEC 61158–2 and IEC 61158–3 standards, which describe the communication aspects of fieldbuses.
- *Process automation applications.* PROFIBUS-PA is widely used in industrial process automation applications, such as refineries and chemical and petrochemical industries, as it enables reliable and efficient communication between field devices and the control system.

11.3.1.1.2 DeviceNet

DeviceNet is a type of industrial network that is part of the fieldbus network family. It was developed by Rockwell Automation (formerly known as Allen–Bradley) and is widely used to connect field devices such as sensors, actuators, and other peripheral devices to a centralized control system. DeviceNet is designed to simplify device installation, configuration, and operation in various industrial applications [15].

DeviceNet features the following characteristics [16]:

- *Topology and protocol.* It uses a linear or bus topology, with devices connected sequentially. It uses the CAN protocol for bidirectional communication between devices and the control system.
- *Plug and play.* Known for its ease of adding devices to the network without complex configurations, simplifying installation and maintenance.
- *Device powering.* It can provide power to devices through the same communication cable, eliminating separate cables and reducing installation complexity.
- *Variety of devices.* It supports various devices, from simple sensors to complex actuators, making it suitable for various industrial applications.
- *Communication speed.* The speed varies (typically from 125 kbps to 500 kbps), making it suitable for applications with moderate data transfer requirements.

The standard that describes DeviceNet is the ODVA (Open DeviceNet Vendors Association) standard, which specifies the technical details of the network, including communication protocols, message formats, and electrical specifications [17].

11.3.1.1.3 Controller area network (CAN)

CAN networks are used in various applications, from automotive to industrial automation. They offer robust communication for distributed control systems, allowing devices to communicate efficiently without overloading the network. The main characteristics of CAN networks include [18]:

- *Reliability*. CAN networks are designed for robust industrial environments with common electrical interference and noise. Their design includes error detection and recovery mechanisms (CRC, or cyclic redundancy check) that ensure communication reliability, even in adverse conditions.
- *Communication speed*. CAN networks support communication at different speeds: 125 kbps, 250 kbps, 500 kbps, and 1 Mbps. Low-speed CAN networks, such as 125 kbps, are often used in automotive applications, while higher speeds, like 500 kbps and 1 Mbps, can be found in industrial and automation applications.
- *Flexible topology*. CAN networks can be configured in various topologies, including linear buses and star networks, offering flexibility in deployment across different industrial environments.
- *Bidirectional communication*. Devices connected to a CAN network can send and receive data because the CAN network is half-duplex, meaning, that communication occurs in both directions, but not simultaneously.
- *Low power consumption*. CAN networks are designed to operate with low power consumption, which is advantageous in systems that require energy efficiency and long battery life.
- *Standards and specifications*. CAN networks are standardized by ISO 11898 and ISO 16845 standards. These standards define the technical aspects of the network, including data transfer rates, frame formats, electrical configurations, and communication protocols.
- *Diverse applications*. CAN networks are used in various industries, including automotive, manufacturing, industrial automation, transportation, etc. They are often employed in vehicle control systems, industrial equipment, and field devices.

In summary, CAN networks provide a reliable and efficient solution for communication in industrial environments. Their ability to operate at high speeds, withstand interference, and adapt to different topologies makes them popular in various industrial applications.

11.3.1.2 Ethernet-based networks

Ethernet-based networks have gained prominence in industrial environments due to their high data rates, scalability, and familiarity (the Ethernet standard is widely used today). They enable seamless communication between devices, controllers, and systems across broader areas within a facility. Popular Ethernet industrial protocols include EtherNet/IP, FOUNDATION Fieldbus, PROFINET, and Modbus TCP.

11.3.1.2.1 EtherNet/IP

This protocol is an extension of the standard Ethernet protocol and is widely used in industrial automation. The main characteristics of EtherNet/IP are as follows [17]:

- *Interoperability.* Interoperability with control systems allows real-time information exchange between devices from different manufacturers.
- *Connection-oriented model.* It utilizes the connection-oriented communication model, meaning, that communication is established and maintained between devices, ensuring reliable data transfers.
- *Real-time communication.* EtherNet/IP supports real-time communication, making it suitable for industrial applications that require synchronization and precision in data exchanges between devices and control systems.
- *Data structure.* It uses a consistent object-based data structure that represents devices and related information. It facilitates the configuration, monitoring, and diagnostics of connected devices.
- *Adaptability to different applications.* EtherNet/IP is flexible and can be used in various applications, from industrial automation to integrating IT (information technology) and OT (operational technology) systems in manufacturing environments.
- *Variable speeds.* It can operate at different data transmission speeds, such as 10 Mbps, 100 Mbps, or 1 Gbps, allowing adaptation to the specific needs of the application.

EtherNet/IP is based on standards and specifications, including IEEE 802.3 Ethernet, TCP/IP (Transmission Control Protocol/Internet Protocol), CIP (Common Industrial Protocol), and the ODVA EtherNet/IP Specification [19].

11.3.1.2.2 FOUNDATION Fieldbus

FOUNDATION Fieldbus is a network technology for process automation that focuses on continuous process control applications, such as in the chemical, petrochemical, and food and beverage industries [20]. In other words, it is used in industrial automation to connect field devices to distributed control systems. FOUNDATION Fieldbus uses an Ethernet-based communication protocol to facilitate communication between field devices, such as transmitters, valves, and distributed control systems.

The standards that define the FOUNDATION Fieldbus standard are the ISA-50 standard (ISA-50.02), which describes the reference model for FOUNDATION Fieldbus and its elements, and the IEC 61158–2 standard (Fieldbus Standard for Use in Industrial Control Systems), which specifies communication aspects, particularly the specification of physical, electrical, and communication characteristics for fieldbus devices.

The FOUNDATION Fieldbus network technology exhibits the following characteristics:

- *Advanced digitalization.* Enables the exchange of digital information among devices, facilitating configuration, monitoring, and diagnostics.
- *Bidirectional communication.* Allows communication between field devices and control systems in both directions.
- *Wire reduction.* Permits devices to share a pair of wires for communication and power, reducing the need for traditional wiring.
- *Variety of devices.* Supports various field devices, catering to different industrial applications.
- *Interoperability.* It is based on open standards, promoting interoperability among devices from different manufacturers.
- *Description standards.* The ISA-50 standard (ISA-50.02) outlines the reference model, while the IEC 61158–2 standard specifies communication aspects.
- *Distributed systems.* Used in distributed systems, enabling efficient communication across different parts of an industrial plant.

In summary, the FOUNDATION Fieldbus technology offers an advanced digital solution for communication between devices and control systems. It reduces wiring, promotes interoperability, and enhances industry efficiency and safety.

11.3.1.2.3 PROFINET

Developed by PROFIBUS & PROFINET International (PI), PROFINET is an advanced Ethernet protocol used in industrial automation and control systems. It offers different communication profiles to meet the specific needs of various industrial applications, ensuring efficiency, reliability, and flexibility. PROFINET offers several specific features:

- *Real-time communication.* PROFINET is designed to provide real-time communication between devices. It is crucial in industrial applications where synchronization and precision are essential.
- *Topology flexibility.* PROFINET supports various network topologies, including star, line, and ring, making the network adaptable to the

specific requirements of each application, facilitating system expansion and maintenance.

- *Determinism.* PROFINET guarantees determinism, allowing field devices to communicate at the right time.
- *Communication profile.* PROFINET offers different communication profiles tailored to devices and industrial applications.
- *Ethernet integration.* PROFINET uses standard Ethernet infrastructure, making integrating into existing industrial environments easy.

PROFINET is based on international standards and specifications to ensure interoperability and compatibility. Standards IEC 61158–2 and IEC 61784–1 describe aspects of fieldbuses communication, and PROFINET's communication profile is defined in standard IEC 61784–2.

11.3.1.2.4 Modbus TCP

Modbus is one of the oldest industrial protocols widely used for communication between field devices and control systems. Modbus TCP extends the traditional Modbus protocol designed to operate over Ethernet. It is a cost-effective option for industrial network implementations, allowing real-time data exchange between devices. The main features of Modbus TCP are as follows [21]:

- *Simplicity and efficiency.* Modbus TCP has become an economical choice for industrial network implementations. It maintains a straightforward message structure, making it easy for devices to exchange information.
- *Real-time communication.* The protocol supports real-time communication, enabling instant data exchange between devices and control systems.
- *Connectivity.* Modbus TCP provides connectivity between devices from various manufacturers, promoting interoperability.
- *Open standard.* It allows broad adoption and avoids dependency on a single vendor.
- *Standards and specifications.* It is based on the specifications of the original Modbus protocol, defining the message structure, the function of each field, and the operation modes. Modbus TCP uses the TCP/IP (Transmission Control Protocol/Internet Protocol) transport protocol.
- *Flexibility and broad adoption.* Modbus TCP is highly flexible and can be adopted in various devices, from industrial controllers to sensors and actuators, making it versatile for multiple applications; its simplicity and broad compatibility allow its use in various industries, such as industrial automation, manufacturing, energy, and more.

This chapter focuses on wired non-IP industrial networks; however, there are many other networks for industrial use, such as wireless (e.g., WirelessHart) and IoT (Internet of Things) networks (e.g., LoRaWAN).

11.4 SERIAL INDUSTRIAL NETWORKS (NON-IP)

Industrial serial networks, aka non-IP industrial networks, created in the 1970s, are specialized communication systems suitable for industrial environments. These networks ensure efficient and reliable data transmission between devices in various industrial processes. Their creation came when key concerns revolved around performance, stability, durability, and interoperability.

11.4.1 Historical background and development

The historical background of industrial serial networks stems from the need for reliable and efficient communication in industrial environments. Initially, these networks relied heavily on wired solutions, providing dedicated and stable communication between industrial systems and devices. However, flexibility, portability, and cost-effectiveness have shifted toward wireless solutions requiring low power consumption. This transition has been further accelerated by the growing need for interoperability and integration with the broader Internet, leading to standards such as 6TiSCH for IP-based industrial communications on the Internet's low-power and lossy networks.

11.4.2 Technical overview

Serial industrial networks, such as the universal asynchronous receiver-transmitter (UART), are designed to transmit and receive data over communication channels. For example, a UART is a hardware or software device that manages asynchronous serial communication between computing devices. The primary purpose of UART design is to facilitate data transmission in industries, and with technological advancements, efforts have been made to improve its energy efficiency [22]. For example, UART implementation on field-programmable gate array (FPGA) using dynamic voltage scaling has been proposed to improve the energy efficiency of communication networks.

11.4.3 Limitations of industrial serial networks

While fundamental to communication in industrial environments, serial industrial networks have inherent limitations that can pose challenges in modern industrial environments. These limitations come from the design philosophy, age, and technological limitations of the era in which they were developed. Following are some of these limitations [23].

- *Technology limitations.* By nature, serial industrial networks have inherent technological limitations that pose challenges, especially when juxtaposed with modern communications systems. These limitations arose from the basic design principles of serial communications and the technological landscape when they were developed.
  - *Sequential data transmission.* One of the primary characteristics of serial communication is the sequential transmission of data. Unlike parallel communication, where multiple bits are transmitted simultaneously, serial communication sends data bit by bit. This sequential nature can lead to bottlenecks, especially when there is a need to transmit large volumes of data or when real-time communication is essential.
  - *Point-to-point communication.* Serial networks are inherently designed for point-to-point communication. This design can limit their scalability and flexibility, especially when there is a need to establish more complex network topologies or integrate with other systems.
  - *Limited bandwidth.* The bandwidth of serial networks is often limited, which can restrict the amount of data transmitted over the network within a given time frame. This limitation becomes particularly pronounced when transmitting high-resolution data or multimedia content is needed.
  - *Vulnerability to noise and interference.* Serial communication can be susceptible to noise and interference, especially in industrial environments with various electronic devices operating simultaneously. This susceptibility can lead to data corruption or loss, impacting communication reliability.
- *Lack of advanced features.* Many serial industrial networks, especially older ones, lack advanced features that are standard in modern communication systems. These may include error correction mechanisms, advanced modulation techniques, or support for higher data rates.
- *Vulnerability to cyber threats.* Although fundamental to communication in industrial environments, serial industrial networks were not designed initially with cybersecurity in mind. As a result, they expose vulnerabilities that can be exploited by cyber threats, especially in the context of increasing integration of operational technologies (OT) and information technology (IT). Serial networks, especially those developed decades ago, often lack the security mechanisms that come standard in modern communications systems. This lack of security features makes them vulnerable to cyberattacks, including eavesdropping, data tampering, and unauthorized access.

- *Lack of modern security mechanisms.* Serial industrial networks, especially those developed in the early days of industrial automation, often lack the advanced security mechanisms standard in modern communications systems. This gap makes them especially vulnerable in today's interconnected and cyber threat–prone environment.
  - *Lack of encryption and authentication.* Many traditional serial networks, such as controller area networks (CANs) and local area interconnect networks (LINs), were not designed with encryption or authentication in mind. For example, the CAN bus, widely used for vehicle communications, inherently lacks security mechanisms, making it vulnerable to various attacks. Without encryption, unauthorized entities can easily intercept and read data transmitted over these networks. Likewise, without authentication, there is no way to verify the message's legitimacy, or the sender.
  - *Limited defense against repeated attacks.* Without modern security features, serial networks can be vulnerable to replay attacks. An attacker can capture legitimate messages and replay them later to create unauthorized actions or disrupt normal network operations.
  - *Inadequate intrusion detection system.* Although there have been many efforts to implement intrusion detection systems (IDS) for serial networks, many operate under limitations, such as low bandwidth, small frame sizes, and limited IT resources; this makes it difficult to detect and mitigate sophisticated cyber threats effectively.
  - *Security modernization challenges.* Due to the legacy nature of many serial networks, retrofitting them with modern security mechanisms can be difficult due to compatibility issues, hardware modifications, and the risk of disrupting critical industrial processes.
  - *There is no standardized security protocol.* Unlike modern communications systems with standardized security protocols, many serial networks do not have a uniform approach to security. Security implementations can vary widely, leading to inconsistencies and potential vulnerabilities.
- *Integration challenges.* During the foundation of many existing industrial systems, serial industrial networks pose particular challenges when integrating with modern systems and technologies. These challenges arise from serial networks' inherent design and technological limitations and the evolving requirements of modern industrial environments. Serial networks, developed in an era before the widespread adoption of digital and IP-based systems, often encounter compatibility problems when interfacing with modern systems. This incompatibility can manifest in data formats, communication protocols, and physical connectors, requiring ports or converters to bridge the gap. The point-to-point nature of serial communication can also limit

scalability, especially in large industrial environments, where many devices must be interconnected.

Extending a serial network can be tedious and require significant reconfiguration. In multi-device environments, ensuring data synchronization between devices can be difficult in a serial network; this is especially true when integrating with systems that operate at higher data rates or require real-time communication. Modern industrial environments often deploy advanced monitoring and diagnostic tools that rely on high-throughput data access and analysis. Integrating these tools with serial networks can be difficult due to limited data rates and a lack of support for advanced communication protocols in serial systems. Due to the legacy nature of many serial networks, modernizing them with modern features or upgrading them to newer standards can be difficult. It may involve hardware modifications, software updates, and potential disruption of ongoing industrial processes.

11.4.4 Differences from TCP/IP-based networks

Serial industrial networks and TCP/IP-based networks represent two distinct communication models, each with its characteristics, advantages, and limitations. While serial networks played a fundamental role in the early days of industrial automation, TCP/IP-based networks have become the de facto standard for modern communications, especially with the rise of the Internet and connection systems.

- Communication model
  - Serial Network: Serial communication is inherently point-to-point, meaning, data is transmitted sequentially from one point to another. It makes direct communication between two devices suitable but can pose challenges in more complex network topologies.
  - TCP/IP network: TCP/IP, or the Internet Protocol suite, is designed for multipoint communication. It uses a set of rules to send and receive messages at the Internet address level, allowing for complex network structures and routing capabilities.
- Data transmission
  - Serial network: Data in a serial network is transmitted sequentially bit by bit. It can lead to potential congestion, especially when transferring large volumes of data.
  - TCP/IP network: TCP/IP networks transmit data in packets, allowing faster and more efficient transmission, especially over long distances.
- Scalability and flexibility
  - Serial network: Due to their point-to-point nature, serial networks can have limitations in scalability and flexibility, especially when integrating with modern systems.

- TCP/IP network: TCP/IP networks are inherently scalable and flexible, allowing many devices and systems to be added without significant reconfiguration.
- Security mechanism
  - Serial network: Older serial networks often lack advanced security mechanisms, making them vulnerable to cyber threats, especially in interconnected environments.
  - TCP/IP network: TCP/IP networks come with protocols and security mechanisms, such as SSL/TLS, to ensure secure communication over the Internet.
- Addressing and routing
  - Serial network: Serial networks often lack advanced routing and addressing capabilities, which can cause problems in multi-device environments.
  - TCP/IP network: TCP/IP networks use IP addresses to identify devices and have complex routing mechanisms to ensure data reaches its intended destination.

In summary, the serial industrial network emerged in the 1970s and played a central role in developing industrial communication systems. These networks, tailored to specific industrial needs, prioritize performance, stability, and interoperability. However, as the technology landscape evolves toward integrating the Internet and IP-based communications, the inherent limitations of serial networks become more apparent. Although they provide point-to-point communication and sequential data transfer, their scalability, security mechanisms, and integration capabilities are often overshadowed by advanced and flexible TCP/IP networks that are more active.

11.5 INDUSTRIAL NETWORK PROTECTION STRATEGIES

Cybersecurity in industrial networks is of utmost importance for the regular and proper functioning of operations in the industry. Attacks can completely disrupt the network's functionality. We will address the main risks and the protective measures that should be implemented to avoid them.

11.5.1 Key security risks in industrial networks

The key security risks that arise in industrial networks include the following [24]:

- *Unauthorized access.* One of the main risks is unauthorized access to the industrial network. It can occur when hackers or malicious employees manage to enter the network and gain access to critical systems. Such a risk can materialize when, for example, an intruder gains unauthorized access to a factory's network through compromised user credentials or by exploiting a vulnerability in an unpatched system. Once inside the network, intruders can disrupt operations, steal confidential data, or damage equipment.

- *Malware and viruses.* Infection by malware, such as viruses, worms, or ransomware, can impact the operation of industrial machines and systems, causing severe disruptions. For instance, an inattentive employee may open a malicious email attachment on an industrial computer, infecting the network with ransomware. As a result, the factory's production systems are encrypted, and the attacker demands a ransom for them to be unlocked. There have been cases where companies were forced to cease their operations due to the high cost of the ransom demanded.
- *Denial-of-service (DDoS) attacks.* DDoS attacks can flood the network with fake traffic, overwhelming it and making it inaccessible. A possible scenario for this case is when, for instance, a well-coordinated group of hackers launches a DDoS attack against the servers of a power plant, flooding them with fake and useless traffic. It overloads the network and prevents operators from accessing control systems, potentially disrupting power distribution.
- *Data interception.* Intercepting sensitive data in transit can lead to disclosing confidential information or unauthorized control of devices. One possible scenario occurs when an attacker manages to intercept communication traffic between an industrial automation system and an unprotected central server. In this case, the attacker can collect sensitive information or alter the sent commands, affecting industrial process control.
- *Social engineering.* Cybercriminals can also use social engineering techniques to deceive employees and obtain information or unauthorized access. In this case, social engineering acts as a facilitator. For example, the attacker poses as a maintenance technician and convinces an employee to disclose confidential information about the network infrastructure. This information is later used to plan a targeted attack.
- *Software and hardware vulnerabilities.* Vulnerabilities in industrial software or hardware can be exploited by attackers to gain unauthorized access or control over industrial systems. For example, a food processing plant uses outdated industrial control software with a known vulnerability. An attacker exploits this vulnerability to access the system and manipulate cooking temperatures, resulting in low-quality products and waste.
- *Lack of monitoring and detection.* The absence of threat monitoring and detection systems leaves industrial networks vulnerable to attacks that can go unnoticed. For instance, consider a nuclear power plant

lacking an effective intruder detection system on its network. It allows attackers to compromise control systems, silently jeopardizing plant and environmental safety.

- *Insecure software development.* Poorly designed or inadequately tested industrial software can contain vulnerabilities that attackers can exploit. Suppose an automobile manufacturer uses an automated control system for robots on its assembly line. A security flaw in the software of these robots allows an attacker to issue remote commands, resulting in workplace accidents and production line damage.
- *Lack of awareness and training.* Employees unaware of cyber threats and security best practices can inadvertently introduce risks into the industrial network. The consequences of careless or malicious actions can be catastrophic. For example, an employee at a chemical plant clicks on a link in a phishing email, leading to a malware attack that compromises the plant's control systems, causing irreparable damage (with potential environmental or health risks), such as chemical leaks and safety hazards.
- *Internal attacks.* Insiders with privileged knowledge can become deliberate internal threats, causing harm to company operations. These cases are more challenging to predict or prevent. Imagine a quality control engineer at an electronics manufacturing plant with privileged access to the plant's control systems. For some reason, they could access the control systems and alter production settings, manufacturing defective products and significant losses.

11.5.2 Security networks in critical environments

Before discussing security networks in critical environments, defining risk levels in industrial settings is essential. The regulation that governs risk levels in labor activities in general, including the industry, is NR4. The risk levels are defined as follows [25]:

- Risk level 1: Low-risk environments, where operations and processes do not pose a significant danger to the safety of workers and the environment.
- Risk level 2: Moderate-risk environments, where incidents that cause minor injuries or limited material damage may occur.
- Risk level 3: High-risk environments, where processes and operations can result in severe injuries and significant material damage.
- Risk level 4: Extremely high-risk environments, where operations pose severe threats to the safety of workers, the environment, and equipment integrity. It includes industries dealing with hazardous, radioactive, or flammable substances, such as nuclear plants, chemical plants, and other critical facilities.

Naturally, not all activities at level 4 require critical digital communication networks (computer networks) in their processes. However, where such communication is necessary, such as in nuclear power plants, steel plants, chemical plants, and others at risk level 4, networks prioritizing critical communication for safety are required. These networks ensure that emergency shutdowns and safety interlocks are transmitted quickly and reliably.

11.5.3 Key protective actions in industrial networks

Cybersecurity in industrial networks is critical to ensuring the integrity of industrial operations and protecting against cyber threats. That is why it is essential to adopt appropriate protective measures. In what follows, in Table 11.1, we present the relationship between security risks and the corresponding protection measures.

It is important to note that cybersecurity is a layered approach, and multiple protective measures may be necessary to mitigate threats on an industrial network effectively. Additionally, it is crucial to keep these measures up to date and in compliance with relevant security standards to ensure the best defense against cyber threats.

Following, we list some cybersecurity practices to mitigate risks and vulnerabilities in critical IT/OT infrastructures [26]:

Table 11.1 Security risks and protection measures

Security risks	*Protection measures*
Unauthorized access	Authentication and access control Network segmentation Awareness and training
Malware and viruses	Antivirus and antimalware Updates and patches
Denial-of-service (DoS) attacks	Firewalls and IDS/IPS Traffic monitoring
Data interception	Cryptography Network segmentation
Social engineering	Social engineering
Software and hardware vulnerabilities	Updates and patches Security tests
Lack of monitoring and detection	IDS/IPS Traffic monitoring
Insecure software development	Security tests
Lack of awareness and training	Awareness and training
Insider attacks	Authentication and access control Traffic monitoring Security and compliance standards

- Use of advanced firewalls to filter unwanted traffic and attacks.
- Use of regular software and system updates to address known vulnerabilities.
- Use of continuous monitoring of suspicious activities and early intrusion detection.
- Use of encryption of sensitive data at rest and in transit for protection against theft.
- Use of multifactor authentication to strengthen system access.
- Secure software development following best practices and rigorous testing.
- Regular employee training should be used to raise awareness of cybersecurity.

These cybersecurity practices help fortify the resilience of critical IT/OT systems against threats and minimize the potential impact of malicious attacks. Critical infrastructures should be connected through demilitarized zones (DMZs) to mitigate these risks, a practice common in large enterprises. A final note is to consider the risks and adopt secure integration practices. Organizations can harness the benefits of interconnecting and integrating critical IT/OT infrastructures while minimizing potential vulnerabilities.

- *Network segmentation.* Isolating critical systems from less-sensitive networks through network segmentation to limit the spread of attacks. Network segmentation involves creating isolated zones or segments to restrict communication between systems. It can be achieved through:
  - *Firewall.* Control traffic between network segments, allowing or blocking specific communications.
  - *VLANs (virtual LANs).* Physically divide a network into logical segments, separating devices into groups based on criteria, such as department or function.
- *Authentication and access control.* Implement multifactor authentication and tightly control access to industrial systems, ensuring that only authorized personnel are allowed access. In addition to MFA (multifactor authentication), other secure authentication techniques include:
  - *Biometrics.* Using fingerprints, facial, or iris recognition to authenticate users.
  - *Digital certificates.* Using cryptographic keys to verify identity.
  - *Hardware tokens.* Physical devices that generate unique codes for authentication.
- *Firewalls and IDS/IPS.* Use firewalls and intrusion detection/prevention systems (IDS/IPS) to monitor and protect network traffic. There are two main types of firewalls:
  - *Application firewall.* Inspects traffic based on specific applications and application rules.

- *Next-generation firewall.* Combines application firewall features with intrusion prevention capabilities.

  IDS (intrusion detection system) monitors traffic for suspicious behavior, while IPS (intrusion prevention system) goes beyond detection to block malicious traffic actively.
- *Updates and patches.* Keep all industrial systems and devices up to date with the latest security fixes to mitigate known vulnerabilities. A real example is the WannaCry ransomware attack in 2017, which exploited a Windows vulnerability to encrypt systems and caused significant damage to companies worldwide due to the lack of patch applications.
- *Traffic monitoring.* Implement traffic monitoring systems to detect unusual behavior and suspicious activity on the network. Monitoring techniques include:
  - *Packet analysis.* Examines the content of network packets.
  - *Log monitoring.* Records network events and activities.

  Traffic monitoring tools include Wireshark, Nagios, SolarWinds, and others.
- *Data backup and recovery.* Perform regular backups of critical data and develop disaster recovery plans to minimize the impact of attacks or failures. Backup strategies include regular backups in secure locations, cloud backup, and differential and incremental backup techniques. Disaster recovery involves creating detailed plans to restore systems and data during an outage.
- *Awareness and training.* Train employees to recognize cyber threats, such as phishing and social engineering, reducing the risk of manipulation.
- *Encryption.* Use encryption to protect the confidentiality of data in transit, making it more difficult for attackers to intercept sensitive information. Encryption techniques include:
  - *AES (Advanced Encryption Standard).* Used to encrypt data in transit and at rest.
  - *RSA (Rivest–Shamir–Adleman).* Used for public key cryptography.
  - *TLS (Transport Layer Security).* Used to protect communications on the web.
- *Security testing.* Conduct regular security testing, such as vulnerability assessments and penetration tests, to identify and remediate vulnerabilities. Vulnerability assessments identify and classify vulnerabilities in systems, while penetration tests simulate attacks to assess a network or system's resistance to intrusion.
- *Security and compliance standards.* Follow recognized security standards, such as ISA/IEC 62443, and ensure regulatory compliance in the industry. In addition to ISA/IEC 62443, other standards include:

- *The NIST Cybersecurity Framework*. Guidelines for improving cybersecurity in organizations.
- *ISO 27001*. Establishes an information security management system.
- *PCI DSS*. Guidelines for protecting payment card information.
- *HIPAA*. Health information security regulation in the USA.

11.6 ADDITIONAL NON-IP INDUSTRIAL NETWORK PROTECTION STRATEGIES

Non-IP networks are generally located in the lower layers, protected by a series of levels of protection that increase from the Internet. The recommendations described in the previous chapter are, without a doubt, essential strategies for protecting non-IP networks. However, if one malware or hacker manages to reach non-IP networks, there will be no protection or detection mechanisms due to the aforementioned limitations.

We know we will have to live with these networks for a while, and simply leaving them unprotected cannot be an option. All recommendations discussed beforehand should be adopted whenever possible, and it is essential to always keep networks as up-to-date as possible with all the protection mechanisms already available and discussed previously, but how can we deal with non-IP legacy networks that cannot be replaced and whose limitations in terms of information security are known?

Serial protocols do not have adequate security mechanisms. Usually, as mentioned earlier, its security is achieved through attempts to isolate and segregate networks, as shown in Figure 11.2. However, this type of strategy does not guarantee that this part of the network will not be affected. That is why it is imperative that all network layers, including the lowest, have layers of protection.

Usually, cybersecurity mechanisms are placed on the industrial IP layer, which uses almost the same protection strategies as a typical business TCP-IP network, such as IPS (intrusion prevention systems), IDS (intrusion detection systems), firewalls, and access control [27]. Firewalls do not make the integration process between non-IP and IP networks but through devices (e.g., gateways, PLC), with network interfaces that work as bridges. Even with all the known limitations in non-IP devices, it is still necessary to add mechanisms that can guarantee a minimum of protection in the event of an invasion; otherwise, if the attacker could access these devices, he will be fully able to control all field equipment, since he will not find any difficulty doing so.

As we have already mentioned, non-IP industrial networks often lack advanced security. These gaps make them especially vulnerable in today's

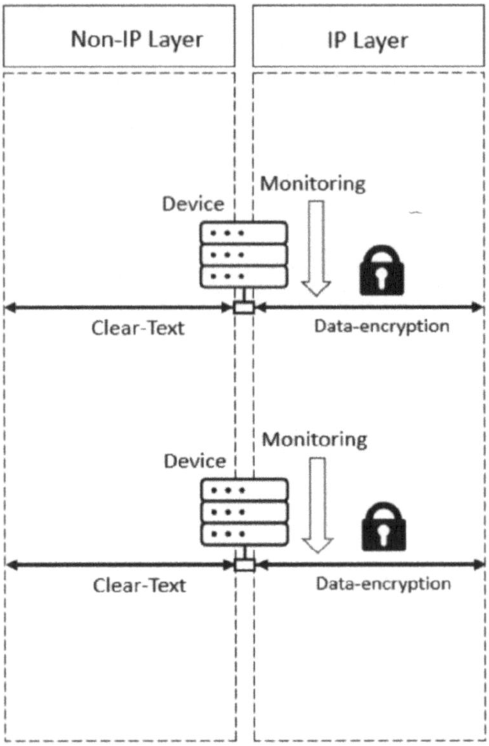

Figure 11.2 Security layers.

interconnected environments. The following gaps should be mitigated with additional actions:

- Lack of encryption and authentication
- Limited defense against repeated attacks
- Inadequate intrusion detection system
- Security modernization challenges
- No standardized security protocol

Strategically, we will separate the security zone components (Figure 11.3) inside the industrial network into parts and work on mitigating actions for each. It can be divided as follows:

- Physical network
- Logical network
- Endpoints
- Devices (zone edge)

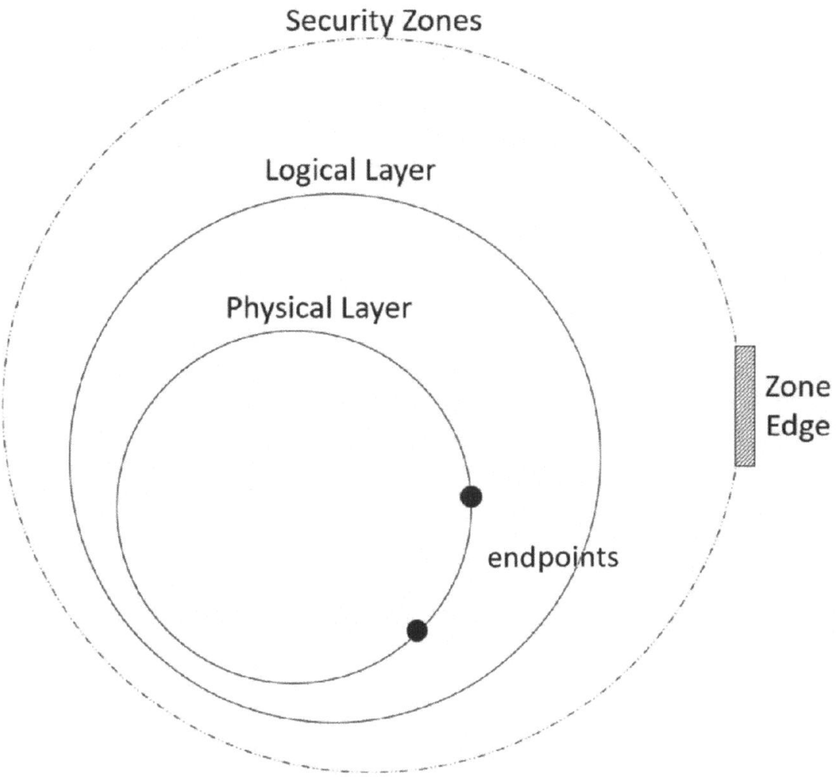

Figure 11.3 Non-IP network components.

11.6.1 Strategy for physical network

Suppose an attacker gains physical access to the network. In that case, he can pretend to be a network entity or sniffer monitoring communications. That is why the physical network must be adequately protected, by not allowing unauthorized persons physical access to network cables and equipment. Property security usually meets this requirement, which naturally prevents people outside the process from approaching the plants or equipment. However, it is recommended that network cables and equipment be physically protected, employing gutters and lockable and monitored cabinets to ensure that only persons with permission can access them. These simple actions can reduce the chance of unauthorized physical access.

Some vendors have specific tools for the physical monitoring of certain protocols. These tools can monitor the health of the network and devices. This monitoring is not aimed at cybersecurity but can indirectly detect abnormal variations in communication (e.g., voltage level) and even changes in the parameterization of devices on the network. The use of these tools can be the first layer of monitoring.

11.6.2 Strategy for logical network

Serial-based industrial protocols represent several security concerns. Most do not require authentication; generally, it is only necessary for a valid address and other parameters, such as function codes. An attacker can easily replicate these messages and defile the communication. All protocols transmit in clear text that can be captured and spoofed due to the lack of encryption.

In this context, it is essential to monitor network traffic through mechanisms that can identify some basic aspects of the protocol, such as error messages, non-existent network addresses, and even control operations not commonly used by the industrial process. This monitoring can be a preventive measure for any attack that aims to impact the plant's operation.

Character count, average data packet size, and communication sequence can also be monitored as indirect mechanisms for detecting communication anomalies. Because they are cyclical and repetitive networks, any slight behavior change can signify a possible attack. This type of monitoring requires in-depth knowledge of network behavior.

11.6.3 Strategy for endpoints

Endpoints are physical devices that connect to and exchange information with a network. Some examples of endpoints are mobile devices, desktop computers, virtual machines, embedded devices, and servers. Endpoints can mean vulnerabilities and should be kept to the smallest number possible.

Only essential devices should be kept on the network. The smaller the number of devices, the smaller the opportunity for attacks. In this sense, only devices that cannot, for some reason, be migrated to more secure infrastructures should be kept on non-IP networks. Those devices that are maintained need to be adequately protected against attacks, especially those arising from contamination by an external agent. All device interfaces must be blocked to prevent them from being used.

Cyberattacks exploit, for example, Universal Serial Bus (USB) interfaces. The USB has been used as an attack vector, and USB mass storage devices containing malware can interface directly with drivers running in the most privileged operating systems. USB interfaces must be blocked.

Remote access to industrial networks must be avoided because it can be vulnerable. In situations where access is unavoidable, it must be made available through access control tools with auditing enabled.

Another action regarding endpoints is related to establishing solid asset management. We cannot protect what we do not know. An inventory of all devices and their configuration is essential to define cybersecurity measures.

Finally, devices on the network must go through a hardening process of mapping threats, mitigating risks, and carrying out corrective activities, with the primary objective of making it prepared to face attack attempts.

11.6.4 Strategy for zone edge

All communication that enters and leaves the security zone of non-IP networks must be controlled and monitored [28]. Only the necessary communication ports must be released. Networks need to be segmented to avoid any improper communication. Any communication must be monitored in real time, and any abnormal event representing an attack must be reported. Upper-network layers must follow the recommendations listed in Section 11.3.

11.7 CONCLUSION

The rapid evolution of industrial networks has been influenced by the revolutionary transformations enabled by the Industry 4.0 paradigm. This technological revolution has reshaped future factory systems' control and communication system requirements. Industrial automation increasingly relies on intelligent and complex distributed measurement and control systems enabled by non-IP industrial networks.

Researchers are applying new technologies, such as artificial intelligence (AI) and machine learning models, to detect anomalies in non-IP network protocols; several studies are available on this subject. Today, technologies exist to monitor these networks passively without interfering in the communication process, by detecting anomalous events. As non-IP networks are cyclical, deterministic, and repeatable, this means any behavior change is easily seen. The challenge today is to make this technology viable from an economic point of view; since they are legacy networks and are decreasing in use, they are not commercially feasible to the leading automation suppliers.

The evolution and adaptation of non-IP industrial networks are critical to ensure their continued relevance and functionality in the modern industrial landscape. Despite their limitations, these networks still play an important role in various industrial environments, providing reliable communication and data transfer between industrial devices and systems.

Network protection strategies are available in the international standards already mentioned, and they give a good overview of actions that can be taken with relative success in protecting networks. Non-IP networks are not typically mentioned in these standards and lack adequate guidance for reducing cyber risks. This chapter addressed the points at which actions can be taken to mitigate the most critical threats, increasing the layers of protection of these networks.

REFERENCES

[1] Abdelghani, T. (2019). Implementation of defense in-depth strategy to secure industrial control system in critical infrastructures. American Journal of Artificial Intelligence, 3(2), 17–22.
[2] Alcaraz, C., & Lopez, J. (2020). Secure interoperability in cyber-physical systems. In Cyber Warfare and Terrorism: Concepts, Methodologies, Tools, and Applications (pp. 521–542). IGI Global.
[3] Aliwa, E., Rana, O., Perera, C., & Burnap, P. (2020). Cyberattacks and countermeasures for in-vehicle networks. arXiv, 2004, 10781.
[4] Andersen, B., Kasparick, M., Ulrich, H., Franke, S., Schlamelcher, J., Rockstroh, M., and Ingenerf, J. (2018). Connecting the clinical IT infrastructure to a service-oriented architecture of medical devices. Biomedical Engineering/Biomedizinische Technik, 63(1), 57–68.
[5] Ani, U. P., Daniel, H., & Tiwari, A. (2017). Review of cybersecurity issues in industrial critical infrastructure: Manufacturing in perspective. Journal of Cyber Security Technology, 1(1), 32–74.
[6] Dadheech, K., Choudhary, A., & Bhatia, G. (2018). Demilitarized zone: A next level to network security. In 2018 International Conference on Intelligent Computing and Control Systems (pp. 595–600). IEEE.
[7] De Bruijne, M., & Van Eeten, M. (2007). Systems that should have failed: Critical infrastructure protection in an institutionally fragmented environment. Journal of Contingencies and Crisis Management, 15(1), 18–29.
[8] de Moura, R. L., Ceotto, L. D. L. F., & Gonzalez, A. (2017). Industrial IoT and advanced analytics framework: An approach for the mining industry. In 2017 International Conference on Computational Science and Computational Intelligence (CSCI). IEEE.
[9] Gardner, J. W., Boyer, K. K., & Gray, J. V. (2015). Operational and strategic information processing: Complementing healthcare IT infrastructure. Journal of Operations Management, 33, 123–139.
[10] de Moura, R. L., Franqueira, V. N. L., & Pessin, G. (2021). Towards safer industrial serial networks: An expert system framework for anomaly detection. In 2021 IEEE 33rd International Conference on Tools with Artificial Intelligence (ICTAI) (pp. 1197–1205). IEEE.
[11] Profibus: Website oficial da Profibus & Profinet International (PI). (s.d.). (2023, September 10). www.profibus.com/
[12] Profinet: Profibus & Profinet International (PI). (2023, September 10). www.profibus.com/
[13] PROFIsafe: Profibus & Profinet International (PI). (2023, September 10). www.profibus.com/
[14] IEC 61158 Standard: International Electrotechnical Commission (IEC). (2023, August 10). www.iec.ch/
[15] DeviceNet: Rockwell Automation. (2023, September 23). www.rockwellautomation.com/
[16] EtherNet/IP: ODVA (Open DeviceNet Vendors Association). (2023, September 10). www.odva.org/
[17] Ethernet Standard (IEEE 802.3): (Institute of Electrical and Electronics Engineers). (2023, September 15). https://standards.ieee.org/

[18] Foundation Fieldbus: Website oficial da FieldComm Group (anteriormente conhecida como Fieldbus Foundation). (s.d.). www.fieldcommgroup.org/
[19] Fedullo, T., Morato, A., Tramarin, F., Rovati, L., & Vitturi, S. (2022). A comprehensive review on time sensitive networks with a special focus on its applicability to industrial smart and distributed measurement systems. Sensors, 22(4), 1638.
[20] Gardner, J. W., Boyer, K. K., & Gray, J. V. (2015). Operational and strategic information processing: Complementing healthcare IT infrastructure. Journal of Operations Management, 33, 123–139.
[21] Modbus TCP: Modbus. (2023, September 10). www.modbus.org/
[22] Haripriya, D., Kumar, K., Shrivastava, A., Al-Khafaji, H. M., Moyal, V., & Singh, S. K. (2022). Energy-efficient UART design on FPGA using dynamic voltage scaling for green communication in industrial sector, Wireless Communications and Mobile Computing v. 2022, (1), 4336647.
[23] Humayed, A., Lin, J., Li, F., & Luo, B. (2017). Cyber-physical systems security—a survey. IEEE Internet of Things Journal, 4(6), 1802–1831.
[24] Kayan, H., Nunes, M., Rana, O., Burnap, P., & Perera, C. (2022). Cybersecurity of industrial cyber-physical systems: A review. ACM Computing Surveys (CSUR), 54(11s), 1–35, https://doi.org/10.1145/3510410.
[25] Lee, E. A. (2008). Cyber physical systems: Design challenges. In Proceedings of the 11th IEEE Symposium on Object Oriented Real-Time Distributed Computing (ISORC) (pp. 363–369). IEEE.
[26] Mantravadi, S., Schnyder, R., Møller, C., & Brunoe, T. D. (2020). Securing IT/OT links for low power IIoT devices: Design considerations for Industry 4.0. IEEE Access, 8, 200305–200321, https://doi.org/10.1109/ACCESS.2020.3035963.
[27] Osei-Kyei, R., et al. (2021). Critical review of the threats affecting the building of critical infrastructure resilience. International Journal of Disaster Risk Reduction, 60, 102316, https://doi.org/10.1016/j.ijdrr.2021.102316.
[28] ZigBee: Zigbee Alliance. (2023, September 17). www.zigbeealliance.org/

Chapter 12

Enhancing the security of firmware over-the-air updates in automotive cyber-physical system

Rachana Y. Patil, Yogesh H. Patil, Asmita Manna, and Manjiri Ranjanikar

LIST OF ABBREVIATIONS

AUTOSAR	automotive open system architecture
CAN	controller area network
DoS	denial-of-service
ECUs	electronic control units
FOTA	firmware over-the-air
HSM	hardware security module
IoT	Internet of Things
ISO/IEC	International Organization for Standardization/International Electrotechnical Commission
OTAP	over-the-air key provisioning
RBAC	role-based access control
SIEM	security information and event management
TPM	trusted platform module

12.1 INTRODUCTION

Over the past decade, the automotive landscape has witnessed a profound transformation as mechanical components in vehicles have progressively given way to their electronic counterparts [1]. This transition has ushered in the era of intelligent vehicles, which boast a sophisticated network of electronic control units (ECUs), sometimes exceeding 50 in number, governing a wide array of vehicle functions, including critical road-assist services and safety systems. Within each ECU resides distinct firmware, serving as the operational software, with a continuous commitment to refinement and enhancement, as is customary with software development [2].

Traditionally, firmware updates in the automotive sector have been administered via physical cables. However, an emerging trend is poised to reshape this landscape of FOTA updates. This innovative approach offers a multitude of advantages [3, 4]. Firstly, it offers vehicle owners a seamless and convenient experience, obviating the need to visit a physical service

 DOI: 10.1201/9781003559993-12

station for updates. Secondly, it ushers in a new era of speed and efficiency, enabling near-instantaneous updates upon the release of new firmware versions, swiftly bringing the latest improvements to the corresponding ECUs. Thirdly, and perhaps most importantly, FOTA updates bolster safety by minimizing a vehicle's duration with potentially flawed firmware, reducing the associated risks to drivers and passengers [5].

FOTA is poised to make significant inroads into the automotive industry, taking cues from its highly successful implementation in the mobile phone sector. The FOTA approach is set to revolutionize how updates are carried out, allowing them to be seamlessly executed at the customer's location rather than necessitating a visit to a dealership [6]. This shift enhances convenience and promises substantial reductions in fleet management costs.

At present, only a select few vehicles are equipped to receive updates, typically focusing on improving infotainment or telematics systems. Consequently, a firmware issue during a vehicle's lifespan typically demands a return to the dealer for resolution. In contrast, our proposed strategy aims to broaden the scope of FOTA services to encompass the entire vehicular system [7]. This expansion offers a holistic and efficient approach to firmware updates, ultimately bolstering the vehicle's overall reliability and sparing vehicle owners from unnecessary disruptions.

Nonetheless, these advantages also bring to the fore a pressing concern—the potential vulnerability to malware and cyberattacks that could compromise real-time vehicle systems [8]. Once a malicious actor gains access through a wireless connection, inadequate security measures can lead to consequences that pose severe threats to safety and human lives in extreme cases [9, 10]. For example, malevolent nodes attempting to execute denial-of-service (DoS) attacks, such as inundating the internal network with excessive traffic, can precipitate the failure of critical vehicle functions [11].

In the contemporary landscape of vehicle security, the predominant concerns have traditionally revolved around physical attacks, such as tampering with brake wires, manipulating lock mechanisms, or engaging in hot-wiring practices. However, with the advent of FOTA capabilities, an entirely novel class of threats emerges, denoted as cyberattacks [12]. These cyberattacks target the fundamental infrastructure encompassing vehicles and their operators, and they possess the ominous potential to be executed on a large scale with minimal effort.

The successful execution of such cyberattacks, exemplified by installing malicious firmware versions, carries catastrophic implications, placing human lives in peril [13]. Furthermore, it is noteworthy that the outcomes of cyberattacks can mimic those of physical attacks, achieving analogous results, such as the incapacitation of brakes, unauthorized door unlocking, or the unauthorized initiation of the vehicle's engine.

In the contemporary automotive industry, there is an unmistakable surge in the complexity of interconnected units and the proliferation of interfaces. This prevailing trend, while facilitating advanced functionalities, concurrently elevates the vulnerability of these networks, rendering unauthorized access a relatively straightforward endeavor [8]. Despite numerous research endeavors aimed at formulating secure solutions for specialized off-chip networks, it is noteworthy that, to the best of our knowledge, none have hitherto ventured to develop specialized solutions commensurate with the ever-advancing technological complexity that underpins these systems [14].

In stark contrast to the past, where wired vehicles were deemed trusted systems, the contemporary scenario presents a significant shift in the paradigm. An intrusion into a single electronic control unit (ECU) could compromise the entire system from within. Furthermore, a prevailing practice within the automotive industry is the tendency to concentrate efforts on individual ECUs in isolation, often neglecting the holistic perspective of the entire system.

12.1.1 Motivation

In the evolving realm of modern vehicles, the significance of FOTA updates is paramount. They serve as the linchpin of innovation, offering an efficient and cost-effective means to augment vehicle capabilities. Much like the dynamic nature of smartphones, FOTA updates enable vehicles to adapt and grow over time, aligning with the demands of today's tech-savvy consumers. However, this technological promise comes with an inherent challenge: the growing concerns surrounding cybersecurity within automotive cyber-physical systems. As vehicles embrace connectivity and autonomy, they become vulnerable to cyber intrusions. These threats compromise vehicle functionality and pose risks to passenger safety. This research is motivated by the convergence of these two critical factors. The objective is to bolster the security and efficiency of FOTA updates, thus propelling advancements in modern vehicles while preemptively tackling cybersecurity issues. This endeavor is aimed at fostering a future in which vehicles seamlessly integrate convenience, connectivity, and security, thereby sustaining innovation in the mobility sector while prioritizing safety as the foremost concern.

12.1.2 Overview of FOTA update processes

FOTA updates are critical to modern software management in various domains, including the automotive industry [15–17]. They enable remote and wireless firmware, software, and configuration updating in electronic control units (ECUs) and devices. Figure 12.1 describes the overall process of FOTA update. A detailed description of each phase is discussed in this section.

12.1.2.1 Preparation and planning

During this phase of FOTA updates, manufacturers and developers meticulously assess update needs, package new firmware, and establish the foundational framework for secure and efficient vehicle updates.

- *Identification of update needs.* Manufacturers and software developers assess the vehicle's current firmware and software to pinpoint areas requiring updates. These needs may arise from various factors, such as security vulnerabilities, performance improvements, bug reports, or the introduction of new features. Prioritization of updates is critical, with high-priority updates addressing critical security issues or safety concerns.
- *Update packaging.* The new firmware version, designed to address identified needs, is carefully compiled into a package suitable for over-the-air transmission. Encryption is frequently used to prevent anyone from altering or accessing the contents of this package. Utilizing compression methods can minimize the time and resources required for the update's data transfer. This approach optimizes efficiency by reducing the size of the data being transferred, resulting in faster transmission times and decreased resource usage.
- *Rollback mechanism.* In this step, a backup plan is in place to roll back to the previous firmware version if the upgrade encounters any problems or is incompatible with the vehicle's hardware. A mechanism for reverting to the previous version is implemented if the update causes unexpected problems or is incompatible with other components. Maintaining reliability and security throughout operations is made possible by rolling back the vehicle's software to a previous version via this approach.

12.1.2.2 Communication infrastructure

In this step of FOTA updates, automobiles use telematics modules and cellular or Wi-Fi connectivity to set up a secure connection with update servers, which will act as the essential conduit for data transfer during the update itself.

- *Telematics module.* The telematics module in today's vehicles facilitates wireless connections between the vehicle and the outside world. These components must first connect the vehicle to the update server for the FOTA process to work.
- *Cellular or Wi-Fi connectivity.* Vehicles use cellular networks or Wi-Fi connections to connect to the update server, depending on the criteria, like manufacturer infrastructure and the update's size and urgency.

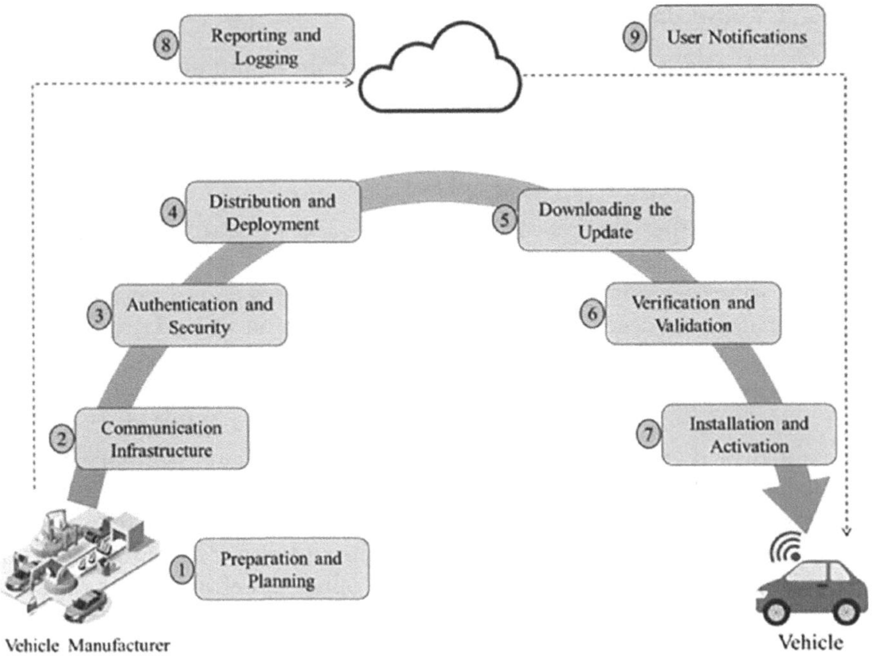

Figure 12.1 Overview of FOTA update process.

12.1.2.3 Authentication and security

In this stage, the rigorous procedures guarantee the integrity of the FOTA update procedure. Encryption protects data integrity from eavesdropping and tampering, while authentication mechanisms confirm the vehicle's authenticity and the update server.

- *Authentication.* Strict measures are used to guarantee that the vehicle and the update server both independently validate the other's identity. The safety and reliability of the update procedure depend on this authentication measure being taken.
- *Encryption.* Data exchanged during FOTA is encrypted, protecting it from eavesdropping and unauthorized changes.

12.1.2.4 Distribution and deployment

At this stage, an update server is managed on a central distribution server, which also acts as a secure repository. Firmware upgrades can be sent out in a controlled and systematic fashion thanks to thoughtful deployment tactics that allow manufacturers to trigger updates remotely or enable vehicles to check for available updates autonomously.

- *Distribution server*. The manufacturer controls a centralized distribution server, where the updates are kept in a safe format. This server stores all the required updated files.
- *Deployment trigger*. The manufacturer can trigger remote updates according to predetermined schedules, user requests, or other conditions. It is possible that vehicles may autonomously check for upgrades regularly.

12.1.2.5 Downloading the update

At this point, the vehicle takes the reins and contacts the distribution server to begin the process. Next, the update files are safely downloaded and stashed in the predetermined location within the car, all set for the installation procedure, guaranteeing a regulated and effective data transmission.

- *Vehicle initiation*. The vehicle initiates contact with the distribution server to obtain the awaiting update.
- *Downloading*. The updates are downloaded from a server and stored in the vehicle's internal memory. There is usually a safe place in the car to keep these documents.

12.1.2.6 Verification and validation

The downloaded update files are checked for corruption, authenticity, and compatibility during this stage. This ensures a safe and dependable update process by checking the data using checksums, verifying the source's legitimacy with digital signatures, and thoroughly assessing compatibility to avoid problems.

- *Checksum verification*. The updated files are checked for integrity by the car using checksums after downloading. This verifies that the data was not corrupted while being sent.
- *Digital signatures*. The updated files' digital signatures are checked to ensure they are legitimate and come from a reliable source.
- *Compatibility checks*. When updating firmware, the FOTA system ensures that it is compatible with the car's current hardware and software by doing comprehensive compatibility tests. This lessens the likelihood of problems due to incompatibility.

12.1.2.7 Installation and activation

This step involves painstakingly updating the vehicle's ECUs (electronic control units) with the latest firmware. A copy of the current firmware is saved in case any problem arises during installation. The upgraded software is activated after a system reboot following installation, guaranteeing optimal performance. After an update is implemented, it is tested thoroughly to ensure it works as intended.

- *Backup*. It is common practice to make a copy of the current firmware or software before installing any updates or patches. If problems emerge during the upgrade process, you can easily revert to the prior version thanks to this protection.
- *Installation*. The updated software is programmed into the vehicle's ECUs and other related electronics. A system reboot is required to ensure the new firmware takes effect after installation.
- *Testing*. Verifying the vehicle's functionality after installing fresh firmware is essential. Problems that may have sprung up during the update are found and fixed during this testing phase.

12.1.2.8 Reporting and logging

During this stage, comprehensive records covering the entire updating procedure are painstakingly kept. These detailed records are useful for auditing and diagnosis; they shed light on whether a certain procedure was successful, which is crucial for establishing responsibility and resolving problems efficiently. All the update history is scrupulously recorded. Each step's success or failure is recorded in these logs. Diagnostics for problem identification and resolution and audits for compliance and accountability are two of the many uses for the data collected in the logs.

12.1.2.9 User notifications

In this phase, users are typically informed about the update process and its outcomes through notifications displayed on the vehicle's dashboard or via dedicated mobile apps. These notifications provide transparency and keep users informed about the status of the update.

The FOTA vehicle updates involve a meticulously orchestrated sequence of steps and stringent security measures to ensure the seamless and secure delivery of updated firmware and software to enhance vehicle functionality and safety.

12.2 LITERATURE REVIEW

This section presents a state-of-the-art review of the existing FOTA mechanisms. Figure 12.2 illustrates the FOTA frameworks and is subsequently elaborated upon in the following subsections of the literature review.

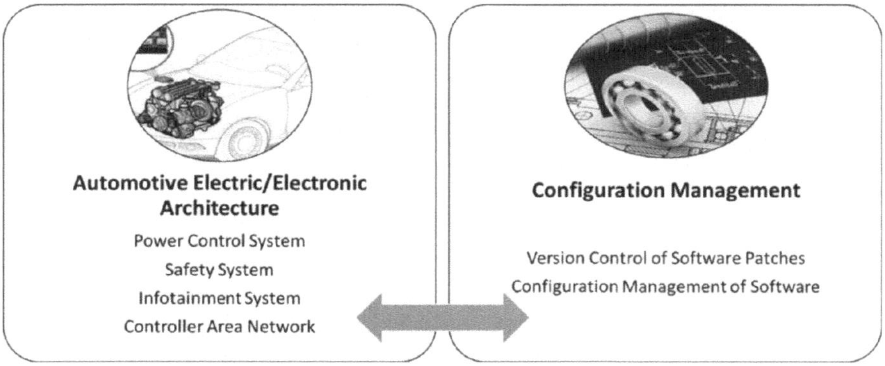

Figure 12.2 FOTA frameworks.

Traditional mechanical automotive devices' design process is more focused on achieving high reliability, performance, and safety. However, with the advent of electronic vehicles, which are majorly controlled by software, data security and privacy are also of utmost importance. These computerized automotive vehicles can connect to external networks to improve user experience. Whenever any automobile is connected to some external network, there is a high susceptibility to being hacked by adversaries. Because of these security and privacy risks, the software used by automobiles must be updated from time to time with the latest security patches [18].

Traditionally, such updates are done manually through a cable connection, which is time-consuming, resource-intensive, and an inefficient approach as a whole. To combat this, automobile companies have developed a special infrastructure called FOTA, which allows automobiles to update their firmware by connecting to the network. With the introduction of FOTA, the need to bring the car to the service station has been removed, the process has become less clumsy, and the updates reach consumers faster. However, with the advent of FOTA, a whole new range of cyberattacks has been introduced, that is, the installation of malicious versions of FOTA by attackers, by which the attackers would be able to perform physical attacks like disabling the brake, starting the engine, unlocking the child safety lock, etc. Eventually, these attacks can have serious consequences, even resulting in injury and death of human beings. Once an attacker installs malicious firmware, virtually (s)he will have complete control over the automotive vehicle. Thus, there is an urgent need to maintain the security of the FOTA updates.

Many researchers have developed different secured, efficient, privacy-aware, and scalable solutions for FOTA updates to make the process more user-friendly [19].

12.2.1 Automotive electric/electronic architecture

Automotive electric/electronic architecture refers to the structured design of electrical and electronic systems in vehicles. It takes care of how control units are arranged, how sensors work, and how controller area network (CAN) architecture is arranged. Understanding this architecture is crucial for FOTA updates because it often relies on high-bandwidth communication channels for transmitting large software packages; FOTA updates often involve multiple ECUs, like safety systems, power control systems, infotainment systems, etc. FOTA updates need to depend on the architecture's diagnostic capabilities to ensure the update's success, to perform rollback in case of failed updates, etc. In short, automotive electric/electronic architecture provides the foundation for seamless and secure FOTA updates in automotive vehicles.

12.2.2 Configuration management

Configuration management is critically important for FOTA updates. It keeps track of the different versions of the software packages, failing which the automotive vehicle may get a wrong software update. Moreover, it helps identify when and how many changes have been made to the firmware, which allows debugging and maintaining the software. It also provides traceability to the deployment history of the firmware. Therefore, it is clear that checking the configuration management for FOTA updates is of utmost importance. Having discussed the technical architecture and the configuration management, let us look into some of the well-known FOTA update standards in the next subsection.

12.2.3 Framework for automotive updates

One of the most common frameworks for automotive FOTA updates is UPTANE [20, 21], an open-source framework for data security and software configurability. In this system, the security of the software updates follows a hierarchical access control structure: access to the vehicle's infrastructure is gained by following multiple levels. Even if one attacker gains access to the vehicular network, (s)he is prevented from causing different harms to the automotive system. Primary-level electronic control units (ECUs) communicate directly with the server for updates, whereas secondary ECUs perform verification by checking the metadata of the primary ECUs. On the server side lies one image repository containing the ECU software components and the corresponding metadata for authentication purposes and an automatically managed director repository that identifies the software images for the update. The director repository is automatically managed and generally encrypts the images of the ECUs, provides different metadata to different primaries, and signs the target metadata using online keys.

Table 12.1 Comparison of FOTA update frameworks

Frameworks	*License type*	*Applicable for*	*Compliant with*
AUTOSAR FOTA framework [24]	Partially open-source	Automotive ECUs	AUTOSAR
eSync FOTA framework [22]	Proprietary	Automotive ECUs	ISO/IEC 26262
GENIVI FOTA framework [23]	Open-source	Automotive ECUs	ISO/IEC 26262
UPTANE [20]	Open-source	IoT and automotives	N/A

Another popular FOTA update platform with data integrity and authentication was proposed by eSync Alliance [22]—the consortium of automotive industries. The heart of the architecture is an orchestrator module that communicates with the server and helps deliver updates to eSync agents on each updatable ECU. The automakers digitally sign all the software updates, or the software developers, and the metadata of the OEMs are also included in the package. This helps the ECUs verify the integrity of the software updates.

Another automotive industry consortium, GENIVI Alliance [23], developed a scalable, secured, privacy-aware, and flexible FOTA framework, providing interoperability among different automakers. To make the system relevant, strategies like digital signature–based authentication methodology, encrypted communication over secure channels, data minimization and anonymization of personal data during update download, and secure bot methodologies are used.

AUTOSAR (automotive open system architecture) also provides a framework for supporting FOTA updates [24, 25]. The framework consists of modules for standardized authentication and encryption protocols, secure boot mechanisms for checking the integrity of software updates during boot time, access control for restricting unauthorized updates, and ECU abstraction. Table 12.1 describes the comparative analysis of various FOTA update frameworks.

12.3 SECURITY CHALLENGES IN FOTA UPDATES

Designing a generic, robust, and complete solution that will be compatible with most automotive vehicles is not easy [26, 27]. The designed solutions must overcome certain security challenges, as mentioned here:

- *Authentication and authorization.* Man-in-the-middle attack is quite common in case of an automatic firmware update over the network. Therefore, it is expected that only authorized entities should be able to initiate and apply FOTA updates. Suppose authentication is not incorporated into the solution. In that case, updates are probably injected with malicious updates, eventually leading to unauthorized access to the automotive vehicles.

- *Data integrity*. Maintaining the integrity of the software update packages while they are available for download is crucial. Suppose the software update packages are infested with malware by some attacker. In that case, the automotive vehicles will download the corrupted version of the software, eventually compromising the safety and security of the automotive vehicles.
- *Secure boot*. Besides ensuring the integrity of the software update, it is also necessary for the automotive vehicles to inspect whether they are running trusted and signed software, especially at the booting time. If the booting process is not secured, attackers could gain control over the car during the booting process itself, and the security would be compromised.
- *Encryption*. When the software update is in transmission, either being deployed by the automakers or being downloaded by automotive vehicles, it is necessary to maintain the update's confidentiality. A strong encryption mechanism is the basic requirement of protecting confidentiality.
- *Securing the FOTA interface and communication channel*. Inherent vulnerabilities in the communication channels used for FOTA updates may lead to side-channel attacks or something similar. Therefore, these vulnerabilities must be plugged in by implementing the security controls, as suggested by various FOTA security standards.
- *Rollback protection*. Sometimes, automakers provide the ability to revert to the older version of the software, and sometimes, the complete software update is rolled back for technical reasons. Ensuring that the attackers cannot force the automotive to go through such rollbacks is challenging.
- *Privacy concerns*. FOTA updates may involve transmitting the vehicle's confidential data, which may lead to privacy concerns. To protect the privacy of the vehicles and adhere to the privacy regulations of the land, incorporating required privacy controls is the need of the hour.
- *Cybersecurity compliance*. Besides the different security and privacy standards, different countries also have their own data privacy laws. It is challenging for the FOTA updates to comply with such varied security standards and laws.

A holistic approach comprising cryptography, secure communication, strong authorization and authentication mechanism, data minimization and anonymization techniques, compliance to best practices from security and privacy standards, and adherence to privacy and security laws of the countries is essential for addressing the identified security challenges. Automakers, security experts, and cyber law experts should come together to mitigate these challenges effectively and ensure the safety and security of FOTA vehicle updates. The security solutions required to overcome challenges, as shown in Figure 12.3, are discussed in further sections of the chapter.

12.4 ENHANCING FOTA UPDATE SECURITY

The previous section discussed the security challenges for FOTA updates. In this section, some security features will be discussed, as shown in Figure 12.4.

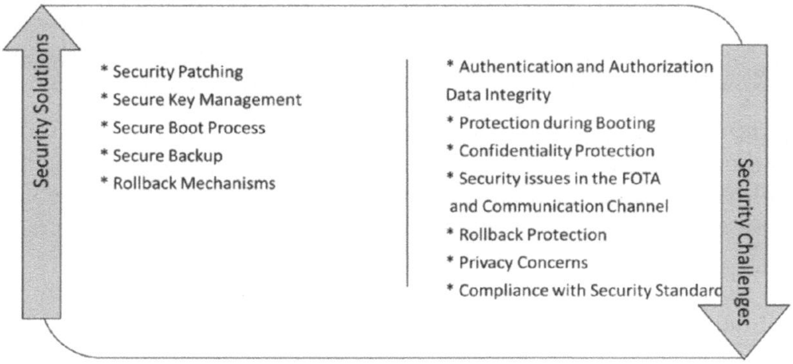

Figure 12.3 Security solutions vs. challenges.

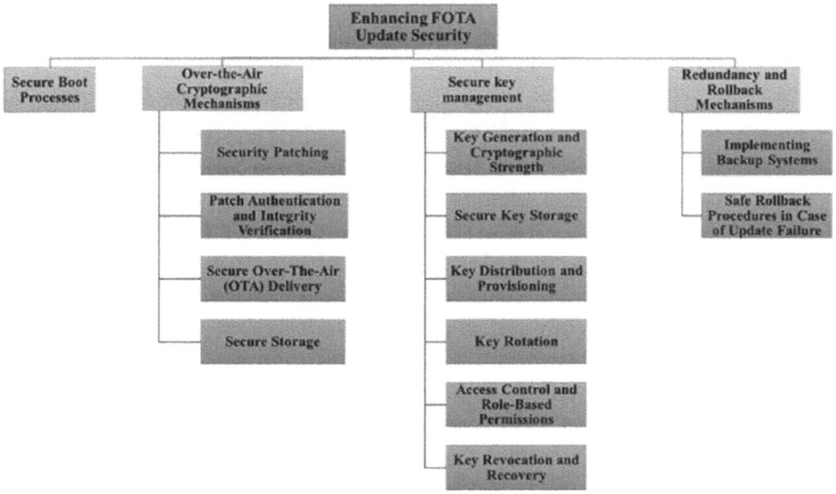

Figure 12.4 Security features of FOTA update.

12.4.1 Implementing secure boot processes

A secure boot process ensures that only the trusted and digitally signed software is run at the boot time, which protects the system against potential threats arising during the update process. When an automotive ECU boots up, secure boot confirms the digital signatures of the bootloader and firmware images and rejects the unsigned and untrusted updates [28]. Eventually, this establishes a chain of trust, starting from the bootloader, firmware, and operating system. The secure boot also ensures that confidential data, like encryption keys and user data, remains as it is without being tampered with. It also protects security against rollback attacks, where the attacker tries to force an old update to be downloaded. The secure boot helps in compliance maintenance and auditing by maintaining records of all updates [29].

12.4.2 Over-the-air cryptographic mechanisms

FOTA updates are particularly critical for vehicles, including automobiles and commercial vehicles, due to their complex and interconnected nature. Implementing strong cryptographic mechanisms for FOTA updates in vehicles is essential for several reasons:

- *Security patching.* Security patching is paramount in the automotive industry as vehicles rely more on intricate software systems encompassing infotainment, navigation, and core vehicle control systems [30]. This growing complexity also brings forth an elevated risk of vulnerabilities that malicious actors could exploit. Such vulnerabilities pose significant safety risks, potentially allowing unauthorized access to critical vehicle functions. However, FOTA updates have emerged as a critical tool for safeguarding vehicles and their occupants. FOTA updates enable manufacturers to respond swiftly to identified vulnerabilities, deploying patches and security updates remotely. This proactive approach not only bolsters the safety of vehicles on the road but also ensures that the automotive industry can adapt to emerging threats in real time, safeguarding the security of connected vehicles and the trust and well-being of their users. In the context of vehicles, cryptography plays a vital role in enhancing security patching processes.
- *Patch authentication and integrity verification.* Cryptographic signatures are used to verify the authenticity and integrity of software patches before they are applied to vehicle systems. When a patch is created, it can be signed with a private key held by the manufacturer or trusted entity. Vehicles, in turn, can verify the signature using the corresponding public key. This ensures that the patch is not tampered with during transmission and comes from a trusted source.

- *Secure over-the-air (OTA) delivery.* Patches delivered from manufacturers or service providers to automobiles must be securely transmitted, which is where cryptography comes in. Over-the-air updates sometimes contain critical software components; encrypting them protects against being intercepted, modified, or accessed by unauthorized parties.
- *Secure storage.* Patches can be kept secure in the vehicle's memory once installed. This prevents any tampering or invasion of privacy. Only authorized processes can decrypt and apply the patch data, since the cryptographic keys are securely handled.

12.4.3 Secure key management

Maintaining the reliability and safety of the updating procedure relies heavily on key management.

- *Key generation and cryptographic strength.* Strong cryptographic keys start with secure key generation, the first step in secure key management. Keys used to sign and encrypt firmware updates should be generated in accordance with industry standards. Keys that are sufficiently random and hence secure against brute force or cryptographic assaults should be generated using cryptographically safe random number generators [31].
- *Secure key storage.* FOTA update keys should be stored safely. It is common practice to store cryptographic keys in a hardware security module (HSM) or a trusted platform module (TPM). These hardware-based solutions provide tamper-resistant storage, which keeps the keys secure even if the hardware is compromised.
- *Key distribution and provisioning.* Secure systems for key distribution play a crucial role in FOTA. In particular, during the initial setup, manufacturers must send automobile encryption and signing keys in a secure manner. Methods for over-the-air key provisioning (OTAP) should be safe, using encrypted communications and protocols to prevent eavesdropping and tampering [32].
- *Rotation.* Key rotation is a fundamental practice in FOTA security. Periodically replacing encryption and signing keys enhances security by limiting the exposure of any single key. Older keys should be securely decommissioned to prevent their accidental or malicious use.
- *Access control and role-based permissions.* Access control mechanisms should be in place to govern who can access and manage cryptographic keys. Role-based access control (RBAC) ensures that only authorized personnel or processes have the necessary privileges to modify, access, or utilize keys [33].

- *Key revocation and recovery.* A clearly established procedure for key revocation and recovery is necessary in case of a security breach or key compromise. In the event of a security breach, this procedure should enable the quick deactivation of compromised keys and the safe installation of fresh, uncompromised keys.

12.4.4 Redundancy and rollback mechanisms

Implementing backup systems. If something goes wrong during a firmware upgrade, the car must still be able to function and be driven safely. The implementation of backup systems during firmware upgrades for vehicles involves strategies such as dual-image systems, parallel systems, data integrity checks, and emergency procedures. These mechanisms ensure seamless transitions between firmware versions, minimize downtime, and enhance overall system resilience, contributing to the safe and continuous operation of the vehicle.

- *Dual-image systems.* There are two copies of the firmware in many vehicles—one is actively running, and another is reserved for updates. During an update, the new firmware is written to the inactive partition [34]. If the update encounters issues or fails, the vehicle can quickly switch back to the previous, known good firmware version located in the active partition. This approach minimizes downtime and ensures that the vehicle can continue to function.
- *Parallel systems.* Some vehicles are equipped with redundant hardware or software systems. This means multiple sets of critical components (e.g., ECUs) can take over if the primary system encounters problems during an update. Redundancy at the hardware level provides an extra layer of resilience, making it less likely for a single failure to disrupt vehicle operation.
- *Data integrity checks.* The vehicle can perform data integrity checks on the new firmware throughout the update process. These checks ensure that the firmware being installed is not corrupted or compromised. If any issues are detected, the system can halt the update and revert to the previous firmware version.
- *Emergency procedures.* Manufacturers often implement procedures that allow the vehicle to revert to a safe operational state if a severe update failure occurs. This might involve disabling non-essential functions or initiating a safe mode to ensure the vehicle remains drivable.

12.4.5 Safe rollback procedures in case of update failure

Safely recovering from a failed update requires the use of a rollback mechanism. To enhance safe rollback procedures, consider implementing a fallback mechanism for unexpected issues, employing incremental rollback approaches to minimize disruption, and incorporating automated recovery procedures for efficient restoration without manual intervention.

- *Verification and validation.* Before a rollback starts, the system verifies and validates that the present firmware is faulty. Error code analysis, system health monitoring, and user feedback are all examples of checks that can be used for this purpose. Verifying the need for a reversal is essential [35].
- *Rollback version.* A stable firmware version is kept as a "rollback" in case something goes wrong. This release undergoes extensive testing and validation to ensure it does not introduce any security holes. It is good to have a safe backup plan.
- *Data preservation.* When performing a rollback, the system keeps your most important files, configurations, and user preferences intact. This ensures that the vehicle may pick up where it left off, data-wise, after a rollback.
- *User notification.* The vehicle's system alerts the driver or owner clearly and straightforwardly of any changes to the vehicle's functioning or features that may occur due to the rewind procedure. Users' faith can be preserved through open channels of communication.
- *Logging and reporting.* Both the update and rollback procedures are logged in detail. These records are useful for determining what went wrong with the update and how to fix it in the future.
- *Testing and validation.* During development, rollback processes are put through extensive testing and validation. This assures their usefulness and safety in practical situations and aids in locating and fixing problems.

12.5 FUTURE DIRECTIONS AND POTENTIAL FURTHER RESEARCH

Enhancing the security of FOTA updates in automotive cyber-physical systems is crucial for ensuring the safety and integrity of modern vehicles. As technology evolves and new threats emerge, there are several future directions and potential areas for further research in this field:

- *Blockchain and distributed ledger technologies.* Explore the use of blockchain and distributed ledger technologies to enhance the security and transparency of FOTA updates. Blockchain can provide a tamper-proof and decentralized ledger for tracking updates and verifying their authenticity.
- *Zero trust architecture.* Develop and implement a zero trust architecture for FOTA updates, which assumes that no device or user should be trusted by default, even if they are within the internal network.

This approach ensures continuous verification and authentication of devices and updates.

- *Machine learning for intrusion detection*. Investigate the use of machine learning algorithms for intrusion detection in automotive systems. ML can help identify anomalous behavior patterns and potential security breaches in real time.
- *Security information and event management (SIEM)*. Integrate SIEM solutions into automotive cyber-physical systems to provide centralized monitoring and real-time response to security events. This can help identify and mitigate security threats promptly.
- *Quantum-safe cryptography*. As quantum computing becomes more powerful, research quantum-safe cryptography algorithms to ensure that FOTA updates remain secure against quantum attacks.
- *Integration of hardware security*. Investigate the integration of hardware-based security solutions, such as hardware security modules (HSMs) or secure enclaves, to enhance the security of FOTA updates.
- *Regulatory compliance and certification*. Research methods to streamline the process of obtaining regulatory compliance and security certifications for FOTA update systems, ensuring that they meet industry standards.

Securing FOTA updates in automotive cyber-physical systems is an ongoing challenge that requires continuous research and adaptation to evolving threats. Collaborative efforts between industry stakeholders, cybersecurity researchers, and policymakers will be essential in addressing these challenges and ensuring the safety and security of connected vehicles.

12.6 CONCLUSION

The core contribution of this chapter lies in providing an extensive overview of FOTA update processes, highlighting their crucial role in modern systems. We thoroughly examined the security challenges inherent in FOTA updates and elucidated various security features aimed at effectively addressing these challenges. Additionally, we explored future trajectories and potential avenues for ongoing research in this important domain. By shedding light on the complexities of FOTA updates, we aim to stimulate discussions and innovations to enhance the security of cyber-physical systems, especially in the automotive industry. Our commitment to advancing understanding and fortifying defenses against cyber threats underscores the significance of our contribution. As we navigate the dynamic landscape of technology and cybersecurity, it is vital to remain vigilant and proactive. Our exploration serves as a foundation for future efforts, guiding stakeholders toward a more secure and resilient future in the realm of FOTA updates and cyber-physical systems.

REFERENCES

[1] Cheng, A., Yin, J., Ma, D. and Dang, X., 2020, August. Application and research of hybrid encryption algorithm in vehicle FOTA system. In *2020 Chinese Control And Decision Conference (CCDC)* (pp. 4988–4993). IEEE.

[2] Nilsson, D. K., Phung, P. H. and Larson, U. E., 2008, May. Vehicle ECU classification based on safety-security characteristics. In *IET Road Transport Information and Control-RTIC 2008 and ITS United Kingdom Members' Conference* (pp. 1–7). IET.

[3] Wang, Z., Han, J. J. and Miao, T., 2019, July. An efficient and dependable FOTA-based upgrade mechanism for in-vehicle systems. In *2019 International Conference on Internet of Things (iThings) and IEEE Green Computing and Communications (GreenCom) and IEEE Cyber, Physical and Social Computing (CPSCom) and IEEE Smart Data (SmartData)* (pp. 196–201). IEEE.

[4] Nilsson, D. K., Sun, L. and Nakajima, T., 2008, November. A framework for self-verification of firmware updates over the air in vehicle ECUs. In *2008 IEEE Globecom Workshops* (pp. 1–5). IEEE.

[5] Qin, G., Dong, X., Yang, L., Wang, W., Xu, Y. and Wang, Y., 2022, October. Research on secure FOTA upgrade method for intelligent connected vehicle based on new domain controller architecture. In *Third international conference on computer communication and network security (CCNS 2022)* (Vol. 12453, pp. 301–307). SPIE.

[6] Vrachkov, D. G. and Todorov, D. G., 2020, September. Research of the systems for Firmware Over The Air (FOTA) and Wireless Diagnostic in the new vehicles. In *2020 XXIX International Scientific Conference Electronics (ET)* (pp. 1–4). IEEE.

[7] Mirfakhraie, T., Vitor, G. and Grogan, K., 2018, July. Applicable protocol for updating firmware of automotive hvac electronic control units (ECUs) over the air. In *2018 IEEE International Conference on Internet of Things (iThings) and IEEE Green Computing and Communications (GreenCom) and IEEE Cyber, Physical and Social Computing (CPSCom) and IEEE Smart Data (SmartData)* (pp. 21–26). IEEE.

[8] Carsten, P., Andel, T. R., Yampolskiy, M. and McDonald, J. T., 2015, April. In-vehicle networks: Attacks, vulnerabilities, and proposed solutions. In *Proceedings of the 10th Annual Cyber and Information Security Research Conference, ACM Digital Library* (pp. 1–8). https://doi.org/10.1145/2746266.2746267.

[9] Luo, F. and Hou, S., 2019. Cyberattacks and countermeasures for intelligent and connected vehicles. *SAE International Journal of Passenger Cars-Electronic and Electrical Systems*, *12*, pp. 55–66.

[10] Nilsson, D. K., Larson, U. E., Picasso, F. and Jonsson, E., 2009. A first simulation of attacks in the automotive network communications protocol flexray. In *Proceedings of the International Workshop on Computational Intelligence in Security for Information Systems CISIS'08* (pp. 84–91). Springer.

[11] Subke, P. and Moshref, M., 2019. Improvement of the resilience of a cyber-physical remote diagnostic communication system against cyber attacks. *SAE International Journal of Advances and Current Practices in Mobility*, *1*, pp. 499–511.

[12] Kukkala, V. K., Thiruloga, S. V. and Pasricha, S., 2022. Roadmap for cybersecurity in autonomous vehicles. *IEEE Consumer Electronics Magazine*, *11*(6), pp. 13–23.

[13] Larson, U. E., Nilsson, D. K. and Jonsson, E., 2008, June. An approach to specification-based attack detection for in-vehicle networks. In *2008 IEEE Intelligent Vehicles Symposium* (pp. 220–225). IEEE.

[14] Borse, M., Shendkar, P., Undre, Y., Mahadik, A. and Patil, R., 2023. Study of hybrid cryptographic techniques for vehicle FOTA system. In *Mobile Computing and Sustainable Informatics: Proceedings of ICMCSI 2023* (pp. 417–430). Springer Nature.

[15] Mayilsamy, K., Ramachandran, N., Moses, B. J. S. and Ravikumar, A., 2022. A hybrid approach to enhance data security in wireless vehicle firmware update process. *Wireless Personal Communications*, *125*(1), pp. 665–684.

[16] Nikic, V., Bortnik, D., Lukic, M. and Mezei, I., 2021, November. Firmware updates over the air using nb-iot wireless technology. In *2021 29th Telecommunications Forum (TELFOR)* (pp. 1–4). IEEE.

[17] El Jaouhari, S. and Bouvet, E., 2022. Secure firmware over-the-air updates for IoT: Survey, challenges, and discussions. *Internet of Things*, *18*, p. 100508.

[18] Mansor, H., Markantonakis, K., Akram, R. N. and Mayes, K., 2015. Let's get mobile: Secure FOTA for automotive system. In *Network and System Security: 9th International Conference, NSS 2015, New York, November 3–5, Proceedings 9* (pp. 503–510). Springer International Publishing.

[19] Sowmya, K., Srinivasan, C., Lakshmy, K. V. and Kumar Bansal, T., 2021. A secure protocol for the delivery of firmware updates over the air in iot devices. In *Soft Computing and Signal Processing: Proceedings of 3rd ICSCSP 2020, Volume 1* (pp. 213–224). Springer.

[20] Kuppusamy, T. K., DeLong, L. A. and Cappos, J., 2018. Uptane: Security and customizability of software updates for vehicles. *IEEE Vehicular Technology Magazine*, *13*(1), pp. 66–73.

[21] Karthik, T., Brown, A., Awwad, S., McCoy, D., Bielawski, R., Mott, C., Lauzon, S., Weimerskirch, A. and Cappos, J., 2016, November. Uptane: Securing software updates for automobiles. In *International Conference on Embedded Security in Car* (pp. 1–11). https://ssl.engineering.nyu.edu/papers/kuppusamy_escar_16.pdf.

[22] Guissouma, H., Hohl, C. P., Lesniak, F., Schindewolf, M., Becker, J. and Sax, E., 2022. Lifecycle management of automotive safety-critical over the air updates: A systems approach. *IEEE Access*, *10*, pp. 57696–57717.

[23] GENIVI, W3C. Vehicle Signal Specification 2019. https://genivi.github.io/vehicle_signal_specification/ (accessed June 23, 2023).

[24] Sivakumar, P., Pavithra, A., Somasundarum, S. K., Somanathan, P. K. and Manimuthu, A., 2022. Role of AUTOSAR in automotive software trends. In *Software Engineering for Automotive Systems: Principles and Applications* (pp. 1–16). CRC Press, Taylor & Francis.

[25] Zerfowski, D. and Crepin, J., 2019. Vehicle computer-automotive-softwareentwicklung neu gedacht. *ATZelektronik*, *14*(7), pp. 36–41.

[26] El Jaouhari, S. and Bouvet, E., 2022. Secure firmware over-the-air updates for IoT: Survey, challenges, and discussions. *Internet of Things*, *18*, p. 100508.

[27] Doddapaneni, K., Lakkundi, R., Rao, S., Kulkarni, S. G. and Bhat, B., 2017, October. Secure fota object for IoT. In *2017 IEEE 42nd Conference on Local Computer Networks Workshops (LCN Workshops)* (pp. 154–159). IEEE.

[28] Siddiqui, A. S., Gui, Y. and Saqib, F., 2020. Secure boot for reconfigurable architectures. *Cryptography*, *4*(4), p. 26.

[29] Gedeon, A. S., Buttyán, L. and Papp, D. F., 2020. *Secure Boot and Firmware Update on a Microcontroller-Based Embedded Board*. Faculty of Electrical Engineering and Informatics, Department of Networked Systems and Services, Budapest University of Technology and Economics.

[30] Cheng, S. M., Chen, P. Y., Lin, C. C. and Hsiao, H. C., 2017. Traffic-aware patching for cyber security in mobile IoT. *IEEE Communications Magazine*, *55*(7), pp. 29–35.

[31] Crowther, K. G., Upadrashta, R. and Ramachandra, G., 2022, November. Securing over-the-air firmware updates (FOTA) for Industrial Internet of Things (IIOT) devices. In *2022 IEEE International Symposium on Technologies for Homeland Security (HST)* (pp. 1–8). IEEE.

[32] Rohini, P. P., 2004. *Over-the-Air Provisioning in CDMA*. Gemplus Technologies.

[33] Nilsson, D. K., Larson, U. E. and Jonsson, E., 2008, June. Low-cost key management for hierarchical wireless vehicle networks. In *2008 IEEE Intelligent Vehicles Symposium* (pp. 476–481). IEEE.

[34] Akiyama, A., Kobayashi, N., Mutoh, E., Kumagai, H., Yamada, H. and Ishii, H., 2010, August. Dual-image guidance system for autonomous vehicle on fast focusing and RGB similarity operation. In *Novel Optical Systems Design and Optimization XIII* (Vol. 7787, pp. 107–114). SPIE.

[35] Hampel, R. and Teleca, S., 2012, September 19. Keeping the connected car current with SOTA/FOTA. *Automative Linux Summit*, The Linux Foundation. https://events.static.linuxfound.org/images/stories/pdf/als2012_hampel.pdf.

Chapter 13

Advanced computational techniques for improving resilience of critical energy infrastructure under cyber-physical attacks

Nawaf Nazir, Sai Pushpak Nandanoori, Thanh Long Vu, Sayak Mukherjee, Soumya Kundu, and Veronica Adetola

LIST OF ABBREVIATIONS

ADMM	alternating direction method of multipliers
API	application programming interface
CPS	cyber-physical system
CVaR	conditional value at risk
DER	distributed energy resources
DRL	deep reinforcement learning
DRO	distributionally robust optimization
FERC	Federal Energy Regulatory Commission
FL	federated learning
GFM	grid-forming inverters
HELICS	hierarchical engine for large-scale infrastructure co-simulation
IEEE	Institute of Electrical and Electronic Engineers
IBR	inverter-based resources
MDP	Markov decision process
MG	microgrid
OPF	optimal power flow
PCC	point of common coupling
P2P	peer-to-peer
RL	reinforcement learning
SAC	soft actor-critic
SDP	semi-definite program
SOS	sum of squares
VaR	value at risk

13.1 INTRODUCTION

In recent years, critical energy infrastructure has been constantly under stress from the increasing disruptions caused by wildfires, hurricanes, other weather-related extreme events, and cyberattacks. Hence, it becomes

DOI: 10.1201/9781003559993-13

paramount to make critical infrastructure resilient to such threats, such as the energy grid. In this regard, the US Federal Energy Regulatory Commission (FERC) has defined *grid resilience* as the "[a]bility to withstand and reduce the magnitude and duration of disruptive events, which includes the capability to anticipate, absorb, adapt to, and rapidly recover from such an event." In this direction, this chapter presents recent advances in optimization techniques and reinforcement learning that can significantly improve the resiliency of energy systems under such threats.

Many classes of such cyberattacks have been discussed in the literature. One type of such attack includes injecting measurement errors, which can be constant offsets or potentially time-varying and, left untreated, may eventually lead to ramp-induced attacks. Another class of attacks includes replay attacks, whereby the measurement output is maliciously changed to reflect the value at a previous timestamp, thus impacting future decisions. A third class of attacks entails corruption of state estimation, whereby the estimated quantities required for future decisions are maliciously changed. Another class of attacks is coordinated attacks, whereby an attacker often has access to multiple critical information of the system and can maliciously pose threats in a coordinated manner, from multiple sources.

The rest of the chapter is organized as follows: Section 13.2 presents several probabilistic optimization techniques to make systems resilient to cyber-physical events. Section 13.3 presents algorithms that enable decision-making agents in cyber-physical networks to act both autonomously and in collaboration to enforce assured resilience across spatiotemporal layers under adversarial scenarios. Section 13.4 presents a reinforcement learning–based method for resilient control under limited system knowledge and limited data sharing. Section 13.5 presents reconfiguration algorithms for optimal interconnection topology that reduces the risks of attacks. Finally, Section 13.6 provides some concluding remarks on the different techniques for resiliency presented in this chapter.

13.2 RESILIENCY THROUGH DISTRIBUTIONAL ROBUST OPTIMIZATION

Cyber-physical events are notoriously hard to predict and numerous in nature. As such, making a system resilient to every possible disruption can quickly become infeasible. Such an approach can also make the system operation overly conservative and impractical. Furthermore, the probability distribution of the occurrence of such events is difficult to predict, and reliable historical data available on such disruptive events is often very sparse. This section presents several optimization techniques that utilize the probability of various cyber-physical events and a distributionally robust optimization (DRO) formulation robust to the sparsity of the available historical data.

13.2.1 Background and significance

Critical energy infrastructure has undergone significant changes in the past decades, which has made these systems more vulnerable to breakdowns and cyberattacks [1, 2]. The resiliency of critical energy infrastructure against various weather-related outages [3] and cyber events [3, 4], including malicious attacks, has been studied in the literature [5, 6]. In the presented optimization methods, this risks associated with various cyber-physical events are assessed, and the system is operated in a resilient manner based on these assessments. This allows us to be resilient against the risks without being overly conservative. Then, to deal with the sparsity of data available on such cyber-physical events, we present a DRO formulation that is robust to a range of disturbance distributions.

Chance constraints constitute a means to provide certain guarantees on constraint satisfaction under uncertainty [7] and have found applications in several domains, including in power system optimization under uncertainty. However, a major drawback of chance constraint formulations is that they only consider the probability of constraint violation and not the impact. In many critical infrastructure systems, minimizing the impact of uncertainties is far more significant. Furthermore, many of these methods require assumptions on the probability distribution of the uncertainty and convex reformulations, which exist only for a very small set of such distributions (e.g., Gaussian) and may not hold in practice.

DRO constraint problems have been well-studied in the literature [8]. In power systems, to deal with distributional robustness, the authors in [9] consider a family of uncertainty distributions and provide convex reformulations for them. They also consider a conditional value at risk (CVaR) approach that accounts for the risk in constraint violation instead of the probability. Similarly, the authors in [10] formulate a tractable problem for log-concave distributions, and in [11], the authors formulate a moment-based ambiguity set for joint chance constraint formulation, whereas in [12], a primal-dual sub-gradient method is considered to optimize risk constrained optimization problems. Even though these methods consider the risk of constraint violations, they can often be overly conservative and, furthermore, still require assumptions on the underlying distribution (e.g., moments) to be accurate.

In case the uncertainty distribution is not known beforehand, sample-based approaches have been used in the literature [13]. However, in rare events, sampling-based approaches fall short as they require an impractically large number of samples to provide acceptable solutions [14], which is often not the case in practice. To overcome this, Wasserstein ambiguity sets have been recently proposed [15]. With Wasserstein ambiguity sets, the chance and risk constraints hold for a family of distributions within a distance (called the Wasserstein distance) from the observed uncertainty realizations.

13.2.2 Modeling and problem formulation

13.2.2.1 Baseline optimization

The baseline optimization problem for a cyber-physical system can be formulated as

$$\underset{x}{\text{Min}} f(x) \tag{13.1a}$$

$$s.t.\ g(x) \leq 0 \tag{13.1b}$$

where x denotes the decision variables (dispatch of resources), $f(x)$ is the problem objective (e.g., minimize operating cost), and $g(x)$ is the system constraint (e.g., system limits). The optimization problem in (1) does not consider uncertainty in system parameters or forecasts, which are critical for practical applications. The next section will consider the optimization problem under uncertainty.

13.2.2.2 Optimization under uncertainty

Uncertainty in the system can arise from various adversarial cyber-physical events. To be resilient against such events, it is important to have sufficient reserves in the dispatchable resources so that the resources can adjust their output values in response to such adversarial events. This section aims to formulate probabilistic methods that determine the amount of reserves required to deal with such adversarial events in a resilient and cost-effective manner. If we consider a bounded set of uncertainty $w \in \Omega_W$, then the stochastic optimization problem can be expressed as

$$\min_{x} f(x, w) \tag{13.2a}$$

$$s.t.\ g(x, w) \leq 0, \forall w \in \Omega_W \tag{13.2b}$$

where x also includes additional decision variables in the form of reserves. However, the preceding optimization problem is infinite-dimensional and cannot be solved scalable. The next section briefly describes some of the traditional probabilistic methods used to solve (2).

13.2.2.3 Chance constraint formulations

In this section, we briefly present various stochastic optimization reformulations to solve (2). One common approach is to make the optimization problem in (2) robust against a specified set of adversarial events. The robust optimization can be formulated as

$$\min_{x} \min_{w} f(x,w) \tag{13.3a}$$

$$s.t.\ g(x,w) \leq 0, \forall w \in \Omega_W \tag{13.3b}$$

The preceding optimization problem can be reformulated using the explicit maximization method, details of which can be found in [16, 17]. However, a robust approach can be overly conservative, especially when the set $w \in \Omega_W$ is large. In such situations, chance constraint–based methods can be employed to avoid constraint violation with a certain probability. A general chance constraint optimization can be expressed as

$$\min_{x} E[f(x,w)] \tag{13.4a}$$

$$s.t. P[g(x,w) \leq 0] \geq 1-\rho \tag{13.4b}$$

where E is the expected value operator, P is the probability operator, and ρ is the pre-specified allowable constraint violation. Chance constraints are closely related to value at risk (VaR), a measure used in the finance industry [18]. Several techniques exist in the literature to obtain tractable reformulations of chance constraints. Most of these methods involve generating functions that produce a family of convex approximations for the probabilistic chance constraint [9]. More details about the chance constraint techniques in power systems and their convex reformulations can also be found in [19].

One of the main drawbacks of chance constraint methods is that they consider the probability of an event and not the magnitude of constraint violation (risk) of the event. To consider the risk of constraint violation, the CVaR method has been employed in the literature [9, 14], especially to deal with low-probability, high-impact events that would otherwise be neglected by the chance constraint formulation, especially for certain types of uncertainty distributions [18]. One such CVaR formulation is obtained by utilizing the Markov generating function. The CVaR-based optimization formulation in that case can be expressed as

$$\min_{x} E[f(x,w)] \tag{13.5a}$$

$$s.t.\ E[g(x,w)+t]_{+} \leq t\rho \tag{13.5b}$$

where t is an optimization variable and $[\,.\,]_+ = max(.,0)$. Further details on the Markov generating functions and other CVaR methods can be found in [9, 19]. Even though CVaR-related methods account for low-probability, high-impact events, they have an underlying assumption on the probability distribution of the uncertainty, that is, the historical data used to represent $w \in \Omega_W$ matches the actual distribution.

13.2.3 Distributionally robust optimization formulation

In this section, we will present the DRO formulation. Previous work in literature [20] has used the Wasserstein ambiguity set method for DRO. In the next section, we will outline this approach, which will then be utilized within the DRO formulation.

13.2.3.1 Wasserstein ambiguity set

Here, we present an overview of the Wasserstein ambiguity set method for DRO. In general, the Wasserstein distance between marginal distributions Q_1 and Q_2 is defined as

$$W(Q_1,Q_2) = \inf_{\pi} \{ \int_{\Xi^2} \|\zeta_1 - \zeta_2\| \pi(d\zeta_1, d\zeta_2) \} \tag{13.6}$$

where π is the joint distribution of ζ_1 and ζ_2, with marginal distributions $Q_1, Q_2 \in P(\Xi)$.

In our case, the Wasserstein distance $W(.,.)$ between the true distribution P and sample probability distribution $\hat{P}_N$ is used to define the ambiguity set $\hat{P}_N$ given by

$$\hat{P}_N = \left\{ P \in P(\Xi) : W\left(P, \hat{P}_N\right) < \epsilon(N) \right\} \tag{13.7}$$

where $P(\Xi)$ represents the set of all probability distributions with support Ξ and where

$$\hat{P}_N = \frac{1}{N} \sum_{k=1}^{N} \delta_{\hat{\zeta}(k)} \tag{13.8}$$

where $\hat{\zeta}(k) \in R^N$ are the samples of distribution and N is the number of such samples available. In literature [20], the value of $\epsilon(N)$ is usually chosen to be

$$\epsilon(N) = C\sqrt{\frac{1}{N}} \log\left(\frac{1}{1-\beta}\right) \tag{13.9}$$

where β is the confidence level of the required constraint satisfaction (e.g., β = 0.05 when the constraints are required to be satisfied with a 95% probability).

In the previous expression, the value of C is calculated as

$$C \approx \inf_{\alpha>0} \sqrt{\frac{1}{2\alpha}\left(1+\log(\frac{1}{N}\sum_{k=1}^{N} e^{\alpha\left\|\hat{\varsigma}(k)-\hat{\mu}\right\|_1^2})\right)} \tag{13.10}$$

where $\hat{\mu}$ is the sample mean. The minimization over α can be solved using the bisection search method. A more detailed description of these expressions and how to obtain the search solutions can be found in [20].

Based on the earlier Wasserstein ambiguity set expression in (7), the stochastic constraint of the form in (4) can be expressed in the DRO form as

$$\inf_{P\in\hat{P}_N} P\left[g(x,w)\leq 0\right] \geq 1-\rho \tag{13.11}$$

Since the preceding constraint is intractable, the goal of this section is to develop a convex approximation of (11) of the form

$$g(x,w)\leq 0, \forall w \in U \tag{13.12}$$

where U is a deterministic uncertainty set such that the robust deterministic constraint in (12) implies satisfaction of the DRO constraint in (11). The rest of the section will focus on finding such a set U that is computationally tractable to obtain.

Now, consider the transformation $v = \hat{\Sigma}^{\frac{1}{2}}(w-\hat{\mu})$, where $\hat{\mu}$ is the sample mean and $\hat{\Sigma}$ is the sample covariance, which implies that

$$U = \hat{\Sigma}^{\frac{1}{2}} V + \hat{\mu} \tag{13.13}$$

Now, based on the set V, and considering Q and Q_N to denote the true distribution and empirical distribution of V, the ambiguity set $\hat{Q}_N$, similar to the expression in (7), can be obtained as

$$\sup_{Q\in\hat{Q}_N} Q\left[v \notin V\right] \leq \rho \tag{13.14}$$

which is now an equivalent representation of the expression in (12), where the set V can now be expressed as

$$V(\sigma) = \{-\sigma 1 < v < \sigma 1\} \tag{13.15}$$

where σ can be found from solving the optimization problem:

$$\min_{0 \le \lambda, 0 \le \sigma \le \sigma_{\max}} \sigma \ s.t.\ h(\sigma, \lambda) \le \rho \tag{13.16}$$

where

$$h(\sigma, \lambda) = \lambda\epsilon + \frac{1}{N}\sum_{k=1}^{N}\left(1 - \lambda\left(\sigma - \left\|\hat{v}^{(k)}\right\|_{\infty}\right)^{+}\right)^{+} \tag{13.17}$$

The preceding optimization problem can again be solved with the bisection search method, as outlined in [20].

Based on the solution of σ, the set $V(\sigma)$ can be expressed as

$$V(\sigma) = Conv(\{\pm\sigma, \pm\sigma\}) \tag{13.18}$$

where $Conv$ represents the convex combination of the terms. Then the set U can be expressed as

$$U = Conv\left(\left\{u^{(1)}, u^{(2)}, \ldots u^{(2m)}\right\}\right) \tag{13.19}$$

$$u^{(i)} = \hat{\Sigma}^{\frac{1}{2}}\hat{v}^{(i)} + \hat{\mu}, \quad 1 \le i \le 2^m \tag{13.20}$$

This results in the following deterministic DRO formulation based on Wasserstein ambiguity set:

$$\min_{x} f(x) \tag{13.21a}$$

$$s.t.\ g\left(x, u^{(i)}\right) \le 0, \quad 1 \le i \le 2^m \tag{13.21b}$$

where f and g are the objective function and the constraint function, respectively, as described earlier in (1). Since the set U can be calculated beforehand, the optimization problem (21) can be solved in a scalable manner using off-the-shelf convex optimization solvers.

Remark 1: It should be noted that the robustness of the DRO formulation can be tuned through the parameter β, in a similar manner to the allowable constraint violation parameter ρ in the case of CVaR formulation. However, the difference between the two is that DRO accounts for the robustness within the distribution itself by considering a family of distributions through a metric.

13.3 RESILIENT AUTONOMOUS DECENTRALIZED AND COORDINATED CONTROLS

In this section, we present and demonstrate novel, adaptive, lightweight algorithms that enable the decision-making agents in a large cyber-physical network to act both autonomously and in collaborative harmony to enforce assured resilience across spatiotemporal layers, even under unforeseen adversarial scenarios (e.g., high-impact, low-probability events). Toward this end, the proposed framework will serve as minimally invasive add-on layers that bridge the existing (faster, reactive) local myopic controls and (slower, predictive) centralized optimization. Importantly, the proposed algorithms will enable the multi-agent network to autonomously and collaboratively enforce resilient operation under no or limited communication environment typical of severe cyber-physical adversarial events

13.3.1 Background and significance

As cyber-physical networks become more complex and larger, and with different parts of the network often operated independently by different stakeholders, there is a strong need for advanced algorithms that allow multiple decision-making agents within the network to act autonomously and collaboratively to enforce resilience under all or most adversarial events. For any given cyber-physical system, there will be state limits that need to be maintained under all circumstances irrespective of the disturbances in the system, and if they are violated, then the objective must be to steer the system to within the specified limits using additional control strategies. Most of the existing works except [21, 22] enforce stability but do not take into account safety. Existing local (primary/device-level) controls within cyber-physical systems today (e.g., inverter controls in a microgrid) are designed and operated in a myopic manner, without any mechanism to actively enforce resilience via local measurement and control action. Furthermore, the current centralized (secondary/system-level) control operates at a slower timescale and is not appropriate for ensuring safety, stability, and operational resilience, especially in the presence of frequent stochastic disturbances or unforeseen adversarial events. Moreover, many adversarial events in critical cyber-physical networks propagate as a cascading failure, which often starts from local disruptions that travel through the network over a short span of time and create a system-wide impact (e.g., a blackout). Therefore, there is a need to develop agile, adaptive, and lightweight control solutions that allow the control agents to autonomously (under no communication) and semi-cooperatively (under limited communication) ensure system-wide resilience under adversarial events. Such solutions, as proposed in this work, should be minimally invasive, that is, need only limited information about the primary/device-level controls, which are often proprietary information and largely model-agnostic, that is, not tied to any specifics of the primary/device-level controls.

This section presents control mechanisms to make multi-agent cyber-physical systems more resilient to known and unknown attacks, even when there is no communication (via local autonomy) or limited communication (via semi-cooperative control). The safety-constrained control actions are designed to take into account the fast and highly nonlinear dynamics of the system and to ensure that the system remains safe during transitions and returns to safety as quickly as possible. Together, the local autonomous and semi-cooperative controls bridge the local myopic primary controls and the centralized dispatcher/coordinator.

13.3.2 Decentralized autonomous controls

In this subsection, we approach decentralized controls with proactive and reactive autonomous control actions.

13.3.2.1 Proactive autonomy

The proactive decentralized autonomous action guarantees operational resilience (and transient safety) to a range of known and likely adversarial events. These proactive autonomous actions will be in the form of (simple) algebraic equations designed offline and easily verified during online (real-time) applications. The proactive autonomous action supports the primary controls by appropriately (and minimally) modifying the local control settings (e.g., inverter power set points in a microgrid) and serves as a bridge between the existing hierarchy of (local/faster/myopic) primary and (centralized/slower) secondary controls.

Mathematically, we will approach the proactive local design as follows: Consider the nonlinear dynamical system of the form $\dot{x} = f(x,u,d)$, where the system states $x \in R^n$, the control inputs $u \in R^m$, and the bounded disturbances $d \in \Delta$. Let the safe region be denoted by C_{safe} (e.g., treat this as the region enclosed by the frequency or voltage limits in a power network). The objective of the proactive phase is to ensure for all deterministic disturbances $d \in \Delta$ that the state evolution is always contained in the safe region, that is, $x(t) \in C_{safe} \forall t > 0$ (as shown in Figure 13.1, proactive phase). The resultant

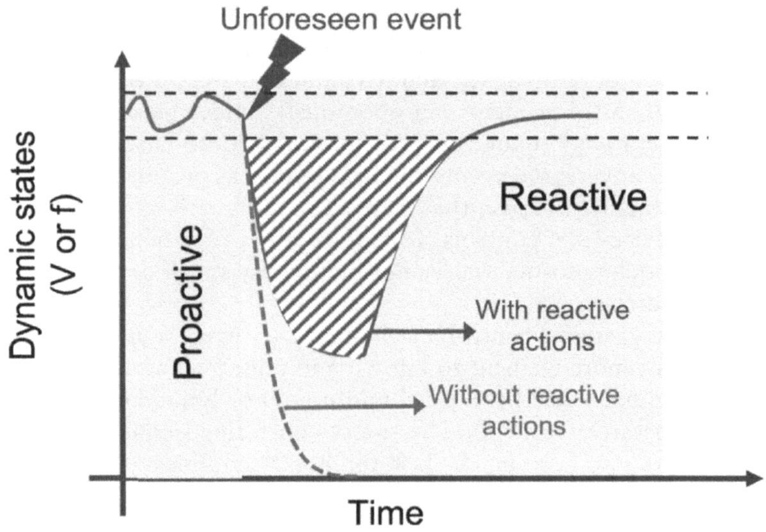

Figure 13.1 Resiliency curve during likely bounded disturbance and unforeseen events.

objective is to identify what are the safe control inputs such that the system evolution will not violate safety constraints, that is, $x(t) \in C_{safe} \forall t > 0$. There are several methods to compute such constrained controls, and here we focus on one such method, Nagumo's theorem [23], which gives the invariance of the closed safe set C_{safe} under $\dot{x} = f(x)$ if and only if for all $x \in C_{safe}, f(x) \in C(x)$, the Bouligand tangent cone to C_{safe} at x. Nagumo's theorem is then adapted for nonautonomous systems, $\dot{x} = f(x,u,d)$ such that the safe set C_{safe} is robust control invariant by $\dot{x} = f(x,u,d)$ if there exists a control law $u(t)$ such that for all $x(0) \in C_{safe}, d \in \Delta, x(t) \in C_{safe}$ e for all $t \geq 0$. We refer to [24] for a detailed treatment of the application of Nagumo's theorem to design proactive decentralized controls and their application to a microgrid network.

While proactive controls can keep the system operating safely within safe limits for bounded disturbances, there may be times when faults or adversarial actions push the system outside of those limits quickly. We next discuss reactive autonomy to recover the system from such a state and return to a safe region.

13.3.2.2 Reactive autonomy

The reactive, decentralized autonomous action guarantees the earliest return to a safe operating region during unforeseen (and large) adversarial disruptions, for example, high-impact, low-probability events. The reactive autonomous actions are computed online and come into action when a high-impact, low-probability event thrusts the system state outside the safe operating regime. The reactive autonomous actions update the myopic control settings to guarantee a robust return to the safe region and work with proactive autonomy once the system state reaches the safe region. In most cases, the system failure is imminent without the reactive control actions (as shown in Figure 13.1, red-dotted line after *an unforeseen event*). The system's operation is maintained by taking reactive actions, and resilience is achieved (as shown in Figure 13.1, brown-striped area in reactive phase).

Let the disturbance be such that $d \notin \Delta$ and it results in $x(t) \notin C_{safe}$. The objective of the reactive decentralized controls is to steer the system to a safe region such that $x(t+\delta) \in C_{safe}$ for a finite positive δ. The design of reactive control involves computing the control inputs in real time such that the system state returns to the safe region. One such constrained control method to identify the reactive control inputs involves efficiently solving for energy functions [25] to ensure a resilient response. The reactive autonomous controls in this work are designed by continuously monitoring the system's critical states, and when they deviate from the safe region, the control set points are updated as a function of the deviated error. Details about the reactive control design are not provided here as it is currently being pursued for a patent.

13.3.3 Coordinated controls

In the onset of adversarial events and other unforeseen events (such as losing an inverter in a microgrid), to ensure system-wide stability guarantees and achieve resilience, we present a constrained cooperative control that appropriately updates the control set points to the device or subsystem controls by transitioning the operating point (set by the centralized slower control). The cooperative control operates relatively at a faster timescale when compared to the supervisory centralized coordinator and is located a level below it but lies above the decentralized autonomous controls among the hierarchy of controls.

To complement the actions of the self-aware (and often myopic) local agents which autonomously act on local detection of adversarial events, the cross-layered coordinated control strategies are introduced to restore the system-wide performance goals (efficiency and stability) by local, safe, and minimal re-alignment of the set points. Real-time distributed coordination schemes, extending algorithms such as projected consensus [26], will be designed for semi-cooperatively tuning (myopic) local controls to enhance system-wide resilience following evasive local actions in response to adversarial events. As preliminary work, we consider the consensus-based cooperative design of controllers at a subset of network locations to mitigate the disturbance and maintain the system-wide objectives [27].

13.3.4 Supporting simulation studies

This subsection demonstrates the proposed decentralized and coordinated controls using a microgrid use case. The 123-bus feeder model with nine grid-forming (GFM) inverters is used. The GFM inverters have device-level controls, such as P-F and Q-V droop controls, which operate quickly to regulate frequency and voltage. There is also a centralized system coordinator that operates more slowly. The GridLAB-D + HELICS + Python co-simulation is used to demonstrate the role of the proposed resilience-promoting controls. To implement the controls, the states, such as the frequency, voltage, and real and reactive powers, are monitored at each inverter every time instance. The decentralized and coordinated controls then modify the real and reactive power set points to the inverters. The decentralized controls require the inverters to know their own droop gains and capacity. The coordinated controls need the communication topology and neighbor droop gains, as well as power measurements. A large disturbance is created while the system is in a steady state, pushing the system state (frequency) outside the safe region. The reactive controls act immediately and modify the real power set points to bring the system frequency back into the safe region.

Figure 13.2 shows how the system behaves without any decentralized controls. After 1 s into the simulation, a large load change is made, which causes the frequency to go outside the safe bounds, which are set to [59.9, 60.1]. Without any safety controls, there is no guarantee that the frequencies will remain inside the safe region. Figure 13.3 shows how the system behaves with reactive autonomous controls at the inverters. When the frequency goes outside the safe bounds, the reactive controls modify the real power set points to bring the frequencies back inside the safe region. The figure also shows how the real powers adjust to the new reference set points to steer the

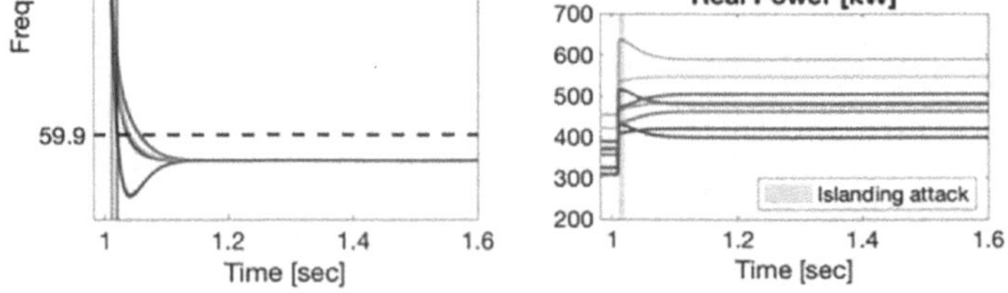

Figure 13.2 System evolution to a large disturbance at $t = 1$ s, and real power as well as real power set points without the proposed controls.

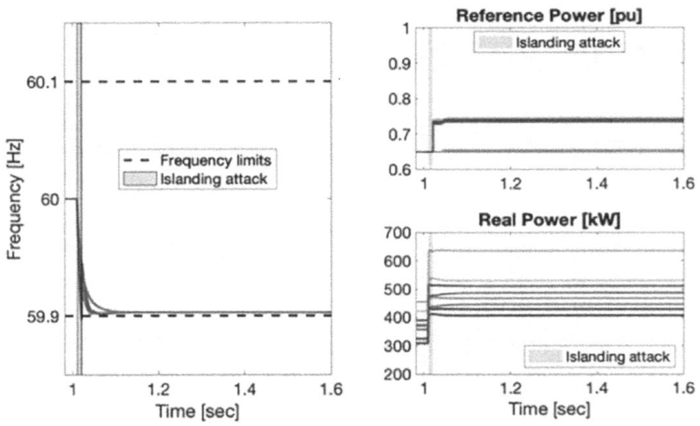

Figure 13.3 System evolution to a large disturbance at $t = 1$ s, and real power as well as real power set points with decentralized autonomous controls.

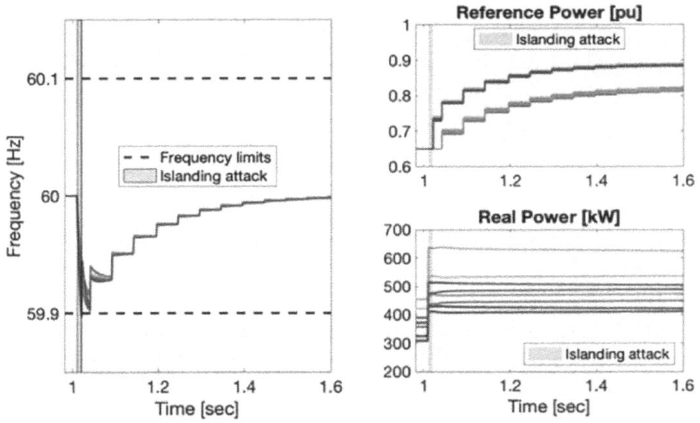

Figure 13.4 System evolution to a large disturbance at $t = 1$ s, and real power as well as real power set points with decentralized autonomous and coordinated controls.

frequencies inside the safe region. It is important to observe that the system frequency, although inside the safe region due to the action of the reactive controls, does not reach the nominal 60 Hz. The centralized coordinator operates slower and adjusts system-wide settings to achieve the desired frequency. However, allowing the system frequency to deviate too far from the desired value is not economical or desirable, so the coordinated controls intervene as soon as possible. Implementing the decentralized autonomous and coordinated controls minimizes frequency deviations outside the safe range and brings the frequency back to the desired value of 60 Hz, as shown in Figure 13.4.

13.4 REINFORCEMENT LEARNING FOR SYSTEM-LEVEL RESILIENT CONTROL

In this section, a new development on using learning-driven controls as an enabler for enhancing the resiliency of cyber-physical systems is described. For the specificity of this section, we present the techniques for the applications to networked microgrids as recently proposed in [28]. The methodology will present an automated decision-making technique that can support recovery in the presence of adversarial actions injected to some of the inverter-based resources.

13.4.1 A networked microgrid perspective

For attaining a net-zero-energy status by the year 2050, networked microgrids emerge as a highly sought-after solution for establishing self-sustaining power grids capable of efficiently integrating renewable energy sources. These resources seamlessly connect with the power grid through power-electronic devices, particularly converter and inverter technologies. Recent advancements in inverter design have brought grid-forming inverters to the forefront of research and development, as highlighted in [29]. GFMs possess the unique capability to function as controllable voltage sources, positioned behind coupling impedance and directly governing the voltage and frequency of the microgrid. For GFM-based microgrid systems, the control architecture comprises multiple tiers, spanning from primary to higher-level control layers, as referenced in [27].

In scenarios where networked microgrids operate under multi-party ownership models, different utilities or operators may own distinct zones within the microgrid network, with restricted data sharing and proprietary information exchange during operational phases. Furthermore, given the escalating complexity of microgrid operations and modelling uncertainties, obtaining precise knowledge of system dynamics becomes a formidable challenge. Consequently, we are interested in answering two critical questions:

1. How can we design higher-level controllers with limited knowledge about the networked microgrid under cyber events, thereby infusing resiliency into microgrid operations?
2. How can we address the issue of limited data sharing across interconnected microgrid networks while accommodating the dynamic electrical couplings that characterize these systems?

Reinforcement learning (RL) using a Markov decision process (MDP)–based framework tries to solve the control tasks with unknown environments or dynamics using interactions. This involves employing interactions with the environment through approaches like value-based or policy-gradient-based methods, or a combination of both, as seen in various studies [30]. Tackling learning control problems, particularly those optimizing across multiple agents within interconnected dynamic environments with distinct action and state spaces, presents notable challenges. Consequently, research has delved into multi-agent reinforcement learning, as evidenced by studies like [31]. RL has seen applications in voltage control [32], control of energy storage in microgrids [33], wide-area damping control [34], volt-VAr control in distribution grids [35], etc. For the learning problem with privacy constraints approaches, such as federated learning (FL), [36] can be very useful, which shares model parameters and gradients between the zones or entities instead of sharing new input data. However, Fed-RL [37] is at the development stage, and a few recent Fed-RL applications in power systems include decentralized volt-var control [38] and energy management for smart homes [39].

This research primarily concentrates on tackling the control design challenge to bolster the overall resilience of a networked microgrid. Our objective is to counteract the adverse impacts of hostile actions directed at the reference signals within the primary control loops of GFMs. We introduce a design framework to implement a "vertically" federated reinforcement learning system within a networked microgrid operating under multi-party ownership. As recently outlined in [28], this design architecture has been implemented and extensively validated by developing a customized training platform, a pivotal element of our research. The validation process involves employing the IEEE-123 bus test feeder (modified) as a benchmark system, featuring three interconnected microgrids.

13.4.2 Resilient reinforcement learning problem

13.4.2.1 Microgrid dynamics

In this subsection, we delve into the dynamics of microgrids. We consider a system composed of r interconnected microgrids, each comprising a total of N GFM inverters and a network with M buses. For the ith GFM inverter, we model it as an AC voltage source with an internal voltage denoted as E_i, and a phase angle represented as δ_i. Mathematically, we express these dynamics as $\delta_i = u_i^{\delta}, E_i = u_i^{V}$. Here, u_i^{δ}, u_i^{V} are the frequency and voltage control input signals to the inverter. The primary control of the GFM inverters constitutes the droop controls as follows:

$$\omega_i^{ref} = \omega_i^{nom} - m_{pi}\left(P_i - P_i^{set}\right) \tag{13.22}$$

$$V_i^{ref} = V_i^{set} - m_{Qi}\left(Q_i - Q_i^{nom}\right) \tag{13.23}$$

where frequency control input u_i^δ is set to the reference frequency ω_i^{ref}, that is, $u_i^\delta = \omega_i^{ref}$, and u_i^V is obtained by passing $V_i^{ref} - V_i$ through a proportional–integral (PI) regulator. Here, V_i, P_i, and Q_i respectively denote voltage, active power, and reactive power, with droop gains as m_{Pi} and m_{Qi}. The primary control intends to perform proportional power sharing and prevent the circulation of reactive power. However, due to its proportional nature, it cannot achieve precise regulation, resulting in steady-state deviations in both frequency and voltage, where secondary controls are required [29].

13.4.2.2 Resilient RL aspects

This section addresses the resilient aspects of RL controls, which act as a supervisory control layer atop existing primary and secondary controls, if applicable. The goal is to train RL agents in the presence of cyber vulnerabilities. We denote the RL outputs as P_i^{res} and V_i^{res}, representing the resilient control inputs. These higher-level control signals are then added to the nominal or pre-specified set points P_{i-nom}^{set} and V_{i-nom}^{set} as follows for the ith GFM inverter:

$$P_i^{set} = P_{i-nom}^{set} + P_i^{res}, V_i^{set} = V_{i-nom}^{set} + V_i^{res} \tag{13.24}$$

The concatenated RL control inputs, denoted as $u^{res} = \left[P_i^{res}, V_i^{res}\right]_{i=1,\ldots,N}$, are designed as a feedback function of the microgrid observations (O), which we will discuss later. During training, an inverter attack is emulated by introducing attack signals at randomly selected i^{th} inverters, perturbing the active power and voltage set points:

$$P_i^{set} = P_{i-nom}^{set} + P_i^{res} + P_i^{attack} \tag{13.25}$$

$$V_i^{set} = V_{i-nom}^{set} + V_i^{res} + V_i^{attack} \tag{13.26}$$

As a result, resilient controllers need to be designed to mitigate the effects of such adversaries. Due to the stochastic nature of the problem, rule-based design of such controllers is impractical. This motivated the use of an RL-based controller design, which was achieved by formulating the resilient microgrid control problem as a partially observable MDP. The MDP is defined by a tuple (S, A, P, r, γ) [40], where the state space (representing microgrid dynamics) $S \subset R^n$ and the action space (GFM inverter set points) $A \subset R^m$ are continuous. $P : S \times A \to S$ is the environment transition function, and $r : S \times A \to R$ is the reward function, and $\gamma \in (0,1)$ is the discount factor.

Observation space. Despite the intricate nature of microgrid dynamics encompassing numerous differential and algebraic variables, our attention is directed toward a specific subset of these variables, contingent upon the underlying problem. Without loss of generality, considering attacks at voltage set points, bus voltage magnitudes $V_i(t)$ (note that $V_i(t)$ is different from inverter voltage set points) are taken as the observation variable O.

Action space. The RL agents implement their actions using the P_i^{res} and V_i^{res} inputs for each individual grid-forming inverter; however, practical set point limiters are implemented to keep the inputs within tolerable bounds.

Rewards. The resilient design needs to keep the quality of service (QoS) variables within desired bounds. For this, the reward $r(t)$ at time t is defined as follows:

$$r(t) = (-cu_{ivld} \; if \; t \le t_a, -\sum_i Q_i \left\| V_i(t) - V_{i,ss} \right\|_2, if \, t > t_a$$
$$and \left\{ V_i(t) < 0.99V_{i,ss} \; or \; V_i(t) > 1.01V_{i,ss} \right\}, 0, t > t_a$$
$$and \left\{ 0.99V_{i,ss} \le V_i(t) \le 1.01V_{i,ss} \right\}$$

where t_a is the instant of the adversarial action, $V_i(t)$ is the voltage magnitude for bus i in the power grid at time t, and $V_{i,ss}$ is the steady-state voltage of bus i before the attack; u_{ivld} is the invalid action penalty if the DRL agent provides action when the network is not attacked. Q_i and c are weights corresponding to voltage deviation and invalid action penalty, respectively.

13.4.3 Resilient RL co-simulation platform for microgrids

The current trends in research within the RL community involve employing benchmarking algorithms found within the OpenAI Gym platform. A simulation engine needs to be integrated into the Python API for operator use to utilize these algorithms. Leveraging the OpenAI Gym interface, we constructed a specialized simulation setup within the GridLAB-D/HELICS co-simulation platform. Our simulation framework relies on GridLAB-D as the microgrid simulation engine. However, our focus on control tasks necessitates utilizing GridLAB-D's subscription/publication architecture to manipulate specific control set points using externally coded Python scripts. This capability is facilitated by the HELICS co-simulation platform [41]. Our architecture uses two main modules: (1) microgrid co-simulation performed by the GridLAB-D/HELICS and (2) the resilient RL algorithmic development using OpenAI Gym. Following the creation of this customized environment, the resilient RL algorithm is developed to introduce ample cyber-resilient behavior. We generate a range of adversarial scenarios during episodic runs, mimicking actuation attacks within the inner-control (primary control) loop of GFMs. Throughout each episodic run, these events are sampled during the dynamic simulation carried out via GridLAB-D/HELICS co-simulation and communicated to the RL agent through the tailored OpenAI Gym interface, as illustrated in Figure 13.5.

13.4.4 Resilient vertical Fed-RL

In a system comprising r interconnected microgrids, the actions of the αth microgrid, executed via GFM inverters, and the observations consisting of concatenated terminal bus voltages are represented as u_{α}^{res} and o_{α}, respectively. Different from conventional federated reinforcement learning, also known as horizontal Fed-RL [37], the αth microgrid's environment is not entirely independent of the βth microgrid's environment due to network coupling. Furthermore, we have the sets $U_{\alpha}u_{\alpha}^{res} = u^{res}$ and $U_{\alpha}o_{\alpha} = O$ for the global networked microgrid environment. In this context, we define r policies of the form $u_{\alpha}^{res} = \pi^{\alpha}(o_{\alpha})$, where α ranges from 1 to r. We opt for deep neural network parametrized policies, denoting them as $\pi_{\theta_{\alpha}}^{\alpha}(.)$, where θ_{α} represents the parameters specific to microgrid α.

To foster federated learning characteristics, we propose employing an actor-critic reinforcement learning architecture, wherein the critic Q-networks for microgrid α are denoted as $Q_{\phi_{\alpha}}^{\alpha}$ and are characterized by neural network parameters ϕ_{α}. We adopt a microgrid-specific approach, utilizing decentralized observation and action spaces for each microgrid. At first, we update the critic networks $Q_{\phi_{\alpha}}^{\alpha}$ with local data within each microgrid. Subsequently, we propose transmitting these critic models to a central coordinator, which can be located at the operator control center managing the networked microgrids. These critic models are then aggregated at the coordinator, blending insights

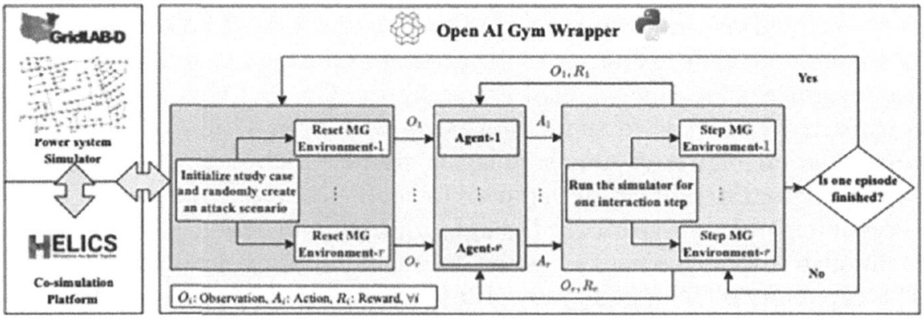

Figure 13.5 Resilient RL co-simulation platform for microgrids.

Source: [28], with IEEE permission.

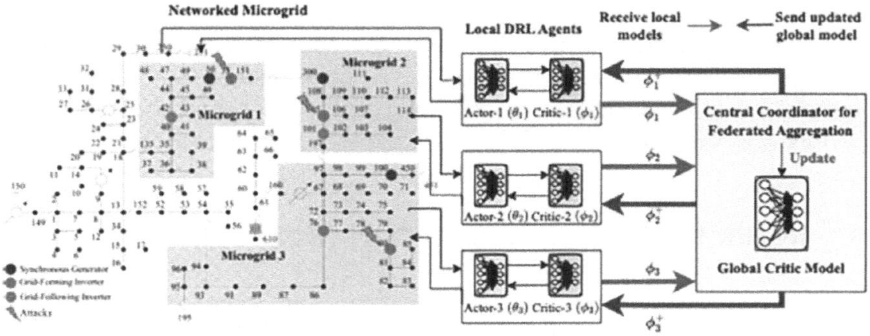

Figure 13.6 Fed-RL framework for networked microgrid.

Source: [28], with IEEE permission.

from the diverse, dynamic behaviors of different microgrids. This combined critique is employed to update the microgrid policies using local data once again. This approach introduces an innovative decentralized architecture for multi-agent systems, wherein the impact of dynamics within interconnected environments is captured by federated averaging of the critic networks. Algorithm 1 delineates the primary steps, and Figure 13.6 presents an overview of the comprehensive framework. In the implementation phase, we extend these concepts to incorporate the state-of-the-art SAC algorithm [30] with entropy regularization. We adapt the standard SAC algorithm from Stable Baselines [42] to integrate it into the federated learning framework. The details can be found in [28].

Algorithm 1: vertical federated resilient RL of networked microgrids [28]

1. **Initialize** critics and policies $Q^{\alpha}_{\phi_\alpha}$, and $\pi^{\alpha}_{\theta_\alpha}$ for the microgrid α.
2. **for** eps = $1, 2, \ldots, n_f$ do
3. **Sample** an adversarial attack scenario from the attack pool.
4. **Generate** episodic simulation data with the GridLAB-D/HELICS-OpenAI Gym emulator.
5. For each of the MG α, **use** o_α, and u^{res}_{α} from the MG α to update the local Q-networks $Q^{\alpha}_{\phi_\alpha}$, $\alpha = 1, \ldots, r$.
6. **Send** critic network $Q^{\alpha}_{\phi_\alpha}$ to the central grid operator.
7. **Perform** aggregation at the coordinator by an averaging operation, and return the aggregated one to each microgrid.
8. **Perform** gradient updates on the policy $\pi^{\alpha}_{\theta_\alpha}$ for each MG using the local observations, actions, and the global critic network model.
9. **end for**

In [28], we have shown a numerical example of our proposed approach using the standardized IEEE 123-bus test feeder system detailed in [27]. Our dynamic simulation utilizes our developed resilient RL co-simulation platform, which incorporates a customized OpenAI Gym interface for RL training.

13.5 INTER-SYSTEM RESILIENT CONTROL

As discussed in the previous sections, microgrids are increasingly considered a promising technology to support the integration of DERs. With an increasing number of microgrids, it is possible to coordinate the operations of networks of microgrids [43]. However, networked microgrids are vulnerable to cyber-physical attacks and faults due to the complex interconnection in both physical and cyber layers. As such, it is necessary to design resilient control systems to support the operations of networked microgrids in responses to cyber-physical attacks and faults.

In [44], networked MGs were investigated for self-healing purpose, in which local MG generation capacities were used to support other MGs when a generation deficiency or fault happens in an MG. Here, the networked MGs are connected/disconnected through a common point of coupling, and both communication and physical network topologies are fixed. Recently, a framework for assembling networked MGs by using consensus algorithms was introduced in [45], yet the communication network was fixed to an all-to-all topology, where each MG can communicate with all other MGs. As such, existing works in [44, 45] considered either fixed physical interconnection networks or fixed communication networks among microgrids.

13.5.1 Threats to distributed decision-making

The following cyber threats can compromise the distributed decision-making process of microgrid controllers:

- *Cyberattacks on controllers*. Generally, there are two types of cyber threats to the MG controllers. First, a malicious controller is investigated, where the attack is considered to be inside the MG control centers. This means that the MG controllers which will be tasked with estimating the supply and demand values of all the individual MGs using P2P architecture will experience malicious perturbation in their estimation program updates [46]. Next, the malicious MG controller sends manipulated measurements of the supply and demand values to their neighbors in the communication graph, also known as byzantine attacks. References such as [47, 48] discuss various aspects of the byzantine attacks and resilient mechanisms.

- *Cyberattacks on communication links.* Generally, this type of attacks occurs on communication links among controllers or communication links between controllers and sensors. [49] presents a description of different types of disruption attacks in the communication infrastructures. These attacks in the communication layers lead to denial-of-service (DoS) or jamming of the communication links.

13.5.2 Reconfiguration problem of networked microgrids

The goal of reconfiguration is for MG controllers to determine the optimal physical interconnection topology of MGs in a distributed manner after they already decided that they should interconnect. While there are several possible objective functions for microgrids, it is desirable that the optimal interconnection topology will correspond with minimal generation cost and a limited number of interconnection lines to reduce the risk of attack to these lines. Accordingly, the objective function of each MG is considered to include the generation cost of generators in that MG plus the loss on lines in that MG and on lines connecting that MG with other MGs in each interconnection topology. Therefore, for MG i we have

$$f_i = \Sigma C_m(g_m) + \lambda\left(\sum_{i:i\to j} z_{ij} l_{ij}\right) \tag{13.27}$$

where the first term stands for the generation cost and the second term stands for the total of line losses.

In the presence of cyberattacks, design framework and algorithms to allow the controllers of $MG_1,\ldots,MG_N$ to distributedly reach optimal interconnection topology corresponding to the smallest total of generation costs and line losses, that is,

$$\min \sum_{i=1}^{N} f_i \tag{13.28}$$

subject to power flow balance constraint, Kirchhoff's voltage law constraint along lines, and voltage limit constraints.

13.5.3 Bi-level optimization approach

A general approach to solve the preceding optimal interconnection topology problem is to search all over the possible switch variable and the local decision variables of all microgrids to find the minimum value of the total objective function $\sum_{i=1}^{N} f_i$, from which the optimal interconnection topology is determined. However, the mixed-integer nature of this problem makes it difficult to solve for large-scale networked microgrids. Moreover, the switch variables are global information decision variables, and each MG controller only knows some of them. As such, it is difficult to solve this problem in a distributed manner by using the general approach.

As such, with the help of the assistant agent, the following bi-level optimization framework, as depicted in Figure 13.7, can be utilized for MG controllers to determine the optimal interconnection topology in a distributed manner, with steps described as follows:

- Step 1: Assistant agent generates multiple interconnection topologies.
- Step 2: Given a topology τ, each controller performs distributed optimization and gets the optimal objective function of its MG. Each controller uses resilient distributed computation to estimate the total of all optimal objective functions of all MGs from exchanging the optimal objective function with other neighboring MGs via the P2P communication network.
- Step 3: Each controller stores all the estimates of the total objective function for all topologies.
- Step 4: Each controller finds the optimal topology corresponding to the smallest total objective function estimate.
- Step 5: If all controllers reach the same optimal topology, then all the MGs interconnect.

Here, two types of distributed calculation are needed:

- Distributed optimal power flow for MG controllers, to determine the optimal objective function consisting of the generation cost and line losses of that MG, together with the losses on lines connecting this MG with other MG.
- Distributed computation for MG controllers, to estimate the total of all optimal objective functions of all MGs from exchanging the optimal objective function with other neighboring MGs via the P2P communication network.

13.5.4 Distributed optimal power flow

In the presence of cyberattacks on controllers, resilient consensus algorithms are utilized to allow each MG controller to estimate the total amount of local optimal values of all MGs in a distributed manner through exchanging its local value with neighbors in the P2P communication network. One resilient consensus algorithm is presented in [46], which states that communication

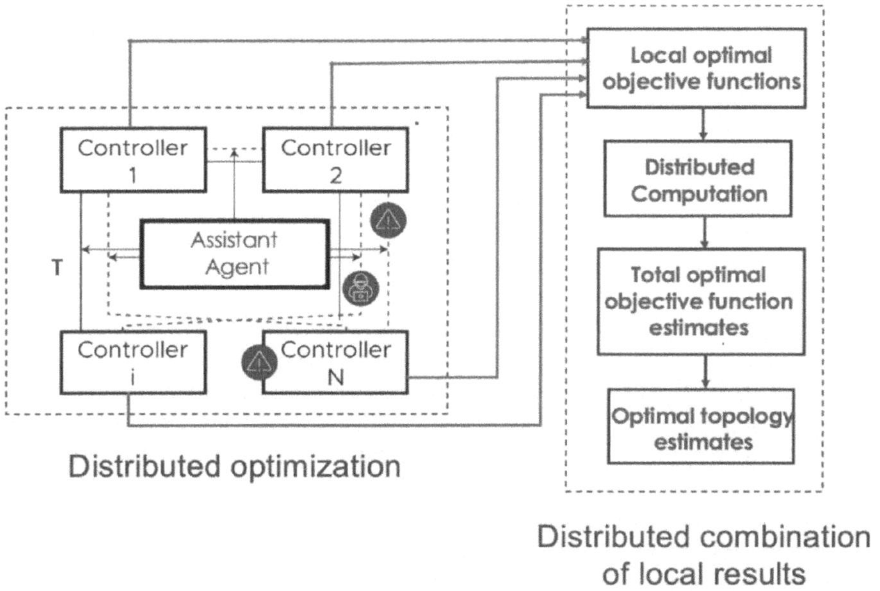

Figure 13.7 Bi-level optimization framework to determine the optimal interconnection topology in a distributed manner.

graph $G=(V,E)$ with N nodes; let f be the maximum number of nodes that use malicious local values during the resilient consensus updates, then all the nodes can recover the local value vectors of all the other MGs if the vertex connectivity of the communication graph is $\geq 2f+1$, then the weight matrix in consensus algorithm W can be constructed with almost any choice of real-valued numbers so that the resilient consensus algorithm converges. Graph theory to construct a communication graph with sufficient vertex connectivity is presented in [50].

Given an interconnection topology sent by the assistant agent, the MG controller in MG i^{th} will exchange information with other MG controllers to solve the distributed optimal power flow problem:

$$\min f_i \tag{13.29}$$

subject to power flow balance constraint, Kirchhoff's voltage law constraint along lines, and voltage limit constraints.

In general, techniques such as the alternating direction method of multipliers (ADMM) [51] and auxiliary problem principle (APP) [52] are used to solve the distributed OPF problem, which is known to have good convergence properties [53]. However, these methods involve a large number of iterations, which limits the applications to small systems [54]. To address this issue, the reduced equivalent network approximation (ENApp) method has been introduced for radially disconnected systems [55]. In this method, the shared variables are exchanged among the interconnected MGs, and distributed optimization is performed on each microgrids until the shared variables converge.

13.6 CONCLUSIONS

This chapter presented various techniques to improve the resilience of critical infrastructure under cyber-physical attacks, including those utilizing risk-based robust optimization, decentralized controls, Fed-RL, and through bi-level optimization of networked microgrids. These methods range from data-driven to model-driven and hybrid techniques and are minimally invasive, easy to implement, fast-acting, and generic enough to be applied to a variety of cyber-physical systems. These methods have been shown to generate superior performance through extensive testing. Future work will examine the conservativeness of these techniques and explore ways to increase operational efficiency. Another avenue of future work will be on transferring developed methodologies onto real-time hardware-in-the-loop settings. This will reduce disparities between the simulation results and the actual implementation.

13.6.1 Acknowledgments

This research is supported by the Resilience through Data-driven Intelligently-Designed Control (RD2C) Initiative, under the Laboratory Directed Research and Development (LDRD) Program at Pacific Northwest National Laboratory (PNNL). PNNL is a multi-program national laboratory operated for the US Department of Energy (DOE) by Battelle Memorial Institute under Contract No. DE-AC05–76RL01830. In this chapter, the authors summarize the collection of works performed by the larger research team from PNNL: Subhrajit Sinha, Alok Kumar Bharti, Priya T Mana, Ramij R. Hossain, Sheik M. Mohiuddin, Yuan Liu, Wei Du, Rohit A. Jinsiwale, Tianzhixi Yin, Qiuhua Huang (presently with Colorado School of Mines), Ankit Singhal (presently with Indian Institute of Technology, Delhi), Kyung-Bin Kwon (PNNL intern, University of Texas at Austin), and their collaborator, Sandip Roy (while he was with Washington State University). We would also like to acknowledge Karan Kalsi and Kevin Schneider from PNNL for their guidance on the research.

13.6.2 Author bios

Nawaf Nazir (senior member, IEEE) is a senior research scientist at Pacific Northwest National Laboratory, Richland, WA. His work is at the intersection of optimization, control systems, and machine learning, with applications to energy systems, transportation systems, and edge devices.

He is the principal investigator (PI) of several multi-million-dollar DOE and ARPA-E projects involving collaboration with multiple universities, national labs, and industry. He has spearheaded and chaired several workshops and panels at venues such as the Transportation Research Board (TRB), PES Grid Edge Conference, American Control Conference (ACC), INFORMS, and PES General Meeting.

Sai Pushpak Nandanoori has been a senior research engineer at PNNL since March 2018. He graduated from Iowa State University, and his research interests lie in stochastic systems, developing system theoretic techniques, and data-driven tools such as transfer operators to tackle challenging problems in the areas of power systems, microgrids, and cyber-physical systems.

Thanh Long Vu (member, IEEE) received his BEng degree in automatic control from the Hanoi University of Technology in 2007, and his PhD degree in electrical engineering from the National University of Singapore in 2012. He is currently a senior controls scientist with Pacific Northwest National Laboratory (PNNL). Prior to joining PNNL, he was a research scientist with the Massachusetts Institute of Technology. His core research interests include the fields of power systems, control systems, and machine learning.

Dr. Sayak Mukherjee is a research scientist in the optimization and control group at the Pacific Northwest National Laboratory. He received his PhD in electrical engineering from North Carolina State University, USA, in 2020 and his BEE from Jadavpur University, India, in 2015, with medal for first-class second position. His areas of expertise include control, reinforcement learning, resilient control designs, large-scale power system stability and control, grid operation with distributed energy resources, etc.

Soumya Kundu received his bachelor's degree in electrical engineering and master's degree in control systems engineering from the Indian Institute of Technology–Kharagpur, Kharagpur, India, in 2009, and his PhD degree in control systems from the University of Michigan–Ann Arbor in 2013. He was a postdoctoral research associate with the Los Alamos National Laboratory before joining the Pacific Northwest National Laboratory in 2016 as a staff research engineer. His research is primarily focused on developing systems theoretic tools to facilitate dynamic security and operational optimality of the future electrical grid.

Veronica Adetola is a control systems scientist with expertise in model predictive control, robust and adaptive control, machine learning for control, dynamical system analysis, and control architecture design. She has more than ten years' experience in research, development, and demonstration of advanced technology solutions for energy-efficient systems, grid-interactive buildings, transport refrigeration, and aerospace air management systems. Before joining PNNL in 2019, Veronica worked at the United Technologies Research Center, where she contributed to and successfully led activities in support of multiple UTC businesses and government-funded research programs.

REFERENCES

[1] E. Zio and G. Sansavini, "Vulnerability of smart grids with variable generation and consumption: A system of systems perspective," IEEE Transactions on Systems, Man, and Cybernetics: Systems, vol. 43, no. 3, pp. 477–487, 2013.
[2] N. Nazir and M. Almassalkhi, "Convex inner approximation of the feeder hosting capacity limits on dispatchable demand," in 2019 IEEE 58th Conference on Decision and Control (CDC). IEEE, 2019, pp. 4858–4864.
[3] N. Ferc and R. Entities, "The February 2021 cold weather outages in Texas and the South Central United States," 2021. https://www.ferc.gov/media/february-2021-cold-weather-outages-texas-and-south-central-united-states-ferc-nerc-and.
[4] 116th Congress, H.r. 360—cyber sense act of 2020, 2020. www.congress.gov/bill/116th-congress/house-bill/360
[5] M. Dehghani, T. Niknam, M. Ghiasi, N. Bayati, and M. Savaghebi, "Cyber-attack detection in dc microgrids based on deep machine learning and wavelet singular values approach," Electronics, vol. 10, no. 16, p. 1914, 2021.
[6] H. Wang, J. Ruan, Z. Ma, B. Zhou, X. Fu, and G. Cao, "Deep learning aided interval state prediction for improving cyber security in energy internet," Energy, vol. 174, pp. 1292–1304, 2019.
[7] N. V. Sahinidis, "Optimization under uncertainty: State-of-the-art and opportunities," Computers & Chemical Engineering, vol. 28, no. 6–7, pp. 971–983, 2004.
[8] G. C. Calafiore and L. E. Ghaoui, "On distributionally robust chance-constrained linear programs," Journal of Optimization Theory and Applications, vol. 130, no. 1, pp. 1–22, 2006.
[9] T. Summers, J. Warrington, M. Morari, and J. Lygeros, "Stochastic optimal power flow based on conditional value at risk and distributional robustness," International Journal of Electrical Power & Energy Systems, vol. 72, pp. 116–125, 2015.
[10] B. Li, J. L. Mathieu, and R. Jiang, "Distributionally robust chance constrained optimal power flow assuming log-concave distributions," in 2018 Power Systems Computation Conference (PSCC). IEEE, 2018, pp. 1–7.
[11] L. Yang, Y. Xu, H. Sun, and W. Wu, "Tractable convex approximations for distributionally robust joint chance-constrained optimal power flow under uncertainty," IEEE Transactions on Power Systems, vol. 37, no. 3, pp. 1927–1941, 2021.
[12] A. N. Madavan and S. Bose, "A stochastic primal-dual method for optimization with conditional value at risk constraints," Journal of Optimization Theory and Applications, vol. 190, no. 2, pp. 428–460, 2021.

[13] J. Cheng, C. Gicquel, and A. Lisser, "Partial sample average approximation method for chance constrained problems," Optimization Letters, vol. 13, no. 4, pp. 657–672, 2019.
[14] J. Barrera, T. Homem-de Mello, E. Moreno, B. K. Pagnoncelli, and G. Canessa, "Chance-constrained problems and rare events: An importance sampling approach," Mathematical Programming, vol. 157, no. 1, pp. 153–189, 2016.
[15] A. Cherukuri and A. R. Hota, "Consistency of distributionally robust risk-and chance-constrained optimization under wasserstein ambiguity sets," IEEE Control Systems Letters, vol. 5, no. 5, pp. 1729–1734, 2020.
[16] N. Nazir, T. Ramachandran, S. Bhattacharya, A. Singhal, S. Kundu and V. Adetola, "Optimization-based resiliency verification in microgrids via maximal adversarial set characterization," 2022 American Control Conference (ACC). Atlanta, GA, 2022, pp. 2214–2220, doi: 10.23919/ACC53348.2022.9867826.
[17] X. Bai, L. Qu, and W. Qiao, "Robust AC optimal power flow for power networks with wind power generation," IEEE Transactions on Power Systems, vol. 31, no. 5, pp. 4163–4164, 2015.
[18] R. T. Rockafellar and S. Uryasev, "Optimization of conditional value-at-risk," Journal of Risk, vol. 2, pp. 21–42, 2000. https://www.ise.ufl.edu/uryasev/files/2011/11/CVaR1_JOR.pdf.
[19] E. Dall'Anese, K. Baker, and T. Summers, "Chance-constrained ac optimal power flow for distribution systems with renewables," IEEE Transactions on Power Systems, vol. 32, no. 5, pp. 3427–3438, 2017.
[20] C. Duan, W. Fang, L. Jiang, L. Yao, and J. Liu, "Distributionally robust chance-constrained approximate ac-OPF with Wasserstein metric," IEEE Transactions on Power Systems, vol. 33, no. 5, pp. 4924–4936, 2018.
[21] S. Kundu, S. Geng, S. P. Nandanoori, I. A. Hiskens, and K. Kalsi, "Distributed barrier certificates for safe operation of inverter-based microgrids," in 2019 American Control Conference (ACC). IEEE, 2019, pp. 1042–1047.
[22] S. Kundu, W. Du, S. P. Nandanoori, F. Tuffner, and K. Schneider, "Identifying parameter space for robust stability in nonlinear networks: A microgrid application," in 2019 American Control Conference (ACC). IEEE, 2019, pp. 3111–3116.
[23] F. Blanchini, "Set invariance in control," Automatica, vol. 35, no. 11, pp. 1747–1767, 1999.
[24] J.-B. Bouvier, S. P. Nandanoori, M. Ornik, and S. Kundu, "Distributed transient safety verification via robust control invariant sets: A microgrid application," in 2022 American Control Conference (ACC). IEEE, 2022, pp. 2202–2207.
[25] D. Panagou, D. M. Stipanović, and P. G. Voulgaris, "Distributed coordination control for multi-robot networks using Lyapunov-like barrier functions," IEEE Transactions on Automatic Control, vol. 61, no. 3, pp. 617–632, 2015.
[26] A. Nedic, A. Ozdaglar, and P. A. Parrilo, "Constrained consensus and optimization in multi-agent networks," IEEE Transactions on Automatic Control, vol. 55, no. 4, pp. 922–938, 2010.
[27] A. Singhal, T. L. Vu, and W. Du, "Consensus control for coordinating grid-forming and grid-following inverters in microgrids," IEEE Transactions on Smart Grid, vol. 13, no. 5, pp. 4123–4133, September 2022, doi: 10.1109/TSG.2022.3158254
[28] S. Mukherjee, R. R. Hossain, Y. Liu, W. Du, V. Adetola, S. M. Mohiuddin, Q. Huang, T. Yin, and A. Singhal, "Enhancing cyber resilience of networked

microgrids using vertical federated reinforcement learning," in *2023 IEEE Power & Energy Society General Meeting (PESGM)*, Orlando, FL, 2023, pp. 1–5, doi: 10.1109/PESGM52003.2023.10252480, also available in arXiv preprint arXiv:2212.08973, 2022.
[29] W. Du, F. K. Tuffner, K. P. Schneider, R. H. Lasseter, J. Xie, Z. Chen, and B. Bhattarai, "Modeling of grid-forming and grid-following inverters for dynamic simulation of large-scale distribution systems," IEEE Transactions on Power Delivery, vol. 36, no. 4, pp. 2035–2045, 2020.
[30] T. Haarnoja, A. Zhou, P. Abbeel, and S. Levine, "Soft actor-critic: Off-policy maximum entropy deep reinforcement learning with a stochastic actor," in International Conference on Machine Learning. PMLR, July 2018, pp. 1861–1870, https://proceedings.mlr.press/v80/haarnoja18b/haarnoja18b.pdf
[31] R. Lowe, Y. I. Wu, A. Tamar, J. Harb, O. P. Abbeel, and I. Mordatch, "Multi-agent actor-critic for mixed cooperative-competitive environments," in NeurIPS, 2017, pp. 6379–6390.
[32] S. Mukherjee, R. Huang, Q. Huang, T. L. Vu, and T. Yin, "Scalable voltage control using structure-driven hierarchical deep reinforcement learning," arXiv preprint arXiv:2102.00077, 2021. https://arxiv.org/abs/2102.00077.
[33] J. Duan, Z. Yi, D. Shi, C. Lin, X. Lu, and Z. Wang, "Reinforcement-learning-based optimal control of hybrid energy storage systems in hybrid ac–dc microgrids," IEEE Transactions on Industrial Informatics, vol. 15, no. 9, pp. 5355–5364, 2019.
[34] S. Mukherjee, A. Chakrabortty, H. Bai, A. Darvishi, and B. Fardanesh, "Scalable designs for reinforcement learning-based wide-area damping control," IEEE Transactions on Smart Grid, vol. 12, no. 3, pp. 2389–2401, 2021.
[35] W. Wang, N. Yu, Y. Gao, and J. Shi, "Safe off-policy deep reinforcement learning algorithm for volt-var control in power distribution systems," IEEE Transactions on Smart Grid, vol. 11, no. 4, pp. 3008–3018, 2020.
[36] K. Bonawitz, H. Eichner, W. Grieskamp, D. Huba, A. Ingerman, V. Ivanov, C. Kiddon, J. Konečný, S. Mazzocchi, B. McMahan, T. Van Overveldt, D. Petrou, D. Ramage, and J. Roselander, "Towards federated learning at scale: System design," Proceedings of Machine Learning and Systems, vol. 1, pp. 374–388, 2019.
[37] X. Wang, C. Wang, X. Li, V. C. Leung, and T. Taleb, "Federated deep reinforcement learning for Internet of Things with decentralized cooperative edge caching," IEEE Internet of Things Journal, vol. 7, no. 10, pp. 9441–9455, 2020.
[38] H. Liu and W. Wu, "Federated reinforcement learning for decentralized voltage control in distribution networks," IEEE Transactions on Smart Grid, vol. 13, no. 5, pp. 3840–3843, September 2022. https://doi.org/10.1109/TSG.2022.3169361.
[39] S. Lee and D.-H. Choi, "Federated reinforcement learning for energy management of multiple smart homes with distributed energy resources," IEEE Transactions on Industrial Informatics, vol. 18, no. 1, pp. 488–497, 2020.
[40] R. Sutton and A. Barto, Reinforcement learning—An introduction. MIT Press, Cambridge, 1998.
[41] B. Palmintier, D. Krishnamurthy, P. Top, S. Smith, J. Daily, and J. Fuller, "Design of the helics high-performance transmission-distribution-communication-market co-simulation framework," in 2017 Workshop on MSCPES. IEEE, 2017, pp. 1–6.

[42] A. Raffin, A. Hill, A. Gleave, A. Kanervisto, M. Ernestus, and N. Dormann, "Stable-baselines3: Reliable reinforcement learning implementations," Journal of Machine Learning Research, vol. 22, no. 268, pp. 1–8, 2021.
[43] Z. Li, M. Shahidehpour, F. Aminifar, A. Alabdulwahab, and Y. Al-Turki, "Networked microgrids for enhancing the power system resilience," Proceedings of the IEEE, vol. 105, no. 7, pp. 1289–1310, 2017.
[44] Z. Wang, B. Chen, J. Wang, and C. Chen, "Networked microgrids for self-healing power systems," IEEE Transactions on Smart Grid, vol. 7, no. 1, pp. 310–319, 2016.
[45] K. P. Schneider, J. Glass, C. Klauber, B. Ollis, M. J. Reno, M. Burck, L. Muhidin, A. Dubey, W. Du, L. Vu, J. Xie, D. Nordy, W. Dawson, J. Hernandez-Alvidrez, A. Bose, D. Ton, and G. Yuan, "A framework for coordinated self-assembly of networked microgrids using consensus algorithms," IEEE Access, vol. 10, pp. 3864–3878, 2022. https://doi.org/10.1109/ACCESS.2021.3132253, 2022.
[46] S. Sundaram and C. N. Hadjicostis, "Distributed function calculation via linear iterative strategies in the presence of malicious agents," IEEE Transactions on Automatic Control, vol. 56, no. 7, pp. 1495–1508, 2010.
[47] H. J. LeBlanc and X. Koutsoukos, "Resilient first-order consensus and weakly stable, higher order synchronization of continuous-time networked multiagent systems," IEEE Transactions on Control of Network Systems, vol. 5, no. 3, pp. 1219–1231, 2017.
[48] H. Zhang, E. Fata, and S. Sundaram, "A notion of robustness in complex networks," IEEE Transactions on Control of Network Systems, vol. 2, no. 3, pp. 310–320, 2015.
[49] A. Gusrialdi and Z. Qu, "Smart grid security: Attacks and defenses," in Smart grid control. Springer, 2019, pp. 199–223.
[50] T. L. Vu, S. Mukherjee, and V. Adetola, "Resilient communication scheme for distributed decision of interconnecting networks of microgrids," in 2023 IEEE Power & Energy Society Innovative Smart Grid Technologies Conference (ISGT). IEEE, 2023, pp. 1–5.
[51] B. D. Biswas, M. S. Hasan, and S. Kamalasadan, "Decentralized distributed convex optimal power flow model for power distribution system based on alternating direction method of multipliers," IEEE Transactions on Industry Applications, vol. 59, no. 1, pp. 627–640, 2023.
[52] H. Du, T. Lin, Q. Li, X. Fu, and X. Xu, "Decentralized optimal power flow based on auxiliary problem principle with an adaptive core," Energy Reports, vol. 8, pp. 755–765. www.sciencedirect.com/science/article/pii/S2352484722015979
[53] D. K. Molzahn, F. Dörfler, H. Sandberg, S. H. Low, S. Chakrabarti, R. Baldick, and J. Lavaei, "A survey of distributed optimization and control algorithms for electric power systems," IEEE Transactions on Smart Grid, vol. 8, no. 6, pp. 2941–2962, 2017.
[54] T. Erseghe, "Distributed optimal power flow using ADMM," IEEE Transactions on Power Systems, vol. 29, no. 5, pp. 2370–2380, 2014.
[55] R. Sadnan and A. Dubey, "Distributed optimization using reduced network equivalents for radial power distribution systems," IEEE Transactions on Power Systems, vol. 36, pp. 3645–3656, 2020. https://api.semanticscholar.org/CorpusID:227305089

Index